With

N

B. Sc. Loyola '53

Chemistry of Peptide Synthesis

N. Leo Benoiton
University of Ottawa
Ottawa, Ontario, Canada

Boca Raton London New York Singapore

A CRC title, part of the Taylor & Francis imprint, a member of the Taylor & Francis Group, the academic division of T&F Informa plc.

Published in 2006 by
CRC Press
Taylor & Francis Group
6000 Broken Sound Parkway NW, Suite 300
Boca Raton, FL 33487-2742

Printed in the United States of America on acid-free paper
10 9 8 7 6 5 4 3 2 1

International Standard Book Number-10: 1-57444-454-9 (Hardcover)
International Standard Book Number-13: 978-1-57444-454-4 (Hardcover)
Library of Congress Card Number 2005005753

Library of Congress Cataloging-in-Publication Data

Benoiton, N. Leo.
Chemistry of peptide synthesis / N. Leo Benoiton.
p. ; cm.
Includes bibliographical references.
ISBN-13: 978-1-57444-454-4 (hardcover : alk. paper)
ISBN-10: 1-57444-454-9 (hardcover : alk. paper)
1. Peptides--Synthesis.
[DNLM: 1. Peptide Biosynthesis. QU 68 B456c 2005] I. Title.

QP552.P4B46 2005
612'.015756--dc22

2005005753

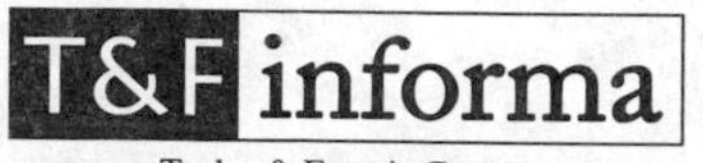

Taylor & Francis Group
is the Academic Division of T&F Informa plc.

Visit the Taylor & Francis Web site at
http://www.taylorandfrancis.com

and the CRC Press Web site at
http://www.crcpress.com

Dedication

This book is dedicated to Rao Makineni, a unique member and benefactor of the peptide community.

Preface

This book has emerged from courses that I taught to biochemistry students at the undergraduate and graduate levels, to persons with a limited knowledge of organic chemistry, to chemists with experience in other fields, and to peptide chemists. It assumes that the reader possesses a minimum knowledge of organic and amino-acid chemistry. It comprises 188 self-standing sections that include 207 figures written in clear language, with limited use of abbreviations. The focus is on understanding how and why reactions and phenomena occur. There are a few tables of illustrative data, but no tables of compounds or reaction conditions. The material is presented progressively, with some repetition, and then with amplification after the basics have been dealt with. The fundamentals of peptide synthesis, with an emphasis on the intermediates that are encountered in aminolysis reactions, are presented initially. The coupling of Nα-protected amino acids and Nα-protected peptides and their tendencies to isomerize are then addressed separately. This allows for easier comprehension of the issues of stereomutation and the applicability of coupling reactions. Protection of functional groups is introduced on the basis of the methods that are employed for removal of the protectors. A chapter is devoted to the question of stereomutation, which is now more complex, following the discovery that Nα-protected amino acids can also give rise to oxazolones. Other chapters are devoted to solid-phase synthesis, side-chain protection and side reactions, amplification on coupling methods, and miscellaneous topics. Points to note are that esters that undergo aminolysis are referred to as activated esters, which is why they react, and not active esters, and that in two cases two abbreviations (Z and Cbz; HOObt and HODhbt) are used haphazardly for one entity because that is the reality of the peptide literature. An effort has been made to convey to the reader a notion of how the field of peptide chemistry has developed. To this end, the references are located at the end of each section and include the titles of articles. Most references have been selected on the basis of the main theme that the chapter addresses. When the relevance of a paper is not obvious from the title, a phrase has been inserted in parentheses. The titles of papers written in German and French have been translated. For obvious reasons the number of references had to be limited. I extend my apologies to anyone who considers his or her work to have been unjustifiably omitted. Some poetic license was exercised in the creation of the manuscript and the reaction schemes. Inclusion of all details and exceptions to statements would have made the whole too unruly.

I am greatly indebted to Dr. Brian Ridge of the School of Chemical and Biological Sciences of the University of Exeter, United Kingdom, for his critical review of the manuscript and for his suggestions that have been incorporated into the manuscript. I solely am responsible for the book's contents. I thank Professor John Coggins of the University of Glasgow for providing the references for Appendix 3, and I am grateful to anyone who might have provided me with information that appears in this book. I am grateful to the University of Ottawa for the office and library services that have been provided to me. I am indebted to Dr. Rao Makineni for generous support provided over the years. I thank the publishers for their patience during the long period when submission of the manuscript was overdue. And most important, I thank my wife Ljuba for her patience and support and express my sincere apologies for having deprived her of the company of her "retired" husband for a period much longer than had been planned.

Table of Contents

1 Fundamentals of Peptide Synthesis

1.1 CHEMICAL AND STEREOCHEMICAL NATURE OF AMINO ACIDS

The building blocks of peptides are amino acids, which are composed of a carbon atom to which are attached a carboxyl group, an amino group, a hydrogen atom, and a so-called side-chain R^2 (Figure 1.1). The simplest amino acid is glycine, for which the side-chain is another hydrogen atom, so there are no stereochemical forms of glycine. Glycine is not a chiral compound, but two configurations or arrangements of substituents around the central α-carbon atom are possible for all other amino acids, so each exists in two stereochemical forms, known as the L-isomers for the amino acids found in proteins and the D-isomers for those with the opposite configurations. The natural amino acids are so designated because they have the same configuration as that of natural glyceraldehyde, which arbitrarily had been designated the L-form. Two isomers of opposite configuration or chirality (handedness) have the relationship of mirror images and are referred to as enantiomers. Enantiomers are identical in all respects except that solutions of the isomers rotate plane-polarized light in opposite directions. The enantiomer deflecting polarized light to the right is said to be dextrorotatory (+), and the enantiomer deflecting polarized light to the left is levorotatory (–). There is no correlation between the direction of this optical rotation and the configuration of the isomer — the direction cannot be predicted from knowledge of the absolute configuration of the compound. According to the Cahn–Ingold–Prelog system of nomenclature, L-amino acids are of the (*S*)-configuration, except for cysteine and its derivatives. In discussion, when the configuration of an amino acid residue is not indicated, it is assumed to be the L-enantiomer.[1]

1. JP Moss. Basic terminology of stereochemistry. *Pure Appl. Chem.* 68, 2193, 1996.

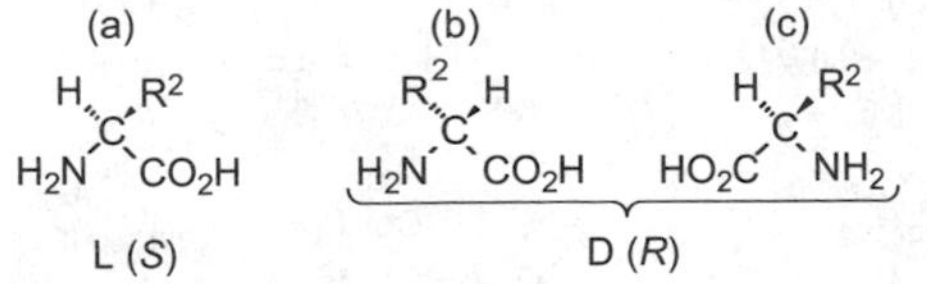

(a) H, R^2, C, H_2N, CO_2H — L (*S*)

(b) R^2, H, C, H_2N, CO_2H (c) H, R^2, C, HO_2C, NH_2 — D (*R*)

FIGURE 1.1 Chemical and stereochemical nature of amino acids. Substituents in (a) and (b) are on opposite sides of the plane N–Cα–C, the bold bond being above the plane. Interchange of any two substituents in (a) changes the configuration. For the Cahn-Ingold-Prelog system of nomenclature, the order of preference NH_2 > COOH > R^2 relative to H is anticlockwise in (a) = (*S*) and clockwise in (c) = (*R*).

1.2 IONIC NATURE OF AMINO ACIDS

Each of the functional groups of the amino acid can exist in the protonated or unprotonated form (Figure 1.2). The ionic state of a functional group is dictated by two parameters: its chemical nature, and the pH of the environment. As the pH changes, the group either picks up or loses a proton. The chemical constitution of the group determines over which relatively small range of pH this occurs. For practical purposes, this range is best defined by the logarithm of the dissociation constant of the group, designated pK_a (the subscript "a" stands for acid — it is often omitted), which corresponds to the pH at which one-half of the molecules are protonated and one-half are not protonated. The p*K*s of functional groups are influenced by adjacent groups and groups in proximity — in effect, the environment. So the p*K* of a group refers to the constant in a particular molecule and is understood to be "apparent," under the circumstances (solvent) in question. As an example, the p*K* of the CO_2H of valine in aqueous solution is 2.3, and the p*K* of the NH_3^+ group is 9.6. Below pH 2.3, greater than half of the carboxyl groups are protonated; above pH 2.3, more of them are deprotonated. According to the Henderson–Hasselbalch equation, pH = p*K* + log $[CO_2^-]/[CO_2H]$, which describes the relationship between pH and the ratio of the two forms; at pH 4.3, the ratio of the two forms is 100. The same holds for the amino group. Above pH 9.6, more of them are unprotonated; below pH 9.6, more of the amino groups are protonated. Note that the functional groups represent two types of acids: an uncharged acid ($-CO_2H$), and a charged acid ($-NH_3^+$). The deprotonated form of each is the conjugate base of the acid, with the stronger base ($-NH_2$) being the conjugate form of the weaker acid. Because the uncharged acid is the first to lose its proton when the two acids are neutralized, the amino acid is a charged molecule at all values of pH. It is a cation at acidic pH, an anion at alkaline pH, and predominantly an ion of both types or zwitter-ion at pHs between the two p*K* values. The amino acids are also zwitter-ions when they crystallize out of solution. A midway point on the pH scale, at which the amino acid does not migrate in an electric field, is referred to as the isoelectric point, or pI.

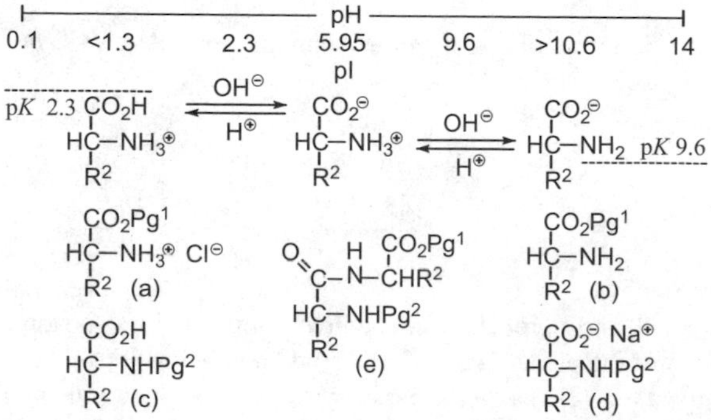

pH: 0.1 | <1.3 | 2.3 | 5.95 (pI) | 9.6 | >10.6 | 14

p*K* 2.3 CO_2H–HC(NH_3^+)–R^2 ⇌ (OH⁻ / H⁺) CO_2^-–HC(NH_3^+)–R^2 ⇌ (OH⁻ / H⁺) CO_2^-–HC(NH_2)–R^2 p*K* 9.6

(a) CO_2Pg^1–HC(NH_3^+ Cl^-)–R^2

(b) CO_2Pg^1–HC(NH_2)–R^2

(c) CO_2H–HC($NHPg^2$)–R^2

(d) CO_2^- Na^+–HC($NHPg^2$)–R^2

(e) O=C(–NH–CHR^2–CO_2Pg^1)–HC($NHPg^2$)–R^2

FIGURE 1.2 Ionic nature of amino acids. Pg = Protecting group. (a) Insoluble in organic solvent and soluble in aqueous acid; (d) insoluble in organic solvent and soluble in aqueous alkali; (b), (c), and (e), soluble in organic solvent.

In practice, a peptide is formed by the combination of two amino acids joined together by the reaction of the carboxyl group of one amino acid with the amino group of a second amino acid. To achieve the coupling as desired, the two functional groups that are not implicated are prevented from reacting by derivatization with temporary protecting groups, which are removed later. Such coupling reactions do not go to completion, and one is able to take advantage of the ionic nature of functional groups to purify the desired product. The protected peptide is soluble in organic solvent and insoluble in water, acid or alkali (Figure 1.2). Unreacted *N*-protected amino acid is also soluble in organic solvent, but it can be made insoluble in organic solvent and soluble in aqueous solution by deprotonation to the anion or salt form by the addition of alkali. Similarly, unreacted amino acid ester is soluble in organic solvent and insoluble in alkali, but it can be made soluble in aqueous solution by protonation to the alkylammonium ion or salt form by the addition of acid. Thus, the desired protected peptide can be obtained free of unreacted starting materials by taking advantage of the ionic nature of the two reactants that can be removed by aqueous washes. This is the simplest method of purification of a coupling product and should be the first step of any purification when it is applicable.

1.3 CHARGED GROUPS IN PEPTIDES AT NEUTRAL PH

The pK_as of carboxyl and ammonium groups of the amino acids are in the 1.89–2.34 and 8.8–9.7 ranges, respectively.[2] These values are considerably lower than those (4.3 and 10.7, respectively) for the same functional groups in a compound such as δ-aminopentanoic acid, in which ionization is unaffected by the presence of neighboring groups. In the α-amino acid, the acidity of the carboxyl group is increased (more readily ionized) by the electron-withdrawing property of the ammonium cation. The explanation for the decreased basicity of the amino group is more complex and is attributed to differential solvation. The zwitter-ionic form is destabilized by the repulsion of dipolar solvent molecules. The anionic form is not destabilized by this effect, so there is a decrease in the concentration of the conjugate acid ($-H_3N^+$). In a peptide, the effect of the nitrogen-containing group has been diminished by its conversion from an ammonium cation to a peptide bond. Thus, the acidity of the α-carboxyl group of a peptide is intermediate (p*K* 3.0–3.4), falling between that of an amino acid and an alkanoic acid. In contrast, incorporation of the carboxyl group of an amino acid into a peptide enhances its effect on the amino group, rendering it even less basic than in the amino acid. Thus, the p*K*s of α-ammonium groups of peptides are lower (7.75–8.3) than those of amino acids. This lower value in a peptide explains the popularity to biochemists over recent decades of glycylglycine as a buffer — it is efficient for controlling the pH of enzymatic reactions requiring a neutral pH. The acidities of the functional groups in *N*-protected amino acids and amino acid esters are similar to those of the functional groups in peptides (Figure 1.3).

Other ionizable groups are found on the side chains of peptides. These include the β-CO_2H of aspartic acid (Asp), the γ-CO_2H of glutamic acid (Glu), the ε-NH_2 of lysine (Lys), and the δ-guanidino of arginine (Arg). The β-CO_2H group is more acidic than the γ-CO_2H group because of its proximity to the peptide chain, but both

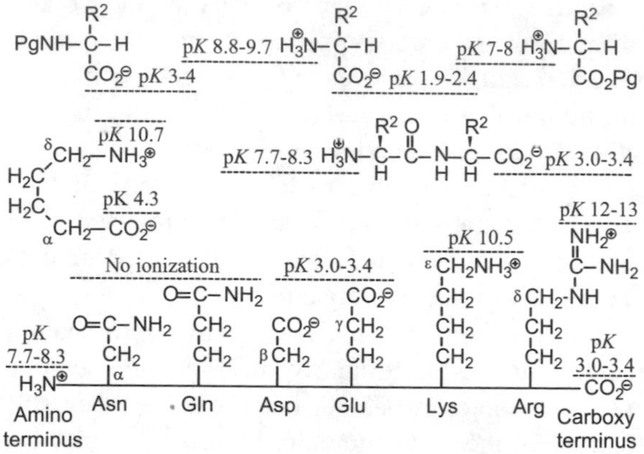

FIGURE 1.3 Charged groups in peptides at neutral pH.

exist as anions at neutral pH. The guanidino group is by nature more basic than the ε-NH_2 group, but both are positively charged at neutral pH. The carboxamido groups of asparagine (Asn) and glutamine (Gln), the amides of aspartic and glutamic acids, are neutral and do not ionize over the normal pH scale. The imidazole of histidine (His) is unique in that it is partially protonated at neutral pH because its p*K* is close to neutrality. The phenolic group of tyrosine (Tyr) and the sulfhydryl of cysteine (Cys) are normally not ionized but can be at mildly alkaline pH. Other functional groups do not pick up or lose a proton under usual conditions. The indole nitrogen of tryptophan (Trp) is so affected by the unsaturated rings that it picks up a proton only at very acidic pH (<2). In summary, p*K*s of carboxyl groups of peptides and *N*-protected amino acids are in the "normal" range; p*K*s of amino groups of peptides and amino acid amides and esters are one or more pH units lower than those of ε-amino groups of lysine.[2]

2. JP Greenstein, and M Winitz. *Chemistry of the Amino Acids*, Wiley, New York, 1961, pp. 486-500.

1.4 SIDE-CHAIN EFFECTS IN OTHER AMINO ACIDS

Glycine (Gly) does not have a side chain, and as a consequence it behaves atypically. Its derivatives are more reactive than those of other amino acids, and it can even undergo reaction at the α-carbon atom. In contrast, valine (Val) and isoleucine (Ile) are less reactive than other amino acids because of hindrance resulting from a methyl substituent on the β-carbon atom of the side-chain (Figure 1.4). Hindrance is manifested primarily at the carboxyl group, and it leads to a greater ease of cyclization once the residue is activated. Threonine (Thr) becomes a hindered amino acid when its secondary hydroxyl group is substituted, as its structure then resembles those of the β-methylamino acids. Leucine (Leu), isoleucine, valine, phenylalanine (Phe), tyrosine (Tyr), tryptophan, and methionine (Met) have hydrophobic side chains. Alanine (Ala) seems anomalous in this regard — a residue imparting hydrophilicity

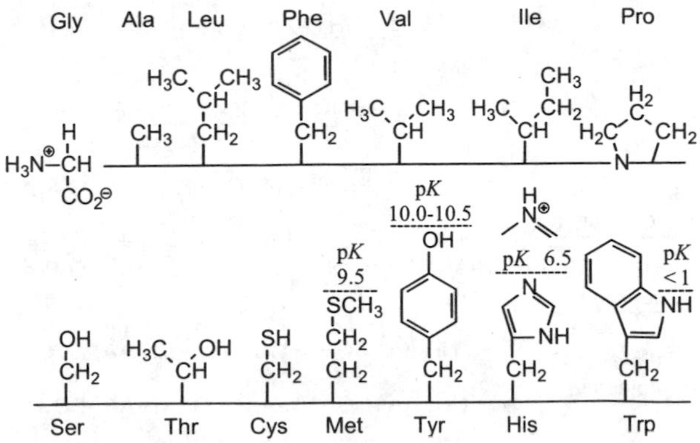

FIGURE 1.4 Side-chain effects in other amino acids.

to a peptide chain. This is evident from reversed-phase, high-performance liquid chromatography of L-alanyl-L-alanine and L-alanyl-L-alanyl-L-alanine, the latter emerging earlier than the dipeptide. The thioether of methionine and the indole ring of tryprophan are sensitive to oxygen, undergoing oxidation during manipulation. Air also oxidizes the sulfhydryl group of cysteine to the disulfide. The alcoholic groups of serine and threonine are not sensitive to oxidation. The propyl side-chain of proline (Pro) is linked to its amino group, making it an imino instead of an amino acid. α-Carbon atoms linked to a peptide bond formed at the carboxyl group of an imino acid adopt the *cis* rather than the usual *trans* relationship. In addition, the cyclic nature of proline prevents the isomerization that amino acids undergo during reactions at their carboxyl groups. Threonine and isoleucine each contain two stereogenic centers (asymmetric carbon atoms). The amino and hydroxyl substituents of threonine are on opposite sides of the carbon chain (*threo*) in the Fischer representation, but the amino and methyl groups of isoleucine are on the same side of the chain (*erythro*). Isomerization at the α-carbon atom of L-threonine generates the D-allothreonine diastereoisomer, with "allo" (other) signifying the isomer that is not found in proteins. The enantiomer or mirror-image of L-threonine is D-threonine.

1.5 GENERAL APPROACH TO PROTECTION AND AMIDE-BOND FORMATION

The initial step in synthesis is suppression of the reactivity of the functional groups in the amino acids that are not intended to be incorporated into the peptide bond. This is usually achieved by the derivatization of the groups, but it may also involve their chelation with a metal ion or conversion into a charged form. It is vital that the modification be reversible. Peptide-bond formation is then effected by abstraction of a molecule of water between the free amino and carboxyl groups in the two amino acid derivatives (Figure 1.5). The next step is liberation of the functional group that is to enter into formation of the second peptide bond. This selective deprotection of one functional group without affecting the protection of the other groups is the critical feature of the synthesis. It is ideally achieved by use of a chemical mechanism

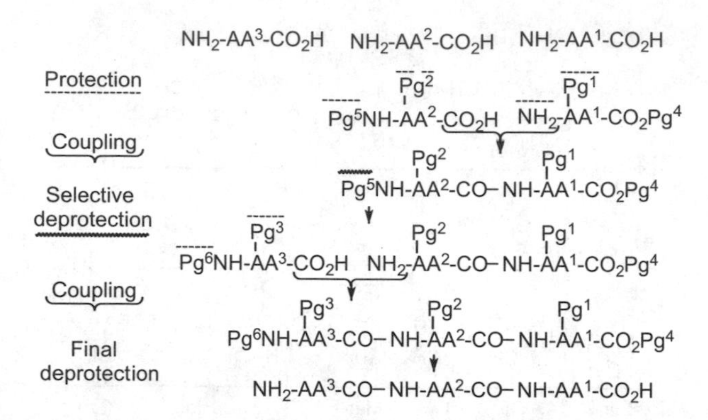

FIGURE 1.5 General approach to protection and amide-bond formation. Pg^1, Pg^2, Pg^3, Pg^4, and Pg^6 may be identical, similar, or different. Pg^5 must be different from Pg^1, Pg^2, and Pg^4. Pg^5 must be removable by a different mechanism (i.e., orthogonal to the other protectors) or be much less stable than the others to the reagent used to remove it.

that is different from that required to deprotect the other groups. In practice, selective deprotection has been accomplished by this approach, as well as by taking advantage of the greater lability to acid of protectors on α-amino groups compared with those on side-chain functional groups. The operations of selective deprotection and coupling are repeated until the desired chain has been assembled. All protecting groups are then removed in one or two steps to give the desired product. In principle, the peptide chain can be assembled starting at the carboxy terminus, with selective deprotection at the amino group, or at the amino terminus, with selective deprotection at the carboxyl group of the growing chain. In either case, the functional groups that are incorporated into the peptide bonds do not participate in subsequent couplings. When two protecting groups require different mechanisms for their removal, they are said to be orthogonal to each other. A set of independent protecting groups, each removable in the presence of the other, in any order, is defined as an orthogonal system. If three different mechanisms are involved in the removal of protecting groups from a peptide, the protectors constitute a tertiary orthogonal system. Some peptides have been synthesized using strategies involving quaternary orthogonal systems.[3]

3. G Barany, RB Merrifield. A new amino protecting group removable by reduction. Chemistry of the dithiasuccinoyl (Dts) function. (orthogonal systems) *J Am Chem Soc* 99, 7363, 1977.

1.6 *N*-ACYL AND URETHANE-FORMING *N*-SUBSTITUENTS

When forming a bond, the nature of the substituent at the carboxyl function of the residue providing the amino group is irrelevant to the reaction; that is, it may be a protector or the nitrogen atom of an amide or peptide bond. In contrast, the nature of the substituent on the amino function of the residue providing the carboxyl group

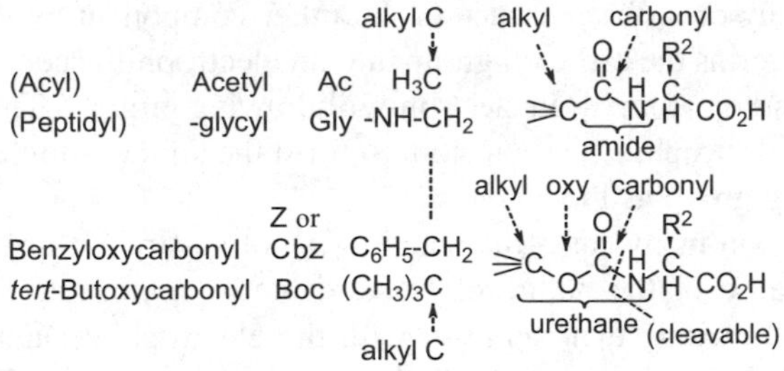

FIGURE 1.6 *N*-Acyl and urethane-forming substituents.

has a dramatic effect on the course of the reaction. Mainly, two types of substituents are at issue. The first is an acyl substituent in which the nitrogen atom is incorporated into an amide bond (Figure 1.6). With rare exceptions, an acyl substituent cannot be removed without affecting the neighboring peptide bond because the sequence of atoms, carbon–carbonyl–nitrogen, is the same as that in a peptide, so acyl substituents are not used as protectors. To introduce reversibility, peptide chemists have inserted an oxygen atom between the alkyl and the carbonyl moieties of the acyl substituent to produce a urethane, in which the *N*-substituent is an alkoxycarbonyl group. Urethanes containing appropriate alkyl groups such as benzyl and *tert*-butyl are readily cleavable at the carbonyl–nitrogen bond, liberating the amino groups. The common alkoxycarbonyl groups are benzyloxycarbonyl (Cbz or Z), *tert*-butoxycarbonyl (Boc), and 9-fluorenylmethoxycarbonyl (Fmoc) (see Section 3.2).

1.7 AMIDE-BOND FORMATION AND THE SIDE REACTION OF OXAZOLONE FORMATION

The two functional groups implicated in a coupling require attention to effect the reaction. The ammonium group of the CO_2H-substituted component must be converted into a nucleophile by deprotonaton (Figure 1.7). This can be done *in situ* by the addition of a tertiary amine to the derivative dissolved in the reaction solvent, or by addition of tertiary amine to the derivative in a two-phase system that allows removal of the salts that are soluble in water. The carboxy-containing component is

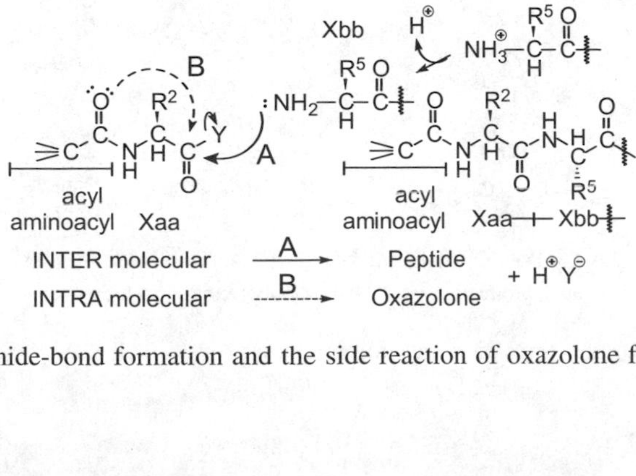

FIGURE 1.7 Amide-bond formation and the side reaction of oxazolone formation.

activated separately or in the presence of the other component by the addition of a reagent that transforms the carboxyl group into an electrophillic center that is created at the carbonyl carbon atom by an electron-withdrawing group Y. The amine nucleophile attacks the electrophilic carbon atom to form the amide, simultaneously expelling the activating group as the anion.

Unfortunately, in many cases the reaction is not so straightforward; it becomes complicated because of the nature of the activated component. There is another nucleophile in the vicinity that can react with the electrophile; namely, the oxygen atom of the carbonyl adjacent to the substituted amino group. This nucleophile competes with the amine nucleophile for the electrophilic center, and when successful, it generates a cyclic compound — the oxazolone. The intermolecular reaction (path A) produces the desired peptide, and the intramolecular reaction (path B) generates the oxazolone. The course of events that follows is dictated by the nature of the atom adjacent to the carbonyl that is implicated in the side reaction.

1.8 OXAZOLONE FORMATION AND NOMENCLATURE

One proton is lost by the activated carboxy component during cyclization to the oxazolone. It is the removal of this proton from the nitrogen atom that initiates the cyclization. Proton abstraction is followed by rearrangement of electrons, shifting the double bond from >C=O to –C=N– with simultaneous attack by the oxygen nucleophile at the electrophilic carbon atom (Figure 1.8). Accordingly, any base that is present promotes cyclization. The nitrogen nucleophile in the coupling is a base, albeit a weak one, so the amino group promotes the side reaction at the same time as it participates in peptide-bond formation. The other component is a good candidate for ring formation because the atoms implicated are separated by the number of atoms required for a five-membered ring. Compounds that have an additional atom separating the pertinent groups such as activated *N*-substituted β-amino acids do not cyclize readily to the corresponding six-membered ring because formation of the latter is energetically less favored.

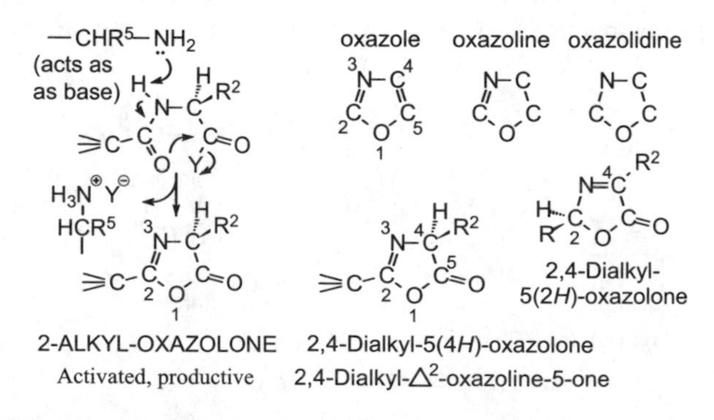

FIGURE 1.8 Oxazolone formation and nomenclature.

The ring compounds in question are internal esters containing a nitrogen atom and were originally referred to as azlactones. They are, in fact, partially reduced oxazoles bearing an oxy group, or more precisely Δ^2-oxazoline-5-ones, with alkyl substituents at positions 2 and 4. The present-day recommended nomenclature is oxazol-5(4*H*)-one or 5(4*H*)-oxazolone, with the parentheses contents indicating the location of the hydrogen atom, and hence the double bond. The alternative structure with the double bond in the 3-position is rare, but it does exist. Such 5(2*H*)-oxazolones are produced when activated *N*-trifluoroacetylamino acids cyclize or when 5(4*H*)-oxazolones from *N*-formylamino acids are left in the presence of tertiary amines. Subsequent discussion relates exclusively to 5(4*H*)-oxazolones.[4,5]

4. F Weygand, A Prox, L Schmidhammer, W König. Gas chromatographic investigation of racemizaton during peptide synthesis. *Angew Chem Int Edn* 2, 183, 1963.
5. FMF Chen, NL Benoiton. 4-Alkyl-5(2*H*)-oxazolones from *N*-formylamino acids. *Int J Pept Prot Res* 38, 285, 1991.

1.9 COUPLING, 2-ALKYL-5(4*H*)-OXAZOLONE FORMATION AND GENERATION OF DIASTEREOISOMERS FROM ACTIVATED PEPTIDES

Aminolysis of the activated component (Figure 1.9, path A) produces the target peptide. The oxazolone (path B) is also an activated form of the substrate, with the same chirality. It undergoes aminolysis at the lactone carbonyl (path E) to produce a peptide with the desired stereochemistry. The stereogenic center of the oxazolone, however, is attached to two double-bonded atoms. Such a bonding arrangement tends to form a conjugated system. The tendency to conjugation is greatest when the carbon atom of the –C=N– is linked to the carbon atom of an aromatic ring, but it is also severe when it is linked to the carbon atom of the *N*-substituent of an activated residue. The ensuing shift of the other double bond or enolization (path G) creates

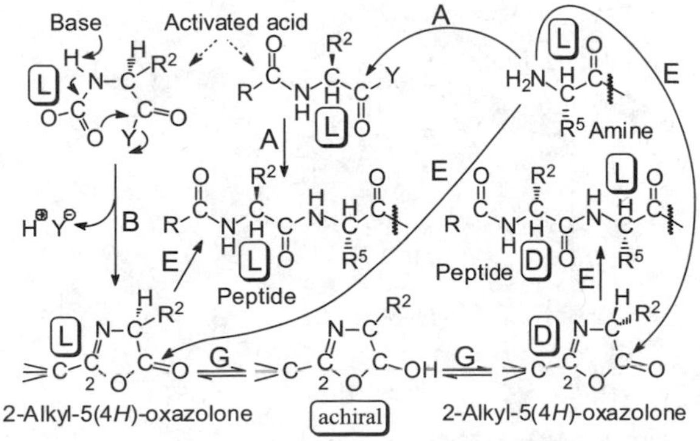

FIGURE 1.9 Coupling, 2-alkyl-5(4*H*)-oxazolone formation and generation of diastereoisomers from activated peptides.

an achiral molecule that has lost its α-proton to the carbonyl function. Reversal of the process (path G), which is promoted by base, generates equal amounts of the two oxazolone enantiomers. Aminolysis of the new isomer produces the undesired diastereoisomer. Thus, the constitution of *N*-acylamino acids and peptides is such that their activation leads to the formation of a productive intermediate, the 2-alkyl-5(4*H*)-oxazolone, that is chirally unstable. The consequence of generation of the 2-alkyl-5(4*H*)-oxazolone is partial enantiomerization of the activated residue, which leads to production of a small or modest amount of epimerized peptide in addition to the desired product.[6–8]

6. M Goodman, KC Stueben. Amino acid active esters. III. Base-catalyzed racemization of peptide active esters. *J Org Chem* 27, 3409, 1962.
7. M Williams, GT Young. Further studies on racemization in peptide synthesis, in GT Young, ed. *Peptides 1962. Proceedings of the 5th European Peptide Symposium*, Pergamon, Oxford, 1963, pp 119-121.
8. I Antanovics, GT Young. Amino-acids and peptides. Part XXV. The mechanism of the base-catalysed racemisation of the *p*-nitrophenyl esters of acylpeptides. *J Chem Soc C* 595, 1967.

1.10 COUPLING OF *N*-ALKOXYCARBONYLAMINO ACIDS WITHOUT GENERATION OF DIASTEREOISOMERS: CHIRALLY STABLE 2-ALKOXY-5(4*H*)-OXAZOLONES

Peptide-bond formation between an *N*-alkoxycarbonylamino acid and an amino-containing component usually proceeds in the same way as described for coupling an *N*-acylamino acid or peptide (see Section 1.9), except for the side reaction (Figure 1.7, path B) of oxazolone formation. Aminolysis of the activated component (Figure 1.10, path A) gives the desired peptide. There are three aspects of the side reaction

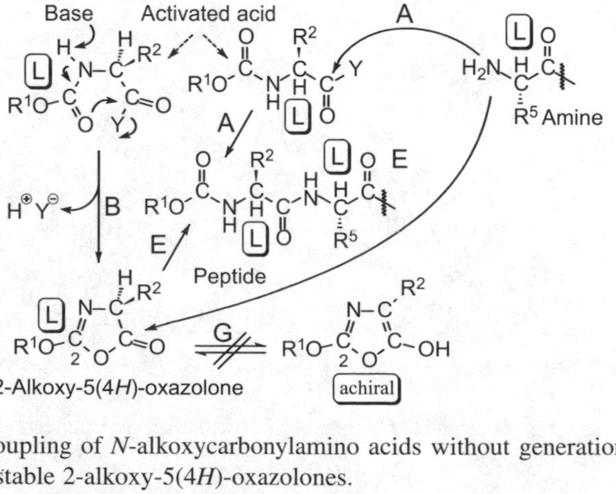

FIGURE 1.10 Coupling of *N*-alkoxycarbonylamino acids without generation of diastereoisomers. Chirally stable 2-alkoxy-5(4*H*)-oxazolones.

that are different, because of the oxygen atom adjacent to the carbonyl group that is implicated in the cyclization. First, the nucleophilicity of the oxygen atom of the carbonyl function has been reduced. The effect is sufficient to suppress cyclization to a large extent, but it is incomplete. 2-Alkoxy-4(5*H*)-oxazolone does form (path B) in some cases. Second, if it does form, it is aminolyzed very quickly (path E) because it is a better electrophile than the 2-alkyl-5(4*H*)-oxazolone. Third, generation of the 2-alkoxy-5(4*H*)-oxazolone is of no consequence because it does not enolize (path G) to give the other enantiomer. It is chirally stable under the usual conditions of operation. So for practical purposes, the situation is the same as if the 2-alkoxy-5(4*H*)-oxazolone did not form. Thus, the constitution of *N*-alkoxycarbonylamino acids is such that their activation and coupling occur without the generation of undesired isomeric forms.

1.11 EFFECTS OF THE NATURE OF THE SUBSTITUENTS ON THE AMINO AND CARBOXYL GROUPS OF THE RESIDUES THAT ARE COUPLED TO PRODUCE A PEPTIDE

When W^a = RC(=O), that is, acyl (Figure 1.11), W^a is not removable without destroying the peptide bond. When W^a = ROC(=O) with the appropriate R, the OC(=O)–NH bond of the urethane is cleavable. When W^b = NHR, W^b is not removable without destroying the peptide bond. When W^b = OR, the O=C–OR bond of the ester is cleavable. During activation and coupling, activated residue Xaa may undergo isomerization, and aminolyzing residue Xbb is not susceptible to isomerization.

When W^a = substituted aminoacyl, that is, when W^a-Xaa is a peptide, there is a strong tendency to form an oxazolone. The 2-alkyl-5(4*H*)-oxazolone that is formed is chirally unstable. Isomerization of the 2-alkyl-5(4*H*)-oxazolone generates diastereomeric products. When W^a = ROC=O, there is a lesser tendency to form an oxazolone. The 2-alkoxy-5(4*H*)-oxazolone that is formed is chirally stable. No isomerization occurs under normal operating conditions. Finally, when W^a = ROC=O, an additional productive intermediate, the symmetrical anhydride, can and often does form.

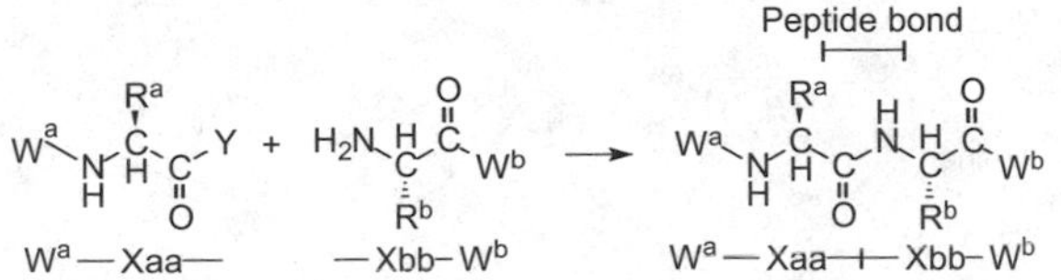

FIGURE 1.11 Effects of the nature of the substituents on the amino and carboxyl groups of the residues that are coupled to produce a peptide.

1.12 INTRODUCTION TO CARBODIIMIDES AND SUBSTITUTED UREAS

Carbodiimides are the most commonly used coupling reagents. Their use frequently gives rise to symmetrical anhyrides, so an examination of their reactions is appropriate at this stage. Previously used in nucleotide synthesis, they were introduced in peptide work by Sheehan and Hess in 1955.[9] Dialkylcarbodiimides (Figure 1.12, designation of the substituents as *N,N′* or 1,3 is superfluous because the structure is unambiguously defined by "carbodiimide") are composed of two alkylamino groups that are joined through double bonds with the same carbon atom. They are in reality dehydrating agents, which abstract a molecule of water from the carboxyl and amino groups of two reactants, with the oxygen atom going to the carbon atom of the carbodiimide, and the hydrogen atoms to the nitrogen atoms, giving an *N,N′*-disubstituted urea (here the designations are required), which is the dialkylamide of carbonic acid. In the process of peptide-bond formation, the carbodiimide serves as a carrier of the acyl group, which may be attached to the nitrogen atom of the urea, giving the *N*-acylurea, or the oxygen atom of the tautomerized or enol form of the urea, giving the *O*-acylisourea. The latter has the double bond at the nitrogen atom, with *O*-substitution necessarily implying the isourea. The most familiar carbodiimide is dicyclohexylcarbodiimide (DCC), which gives rise to the very insoluble *N,N′*-dicyclohexylurea, the *N*-acyl-*N,N′*-dicyclohexylurea, and the *O*-acyl-*N,N′*-dicyclohexylisourea.[9]

9. JC Sheehan, GP Hess. A new method of forming peptide bonds. (carbodiimide) *J Am Chem Soc* 77, 1067, 1955.

1.13 CARBODIIMIDE-MEDIATED REACTIONS OF *N*-ALKOXYCARBONYLAMINO ACIDS

The first step in carbodiimide-mediated reactions of *N*-alkoxycarbonylamino acids is the addition of the reagent to the carboxyl group to give the *O*-acylisourea, which is a transient intermediate (Figure 1.13). The *O*-acylisourea is highly activated, reacting with an amino acid ester (path A) to give dialkylurea and the protected

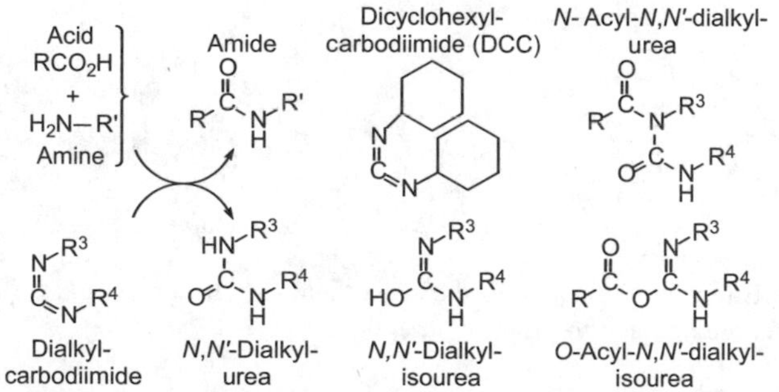

FIGURE 1.12 Introduction to carbodiimides and substituted ureas.[9]

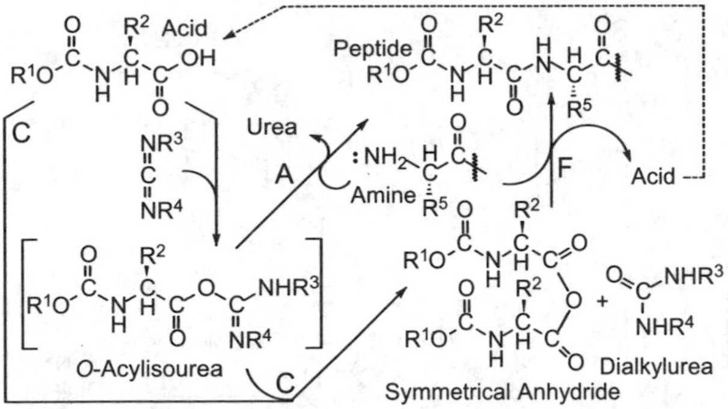

FIGURE 1.13 Carbodiimide-mediated reactions of *N*-alkoxycarbonylamino acids.

dipeptide. The reaction is the same when the nitrogen nucleophile is a peptide. The competing intramolecular cyclization reaction (Figure 1.7, path B) may or may not occur; its occurrence is inconsequential under normal operating conditions. Regardless, another intermolecular reaction may occur as a result of competition between a second molecule of the starting acid and the amino-acid ester for the *O*-acylisourea. The product formed from this reaction (Figure 1.13, path C) contains two *N*-alkoxycarbonylaminoacyl moieties linked to an oxygen atom and is referred to as the symmetrical anhydride. The anhydride is an activated form of the acid that undergoes aminolysis; nucleophilic attack at either carbonyl (path F) gives the desired peptide and an equivalent of the starting acid that is recycled. The stereochemistry is preserved at all stages of the coupling. For practical purposes, whether the symmetrical anhydride is formed or not is immaterial. Thus, there is one necessary intermediate, the *O*-acylisourea, and there are two possible intermediates, the 2-alkoxy-5(4*H*)-oxazolone and the symmetrical anhydride, in carbodiimide-mediated reactions of *N*-alkoxycarbonylamino acids (see Section 2.2 for further details)

1.14 CARBODIIMIDE-MEDIATED REACTIONS OF *N*-ACYLAMINO ACIDS AND PEPTIDES

The first step in carbodiimide-mediated reactions of *N*-acylamino acids and peptides is the same as that for couplings of *N*-alkoxycarbonylamino acids. The *O*-acylisourea is formed (Figure 1.14) and is then aminolyzed to give the peptide (path A). Critical differences arise, however, in terms of the possible side reactions. The competing intramolecular cyclization reaction giving the chirally labile 2-alkyl-5(4*H*)-oxazolone (path B) is much more likely to take place. In fact, the tendency is so strong that vital attention must be devoted to trying to minimize its occurrence. In contrast, the competing intermolecular reaction (path C) giving the symmetrical anhydride of a peptide is not known to occur. The most that can be said is that the latter can exist as transient intermediates; if they do form, they immediately fragment (path H) to the oxazolone and the acid. Thus, in contrast to the case of *N*-alkoxycarbonylamino acids, there is only one necessary intermediate, the *O*-acylisourea, and there is only one possible intermediate, the 2-alkyl-5(4*H*)-oxazolone in

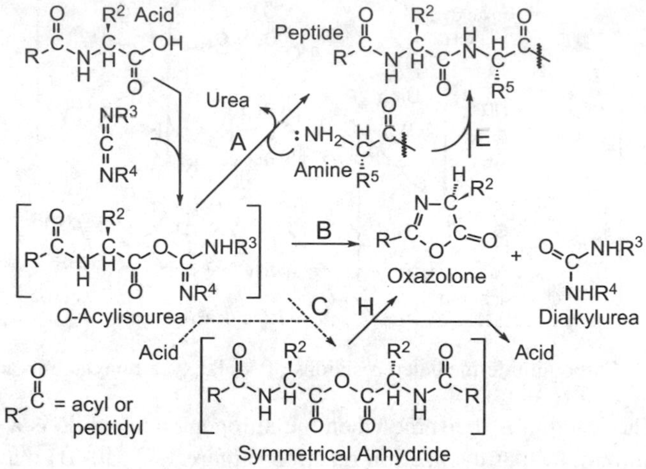

FIGURE 1.14 Carbodiimide-mediated reactions of *N*-acylamino acids and peptides.

carbodiimide-mediated reactions of *N*-acylamino acids and peptides (see Section 2.2 for further details).[10]

10. FMF Chen, NL Benoiton. Do acylamino acid and peptide anhydrides exist? in K Blaha, P Malon, eds. *Peptides 1982. Proceedings of the 17th European Peptide Symposium.* Walter de Gruyter, Berlin, 1983, pp 67-70.

1.15 PREFORMED SYMMETRICAL ANHYDRIDES OF *N*-ALKOXYCARBONYLAMINO ACIDS

A carbodiimide-mediated reaction is usually carried out by adding the coupling reagent to a solution or mixture of the two compounds to be combined. A modified protocol involves addition of the carbodiimide to an *N*-alkoxycarbonylamino acid in the absence of the amino-containing component (Figure 1.15). As a result, the

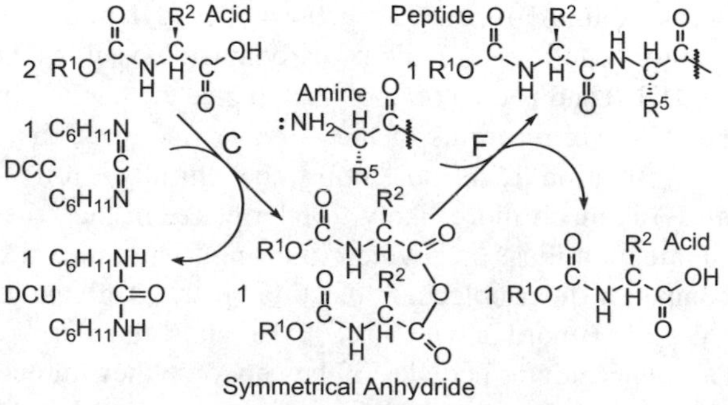

FIGURE 1.15 Preformed symmetrical anhydrides of *N*-alkoxycarbonylamino acids.[13] The reaction is effected in dichloromethane. The *N*,*N*′-dicyclohexylurea is removed by filtration. The symmetrical anhydride is not isolated.

O-acylisourea reacts with the parent acid to give the symmetrical anhydride (path C). Only half of an equivalent of carbodiimde is employed. In principle, this stoichiometry of reactants should force the reaction to completion. The amino-containing component is added several minutes later, sometimes after removal by filtration of the dicyclohexylurea that precipitates if the reagent is dicyclohexylcarbodiimide. The term symmetrical anhydride implies that the parent acid is an *N*-alkoxycarbonylamino acid. This approach to synthesis was introduced in the early 1970s on the basis of knowledge of the properties of symmetrical anhydrides that was gleaned from studies on carbodiimides (see Section 2.2) and mixed anhydrides (see Section 2.5). It obviously follows that the use of preformed symmetrical anhydrides is applicable to synthesis by single-residue addition only (see Section 2.4 for further details).[11–13]

11. H Schüssler, H Zahn. Contribution on the course of reaction of carbobenzoxyamino acids with dicyclohexylcarbodimide. *Chem Ber* 95, 1076, 1962.
12. F Weygand, P Huber, K Weiss. Peptide synthesis with symmetrical anhydrides I. *Z Naturforsch* 22B, 1084, 1967.
13. H Hagenmeier, H Frank. Increased coupling yields in solid phase peptide synthesis with a modified carbodiimide coupling procedure. *Hoppe-Seyler's Z Physiol Chem* 353, 1973, 1972.

1.16 PURIFIED SYMMETRICAL ANHYDRIDES OF *N*-ALKOXYCARBONYLAMINO ACIDS OBTAINED USING A SOLUBLE CARBODIIMIDE

The alternative approach to synthesis by incremental addition using carbodiimides (see Section 1.13) is preparation of the *N*-alkoxycarbonylamino-acid anhydride (see Section 1.15) in dichloromethane, followed by admixture of the anhydride with the amine nucleophile after removal of the dicyclohexylurea by filtration. Filtration of the mixture does not, however, remove all the dialkylurea, and some *N*-acylurea (see Section 1.12) may remain (see Section 2.2). A simple variant giving access to symmetrical anhydride that is free from contaminants is the use of a soluble carbodiimide (Figure 1.16). In this case, soluble means that both the reagent and the substituted ureas produced by its reaction are soluble in water. The latter are soluble because one of the alkyl groups of the carbodiimide is a dialkylaminoalkyl group that is positively charged at neutral and lower pHs. The common soluble carbodiimide is ethyl-(3-dimethylaminopropyl)-carbodiimide hydrochloride (EDC), which has been available since 1961. Its use instead of DCC, followed by washing the solution of the anhydride in dichloromethane with water, yields a solution that is free of substituted-urea contaminants. The idea of using a soluble carbodiimide to prepare purified symmetrical anhydrides (1978) issued from the observation that a symmetrical anhydride in solution had been stable enough to survive washing with aqueous solutions. Pure symmetrical anhydrides of Boc-, *Z*-, and Fmoc-amino acids (see Section 3.2) are obtainable by this procedure, but they are not completely stable on storage. This fact, and the additional effort required to secure these anhydrides,

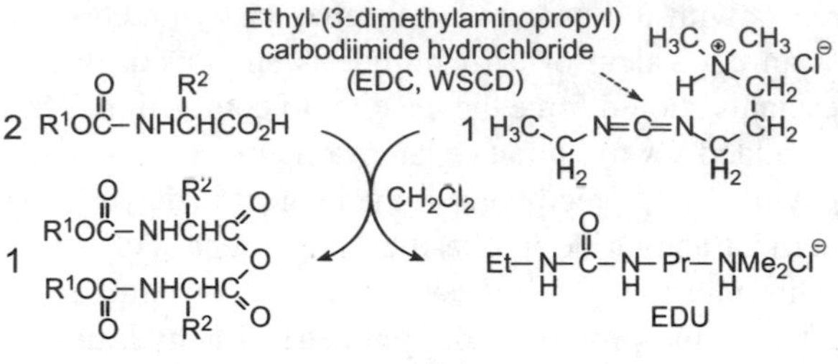

FIGURE 1.16 Purified symmetrical anhydrides of *N*-alkoxycarbonylamino acids obtained using a soluble carbodiimide.[15] The reagent ethyl-(3-dimethylaminopropyl)-carbodiimide hydrochloride,[14] also known as WSCD (water-soluble carbodiimide), the *N,N'*-dialkylurea, and the *N*-acyl-*N,N'*-dialkylurea are soluble in water and thus can be removed from a reaction mixture by washing it with water.

combined with the higher cost of EDC relative to DCC, have diminished the appeal of purified symmetrical anhydrides.[14–17]

14. JC Sheehan, PA Cruickshank, GL Boshart. Convenient synthesis of water-soluble carbodiimides. *J Org Chem* 26, 2525, 1961.
15. FMF Chen, K Kuroda, NL Benoiton. A simple preparation of symmetrical anhydrides of *N*-alkoxycarbonylamino acids. *Synthesis* 928, 1978.
16. EJ Heimer, C Chang, T Lambros, J Meienhofer. Stable isolated symmetrical anhydrides of Nα-9-fluorenylmethyloxycarbonylamino acids in solid-phase peptide synthesis. *Int J Pept Prot Res* 18, 237, 1981.
17. D Yamashiro. Preparation and properties of some crystalline symmetrical anhydrides of N^{α}*tert.*-butyloxycarbonyl-amino acids. *Int J Pept Prot Res* 30, 9, 1987.

1.17 PURIFIED 2-ALKYL-5(4*H*)-OXAZOLONES FROM *N*-ACYLAMINO AND *N*-PROTECTED GLYCYLAMINO ACIDS

Reaction of an *N*-acylamino acid or peptide with a carbodiimide gives the very reactive *O*-acylisourea, which has an inherent tendency to cyclize to the 2-alkyl-5(4*H*)-oxazolone (Figure 1.14, path B). When there is no amine nucleophile present, as generation of the symmetrical anhydride is not pertinent (see Section 1.14), the product is the 2-alkyl-5(4*H*)-oxazolone. Solutions of chemically and enantiomerically pure oxazolones can be obtained by use of the soluble carbodiimide EDC in dichloromethane, followed by removal of water-soluble components by aqueous extraction (Figure 1.17). Elimination of the solvent gives the pure oxazolones. Products generated using DCC require purification by distillation or recrystallization, which results in major losses and partial isomerization. Oxazolones obtained using EDC are prevented from enolizing by the acidic nature of the reagent. With rare exceptions, 2-alkyl-5(4*H*)-oxazolones are of little use for synthesis, but they have been valuable for research purposes. Knowledge of their properties has contributed to our understanding of the side reaction of epimerization that occurs during coupling.[18,19]

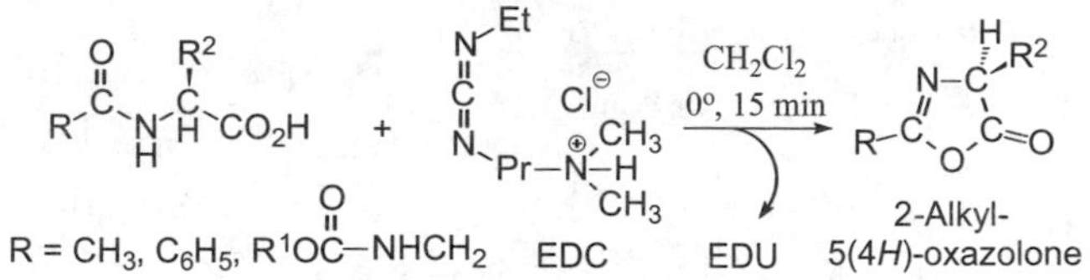

$$R\text{-}C(=O)\text{-}NH\text{-}CH(R^2)\text{-}CO_2H + EDC \xrightarrow[0°,\ 15\ min]{CH_2Cl_2} \text{2-Alkyl-5(4H)-oxazolone} + EDU$$

R = CH_3, C_6H_5, $R^1OC(=O)\text{-}NHCH_2$

FIGURE 1.17 Purified 2-alkyl-5(4*H*)-oxazolones from *N*-acyl and *N*-protected glycylamino acids.[19] The reaction mixture is washed with cold aqueous $NaHCO_3$, after which the dried solvent is removed by evaporation. 2-Alkyl-5(4*H*)-oxazolones had been identified in the 1960s in the laboratories of Goodman in San Diego and Simeon in Wroclaw and Young in Cambridge. The use of ethyl-(3-dimethylaminopropyl)-carbodiimide hydrochloride is the only general method of synthesis that gives enantiomerically pure 2-alkyl-5(4*H*)-oxazolones. The slight acidity of the soluble carbodiimide is sufficient to prevent the oxazolone from tautomerizing.

18. IZ Siemion, K Nowak. New method of synthesis of 2-phenyl-4-alkyl-oxazolones-5. *Rocz Chem* 43, 1479, 1960.
19. FMF Chen, K Kuroda, NL Benoiton. A simple preparation of 5-oxo-4,5-dihydro-1,3-oxazoles (oxazolones). *Synthesis* 230, 1979.

1.18 2-ALKOXY-5(4*H*)-OXAZOLONES AS INTERMEDIATES IN REACTIONS OF *N*-ALKOXYCARBONYLAMINO ACIDS

The reaction of Boc-amino acids with half an equivalent of EDC (see Section 1.16) gives good yields of symmetrical anhydride except for $(Boc\text{-}Val)_2O$ and $(Boc\text{-}Ile)_2O$, for which the yields are 50–60%. This is consistent with the lesser reactivity of β-methylamino acids (see Section 1.4). $(Boc\text{-}Val)_2O$ is crystalline and insoluble in petroleum ether. In an experiment with Boc-valine in 1980 (Figure 1.18), after collection of the anhydride by filtration and evaporation of the solvent, Chen and Benoiton found as residue an activated form of Boc-valine that appeared to be a new compound. Its infrared spectrum showed the absence of the N–H and carbonyl bands of a urethane and instead the two bands characteristic of a 5(4*H*)-oxazolone. Its nuclear magnetic resonance profile showed a sharp doublet for the Cα-proton instead of the two overlapping doublets that are seen for this proton in the spectrum of *N*-substituted-valine derivatives, and a singlet for the *tert*-butoxy protons that was shifted 0.13 ppm downfield from that of the protons of a *tert*-butoxycarbonyl or *tert*-butyl ester group. The product was 2-*tert*-butoxy-4-isopropyl-5(4*H*)-oxazolone — the oxazolone from Boc-valine. The same compound was isolated in 7% yield from an EDC-mediated reaction of Boc-valine with an amino-acid ester that had been terminated after 3 minutes. This was the first demonstration of the formation of a 2-alkoxy-5(4*H*)-oxazolone during the coupling of an *N*-alkoxycarbonylamino acid. The existence of 2-alkoxy-5(4*H*)-oxazolones had been established a few years earlier. They had been identified as the products of the reaction of triethylamine with *N*-benzyloxycarbonylamino acids previously treated with acid halide-forming reagents. Beforehand, it had been believed that cyclization of an activated

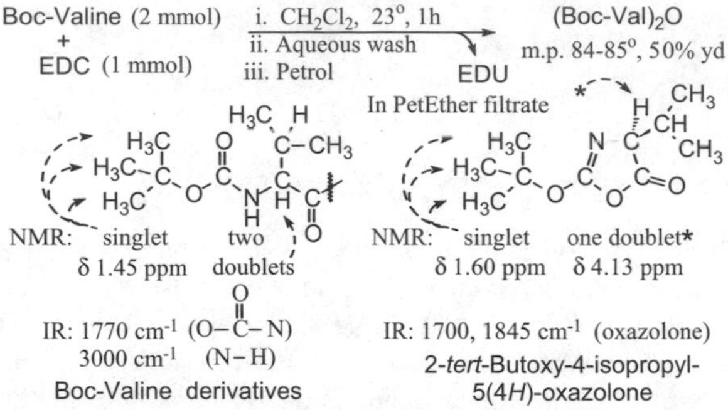

Boc-Valine (2 mmol) + EDC (1 mmol) — i. CH_2Cl_2, 23°, 1h; ii. Aqueous wash; iii. Petrol → (Boc-Val)$_2$O m.p. 84-85°, 50% yd; EDU

In PetEther filtrate

NMR: singlet δ 1.45 ppm; two doublets

IR: 1770 cm^{-1} (O–C(=O)–N); 3000 cm^{-1} (N–H)

Boc-Valine derivatives

NMR: singlet δ 1.60 ppm; one doublet* δ 4.13 ppm

IR: 1700, 1845 cm^{-1} (oxazolone)

2-*tert*-Butoxy-4-isopropyl-5(4*H*)-oxazolone

FIGURE 1.18 2-Alkoxy-5(4*H*)-oxazolones as intermediates in reactions of *N*-alkoxycarbonylamino acids.[22] After removal of the symmetrical anhydride from a reaction mixture containing Boc-valine and ethyl-(3-dimethylaminopropyl)-carbodiimide hydrochloride, the filtrate contained a novel activated form of Boc-valine (20% yield) that was established to be the 2-alkoxy-5(4*H*)-oxazolone. Slow addition of Boc-valine to ethyl-(3-dimethylamino-propyl)-carbodiimide hydrochloride in dilute solution gave a 55% yield. Petrol = petroleum ether, bp 40–60°.

N-alkoxycarbonylamino acid (Figure 1.10, path B) did not occur without immediate expulsion of the alkyl group, giving the amino-acid *N*-carboxyanhydride (see Section 7.13). 2-Alkoxy-5(4*H*)-oxazolones are now recognized as intermediates in coupling reactions and are products that are generated by the action of tertiary amines on activated *N*-alkoxycarbonylamino acids (see Section 4.16).[20–22]

20. M Miyoshi. Peptide synthesis via *N*-acylated aziridinone. I. The synthesis of 3-substituted-1-benzylcarbonylaziridin-2-ones and related compounds. *Bull Chem Soc Jpn* 46, 212, 1973.
21. JH Jones, MJ Witty. The formation of 2-benzyloxyoxazol-5(4*H*)-ones from benzyloxycarbonylamino-acids. *J Chem Soc Perkin Trans 1* 3203, 1979.
22. NL Benoiton, FMF Chen. 2-Alkoxy-5(4*H*)-oxazolones from *N*-alkoxycarbonylamino acids and their implication in carbodiimide-mediated reactions in peptide synthesis. *Can J Chem* 59, 384, 1981.

1.19 REVISION OF THE CENTRAL TENET OF PEPTIDE SYNTHESIS

Experience with synthesis over several decades revealed that enantiomerically pure peptides could be assembled by the successive addition of single residues as the *N*-alkoxycarbonylamino acids and that products constructed by combining two peptides often were not chirally pure. Concurrent developments in the understanding of the chemistry of coupling reactions led to the conclusion that the epimerization that occurred in the latter cases resulted from the formation of the 2-alkyl-5(4*H*)-oxazolones. Oxazolones from *N*-alkoxycarbonylamino acids had not yet been detected, so there emerged a simple rationalization of the question of loss or retention of chiral integrity during coupling that became accepted as a central tenet of peptide synthesis. With this explanation, it was easy for a novice to grasp the rationale underlying the strategies recommended for the successful synthesis of peptides. The reasoning was as follows: First, stereoisomerization can and does occur when an *N*-acylamino acid or peptide is coupled; second, the isomerization results because of formation of the 5(4*H*)-oxazolone; third, no isomerization occurs when Boc- and Z-amino acids are coupled; and fourth, therefore Boc- and Z-amino acids do not isomerize because they don't form 5(4*H*)-oxazolones.

The issue turned out to be much more complicated, however. The conclusion arrived at was demonstrated to be false by the discovery that 2-alkoxy-5(4*H*)-oxazolones exist and are intermediates in coupling reactions (see Section 1.18). It transpired that the reason why Boc- and Z-amino acids do not enantiomerize during coupling is that the 2-alkoxy-5(4*H*)-oxazolones are not chirally labile under normal operating conditions. So the fourth point above had to be revised to: Isomerization does not occur because the oxazolone formed does not isomerize. The new information initially seemed of little practical consequence, but a very disturbing fact emerged. It was realized that there is a danger of isomerization when *N*-alkoxycarbonylamino acids are aminolyzed in the presence of a strong base (see Section 4.17). It will remain intriguing for a long time as to why the erroneous deduction had not been challenged previously, as it is so obvious now that it was fallacious.

1.20 STRATEGIES FOR THE SYNTHESIS OF ENANTIOMERICALLY PURE PEPTIDES

A peptide is constructed by coupling protected amino acids followed by selective deprotection and repetition of these operations (see Section 1.5). An additional critical feature in addition to selectivity in deprotection for successful synthesis is preservation of the chirality of the amino acid residues. This is achieved by employing *N*-alkoxycarbonylamino acids that do not isomerize during aminolysis (see Section 1.10), which implies beginning chain assembly at the carboxy terminus of the peptide (Figure 1.19, right-hand side). This is the only way that peptides are constructed. They are never constructed starting from the amino terminus of the peptide (Figure 1.19, left-hand side) because there is danger of epimerization at the activated residue for every coupling (see Section 1.9) except the first one. A further option is available if the target peptide contains glycine or proline. Glycine is not a

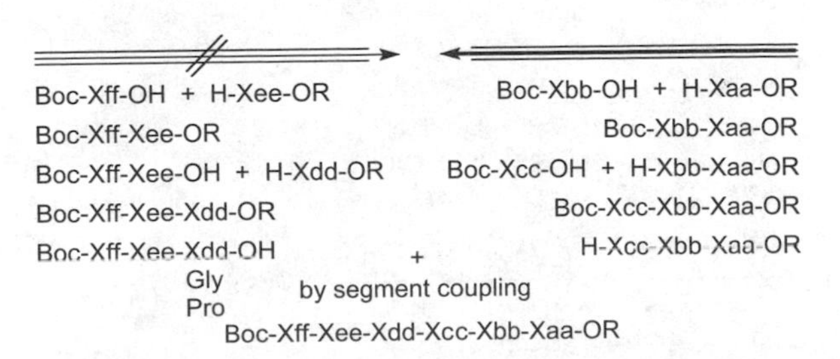

FIGURE 1.19 Strategies for the synthesis of enantiomerically pure peptides. Peptides are always synthesized starting from the carboxy-terminal residue.

chiral amino acid, so activation of a peptide segment at glycyl cannot lead to epimerized products by the oxazolone mechanism (see Section 1.9). Proline is a cyclic amino acid that resists the tendency to form an oxazolone because it would involve two contiguous rings, so activation of a peptide segment at prolyl does not lead to epimerized products. Peptides are, therefore, assembled by single residue addition, starting from the carboxy terminus, using *N*-alkoxycarbonylamino acids. This is complemented by segment coupling at glycyl or prolyl, depending on the constitution of the peptide and the technology employed. When a peptide is constructed by the coupling of segments, the process is referred to as convergent synthesis.

1.21 ABBREVIATED DESIGNATIONS OF SUBSTITUTED AMINO ACIDS AND PEPTIDES

For the purpose of facilitating and simplifying communication, committees of scientists have devised abbreviated designations, the use of which is recommended for representing the structures of derivatized amino acids and peptides. Adherence to the recommendations guarantees unambiguity and quicker understanding by the viewer. Amino acid residues are represented by three-letter abbreviations or symbols, in most cases the first three letters of the name of the amino acid (Figure 1.20). Exceptions are the use of Trp and not Try for tryptophan to avoid confusion with Tyr for tyrosine, and Asn and Gln for asparagine and glutamine to distinguish them from the parent acids. The first letter only of the abbreviation is in upper case. Unspecified residues are indicated by Xaa, Xbb, and so forth. A dash appears at each side of the symbol. The dash to the left indicates removal of H from the α-amino group, and the dash to the right indicates removal of OH from the α-carboxyl group. Protecting groups are placed next to the dashes, indicating their location. An

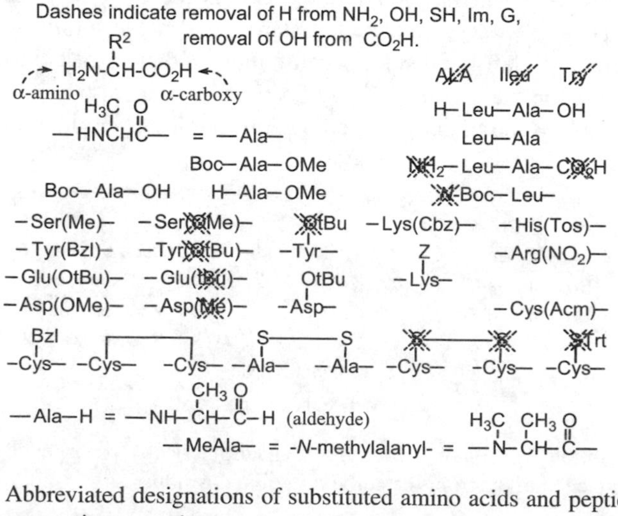

Dashes indicate removal of H from NH_2, OH, SH, Im, G,
removal of OH from CO_2H.
H_2N-CH(R^2)-CO_2H (α-amino, α-carboxy)
[ALA] [Ileu] [Try]
—HNCH(CH_3)C(=O)— = —Ala—
H—Leu—Ala—OH
Leu—Ala
Boc—Ala—OMe
[NH_2]—Leu—Ala—[CO_2H]
Boc—Ala—OH H—Ala—OMe
[N]-Boc—Leu—
—Ser(Me)— —Ser[O](Me)— [O]tBu —Lys(Cbz)— —His(Tos)—
—Tyr(Bzl)— —Tyr[O](tBu)— —Tyr— Z —Arg(NO_2)—
—Glu(OtBu)— —Glu([tBu])— OtBu —Lys—
—Asp(OMe)— —Asp([Me])— —Asp— —Cys(Acm)—
Bzl S——S [S]——[S] [S]Trt
—Cys— —Cys— —Cys— —Ala— —Ala— —Cys— —Cys— —Cys—
—Ala—H = —NH—CH(CH_3)—C(=O)—H (aldehyde)
—MeAla— = -*N*-methylalanyl- = —N(CH_3)—CH(CH_3)—C(=O)—

FIGURE 1.20 Abbreviated designations of substituted amino acids and peptides. Examples of incorrect representations are given.

esterified carboxyl group is indicated by OR. Unsubstituted amino and carboxyl groups are indicated by an H to the left and an OH to the right of the symbols, respectively. All functional groups of the residue are implied in the symbol, so NH_2 or CO_2H should not be added to the abbreviation. It follows that the symbols alone are not meant to be used to represent underivatized amino acids. In cases in which the focus is not on synthetic considerations, the H and OH indicating the terminal groups of peptides may be omitted for the sake of simplicity. For convenience, D-residues may be indicated in lowercase (pro, ala) or with the prefix in italics without the space (*D*Pro).

Substitution on the side chain is indicated by a vertical dash above or below the symbol, or in parentheses to the right, and must be consistent with the above. Alkoxy groups of ω-esters of glutamic and aspartic acids must appear as OR and not R. All other side-chain substituents must appear as tBu, Cbz, and so on, without the substituted atom, and not as OtBu, SBzl, NCbz, and so forth. Consistent with this is correct designation of the disulfide linkage between two cysteine residues by a line and not by –S–S–. An N^{α}-methyl substituent should appear as Me before and adjacent to the symbol without a dash or in parentheses, and without the *N* that is implied. Adherence to the rules provides for a presentation whose meaning is unambiguous and easy to grasp.[23]

23. IUPAC-IUB Joint Commission on Biochemical Nomenclature (JCBN) Nomenclature and Symbolism for Amino Acids and Peptides Recommendations 1983. *Eur J Biochem* 138, 9, 1984.

1.22 LITERATURE ON PEPTIDE SYNTHESIS

The best way to keep abreast of developments in peptide synthesis is to consult the proceedings of the annual symposia of the two major peptide societies. The symposia are held in alternate years — the European Peptide Symposia are held in the even-numbered years, and the American Peptide Symposia are held in the odd-numbered years. The European Peptide Symposia proceedings bear the name Peptides 19xx or 200x, and the American Peptide Symposia proceedings bear the name Peptides followed by various qualifications. Peptide symposia have been held in Japan for many years, but the proceedings began to appear in English only in recent years. Full papers on peptide synthesis appear in organic chemistry journals and journals dedicated to peptide research. A section of chemical abstracts on amino acids, peptides, and proteins is available separately as CA Selects. An annual summary of progress in the field is published by The Royal Society of Chemistry (UK). These publications and the books available on peptide synthesis are catalogued below.

Amino Acids, Peptides, and Proteins. Specialist Periodical Reports:
Royal Society of Chemistry (UK). Vol 33 (2002), literature of 2000; Vol 34 (2004), literature of 2001

Chemical Abstracts Selects: Amino Acids, Peptides, and Proteins.

Dedicated Journals:

Biopolymers (*Peptide Science*) (1995, M. Goodman; 1999, C.M. Deber; 2004, L. Gierash, Eds.), official journal of the American Peptide Society, 2003–.

Journal of Peptide Research (1997, V.J. Hruby, Ed.), created by merger of *International Journal of Peptide and Protein Research* (1973, C.H. Li; 1988, V.J. Hruby, Eds.) and *Peptide Research* (1988, R.A. Houghten, Ed.), official journal of the American Peptide Society –2003.

Journal of Peptide Science (1995, C.H. Schneider; 1999, J. Jones, Eds.), official journal of the European Peptide Society.

Organic Chemistry Journals:

Angew. Chem. Intl. Edn. Engl.; *Chem. Commun.*; *Eur. J. Org. Chem.*, created (1998) by merger of several journals; *J. Am. Chem. Soc.*; *J. Org. Chem.*; *Org. Biomol. Chem.*, created (2003) by merger of *J. Chem. Soc. Perkin Trans.* 1 and 2; *Org. Lett.*; *Synthesis*; *Tetrahedron Lett.*

Proceedings of the American Peptide Symposia:

APS-1 (1968) through APS-17 (2001), APS-18 (2003), APS-19 (2005).

Proceedings of the European Peptide Symposia:

EPS-1 (1958); through EPS-27, Peptides 2002; EPS-28, Peptides 2004.

Proceedings of the Japanese Peptide Symposia:

JPS-34, Peptide Chemistry 1996; JPS-36, Peptide Science, 1999; JPS-40, Peptide Science 2003.

Proceedings of the International Peptide Symposia:

IPS-1 (JPS) (1997); IPS-2 (APS-17) (2001); IPS-3 (EPS-28) (2004).

Atherton, E. and Sheppard, R.C. (1989) Solid Phase Peptide Synthesis, a Practical Approach. 203pp. IPR Press, UK. A working handbook focussing on polyacrylamide resins and Fmoc-chemistry.

Bodanszky, M. (1993) Principles of Peptide Synthesis. 319pp. Springer-Verlag. An authoritative detailed account with >700 references.

Bodanszky, M. (1990) Peptide Chemistry, a Practical Textbook. 198pp. Springer-Verlag. Recipes of procedures.

Chan, W.C. and White, P.D. (Eds.) (2000) Fmoc Solid Phase Peptide Synthesis. A Practical Approach. 368pp. Oxford University Press. Essential procedures and advanced techniques.

Fields, G.B. (Ed.) (1997) Methods in Enzymology. Vol 289. Solid-phase Peptide Synthesis. 710pp. Academic Press. Includes analytical techniques.

Goodman, M., Felix, A., Moroder, M and Toniolo, C. (Eds.) (2002). Houben-Weyl Methods of Organic Chemistry, Vol E22, Synthesis of Peptides and peptidomimetics. Vol E22a, The synthesis of peptides, 901pp. Goerg Thieme Verlag Methods with experimental procedures.

Greenstein, J.P. and Winitz, M. (1961) Chemistry of the Amino Acids. pp 763-1295. John Wiley and Sons.

Gross, J. and Meienhofer, J. (Eds.) (1979-1983) The Peptides: Analysis, Synthesis, Biology. Vols 1-5, 9. Academic Press. Advanced-level reviews.

Jones, J. (1994) The Chemical Synthesis of Peptides. 230pp. Clarenden Press, Oxford. An authoritative account of peptide synthesis.

Kates, S.A. and Albericio, F. (2000) Solid-Phase Synthesis A Practical Guide. 848pp. Marcel Dekker, Inc. Reviews by various authors, >2400 references.

Lloyd-Williams, P., Albericio, F. and Giralt, E. (1997) Chemical Approaches to the Synthesis of Peptides and Proteins. 367pp. CRC Press. The main focus is on large molecules, with 1343 references.

Pennington, M.W. and Dunn, B.M. (Eds.) (1994) Peptide Synthesis Protocols. 350pp. Humana Press. (Also Peptide Analysis Protocols).

Sewald, N. and Jakubke, H-D. (2000) Peptides: Chemistry and Biology. 450pp. Wiley. An overview for newcomers to the field.

Stewart, J.M. and Young, J.D. (1984) Solid Phase Peptide Synthesis, 2nd ed. 184pp. Pierce Chemical Company. The classical working handbook.

Wieland, T., and Bodanszky, M. (1991) The world of peptides: a brief history of peptide chemistry. 298pp. Springer-Verlag.

Wünsch, E. (1974) in Houben-Weyl, Methoden der Organische Chemie, 15/1, 15/2. Synthesen von Peptiden. 1812pp. Georg Thieme Verlag. A two-volume dictionary of methods and compounds, in German.

2 Methods for the Formation of Peptide Bonds

2.1 COUPLING REAGENTS AND METHODS AND ACTIVATED FORMS

The procedures (see Section 1.7) used to combine two amino acid residues to form a peptide are referred to as coupling methods. Coupling involves nucleophilic attack by the amino group of one residue at the electrophilic carbonyl carbon atom of the carboxy-containing component that has been activated by the introduction of an electron-withdrawing group Y. Activation may be carried out either in the presence of the *N*-nucleophile or in the absence of the *N*-nucleophile, which may be by choice or by necessity. Activation in the absence of the nucleophile is referred to as preactivation. When a coupling is effected by the addition of a single compound to a mixture of the two reactants, the compound is referred to as a coupling reagent. In some cases, the coupling reagent requires a subsequent deprotonation of one of the reactants to effect the reaction. The common activated forms of the acid appear in Figure 2.1 in the order of increasing complexity, which also corresponds — with the exception of the mixed anhydride — to the order in which the methods became available. The activating moiety Y is composed of either a halide atom or an azide group or an oxygen atom linked to a double-bonded carbon atom (O–C=), a cationic

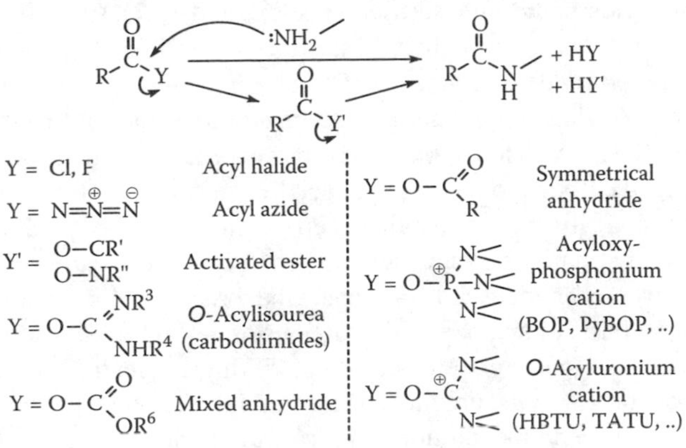

$R-C(=O)-Y$ + $:NH_2$ → $R-C(=O)-NH$ + HY
$R-C(=O)-Y$ → $R-C(=O)-Y'$ → $R-C(=O)-NH$ + HY'

Y = Cl, F — Acyl halide
Y = $N{=}\overset{\oplus}{N}{=}\overset{\ominus}{N}$ — Acyl azide
Y' = O–CR' / O–NR" — Activated ester
Y = O–C(=NR^3)NHR^4 — *O*-Acylisourea (carbodiimides)
Y = O–C(=O)OR^6 — Mixed anhydride
Y = O–C(=O)R — Symmetrical anhydride
Y = O–P$^{\oplus}$(N<)$_3$ — Acyloxy-phosphonium cation (BOP, PyBOP, ..)
Y = O–C$^{\oplus}$(N<)$_2$ — *O*-Acyluronium cation (HBTU, TATU, ..)

FIGURE 2.1 Coupling methods and activated forms.

carbon (O–C$^+$) or phosphorus (O–P$^+$) atom, or a nitrogen atom adjacent to a double-bonded atom (O–N–X=). Some activated forms are much more stable than others. Three different types can be distinguished. The activated form may be a shelf-stable reagent such as an activated ester, a compound of intermediate stability such as an acyl halide or azide or a mixed or symmetrical anhydride that may or may not be isolated, or a transient intermediate, indicated in Figure 2.1 by brackets, that is neither isolable nor detectable. The latter immediately undergoes aminolysis to give the peptide, or it may react with a second nucleophile that originates from the reactants or was added for the purpose, to give the more stable activated ester or symmetrical anhydride R–C(=O)–Y′, whose aminolysis then generates the peptide.

It is important to remember that there are two different types of acyl groups involved in couplings: those originating from an *N*-alkoxycarbonylamino acid and those originating from a peptide. All coupling reagents and methods are applicable to the coupling of *N*-protected amino acids, but not all are applicable to the coupling of peptides. Some methods such as the acyl halide and symmetrical anhydride methods cannot be used for coupling peptides. In addition, the protocols used for coupling may not be the same for the two types of substrates. For these and other reasons, the methods are discussed first in relation to peptide-bond formation from *N*-alkoxycarbonylamino acids (Sections 2.1–2.21). Peptide-bond formation from activated peptides is then addressed separately. In addition, the methods are presented not in the order given in Figure 2.1 but roughly in the order of frequency of usage.

2.2 PEPTIDE-BOND FORMATION FROM CARBODIIMIDE-MEDIATED REACTIONS OF *N*-ALKOXYCARBONYLAMINO ACIDS

The most popular method of forming peptide bonds is the carbodiimide method, using dicyclohexylcarbodiimide (see Sections 1.12 and 1.13). Carbodiimides contain two nitrogen atoms that are slightly basic; this is sufficient to trigger a reaction between the carbodiimide and an acid. The base removes a proton from the acid, generating a carboxylate anion and a quaternized nitrogen atom bearing a positive charge (Figure 2.2). Delocalization of the protonated form to a molecule with a positively charged carbon atom induces attack by the carboxylate on the carbocation generating the *O*-acylisourea. The first step is thus a carboxy-addition reaction, initiated by protonation. The *O*-acylisourea from an *N*-alkoxycarbonylamino acid or peptide has never been detected — hence the brackets. Its existence has been postulated on the basis of analogy with reactions of carbodiimides with phenols that give *O*-alkylisoureas that are well-known esterifying reagents. The normal course of events is for the *O*-acylisourea to undergo aminolysis to give the peptide (path A). However, under certain conditions, some of the *O*-acylisourea undergoes attack by a second molecule of the acid to give the symmetrical anhydride (path C; see Section 2.5). The latter is then aminolyzed to give the peptide and an equivalent of the acid (path F) that is recycled. A third option is that some *O*-acylisourea cyclizes to the oxazolone (see Section 1.10, path B; not shown in Figure 2.2; see Section 2.4) that also gives peptide by aminolysis. Regardless of the path by which the

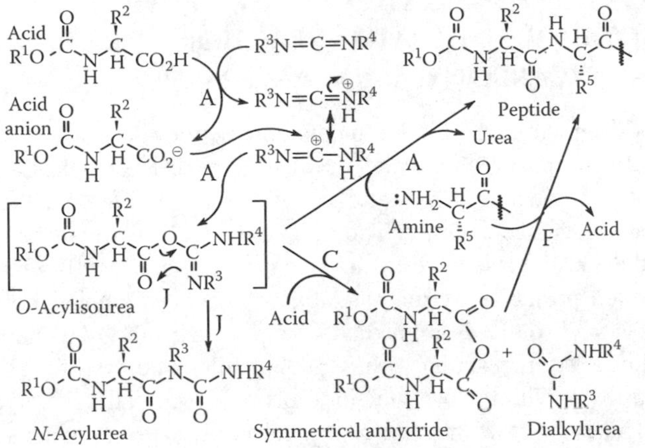

FIGURE 2.2 Peptide-bond formation from carbodiimide-mediated reactions of *N*-alkoxycarbonylamino acids (see Section 1.13).

O-acylisourea generates peptide, the theoretical yield of peptide is one equivalent and one equivalent of *N,N'*-dialkylurea is liberated. However, a fourth and undesirable course of action is possible because of the nature the *O*-acylisourea. The latter contains a basic nitrogen atom ($C{=}NR^3$) in proximity to the activated carbonyl. This atom can act as a nucleophile, giving rise to a rearrangement (path J) that produces the *N*-acylurea (see Section 1.12) that is a stable inert form of the acid. This reaction is irreversible and consumes starting acid without generating peptide. The exact fate of the *O*-acylisourea in any synthesis depends on a multitude of factors; this is addressed in Section 2.3.

A copious precipitate of *N,N'*-dicyclohexylurea separates within a few minutes in any reaction using dicyclohexylcarbodiimide. This allows for its removal; however, it and the *N*-acyl-*N,N'*-dicyclohexylurea are partially soluble in organic solvents used for synthesis and insoluble in aqueous solutions, so they are not easy to remove completely from the products of a reaction. In fact, removal of final traces of these secondary products is often extremely frustrating. It is for this reason that soluble carbodiimides (see Section 1.16) were introduced. The corresponding ureas from these are soluble in aqueous acid, and their removal from a product can be achieved simply by washing a water-immiscible solution of the compounds with aqueous acid. *N,N'*-Dicyclohexylurea presents a problem in solid-phase synthesis because it cannot be removed by filtration. This has led to its replacement by diisopropylcarbodiimide, which gives a urea that is soluble in organic solvents (see Sections 5.14 and 7.1). The reaction of a carbodiimide with a carboxylic acid begins by protonation. This is the first of many reactions to be encountered that are initiated by protonation or deprotonation.[1,2]

1. JC Sheehan, GP Hess. A new method of forming peptide bonds. *J Am Chem Soc* 77, 1067, 1955.
2. NL Benoiton, FMF Chen. Not the alkoxycarbonylamino-acid *O*-acylisourea. *J Chem Soc Chem Commun* 543, 1981.

2.3 FACTORS AFFECTING THE COURSE OF EVENTS IN CARBODIIMIDE-MEDIATED REACTIONS OF *N*-ALKOXYCARBONYLAMINO ACIDS

The course of events in carbodiimide-mediated reactions depends on a multitude of factors, including both the nature and stoichiometry of the reactants, the nature of the solvent, the temperature, the presence of tertiary amine, and the presence of additives or auxiliary nucleophiles. The role of the latter is treated in Section 2.11. Regardless, the first intermediate in the reaction is the *O*-acylisourea, the fate of which is dictated primarily by the availability of *N*-nucleophile. For reactions in solution, the *O*-acylisourea is immediately captured by the *N*-nucleophile, giving the target peptide, but if reaction with *N*-nucleophile is delayed for any reason, the *O*-acylisourea reacts with the parent acid to give the symmetrical anhydride (Figure 2.3, path C). This is the case if there is no *N*-nucleophile, such as when a preactivation is effected or when the intention is to prepare the symmetrical anhydride. The same result is obtained when the reactant is an *O*-nucleophile that has been added to produce an ester. Delays in consumption of *O*-acylisourea can also be a result of the slower approach of an *N*-nucleophile that is insoluble in the reaction medium. Such is the case in solid-phase synthesis; here a portion of the peptide emanates from the symmetrical anhydride. There is a second implication if the reaction with *N*-nucleophile is delayed: the *O*-acylisourea has a tendency to cyclize to the 2-alkoxy-5(4*H*)-oxazolone (path B) or rearrange to the *N*-acylurea (path J). Oxazolone formation is inconsequential; however, *N*-acylurea formation consumes *O*-acylisourea without producing peptide. More *N*-acylurea is generated at a higher temperature — that is why carbodiimide-mediated reactions are always carried out at a low temperature (0° or 5°).

More *N*-acylurea is generated if tertiary amine is present because the latter removes any protons that might prevent the rearrangement (see Section 2.12). The two intramolecular reactions also occur to a greater extent when interaction between the *O*-acylisourea and the *N*-nucleophile is impeded by the side chain of the activated residue. This means that more 2-alkoxy-5(4*H*)-oxazolone and *N*-acylurea are generated when the activated residues are hindered (see Section 1.4). A corollary of the above is that the best way to prepare an *N*-acylurea, should it be needed, is to heat

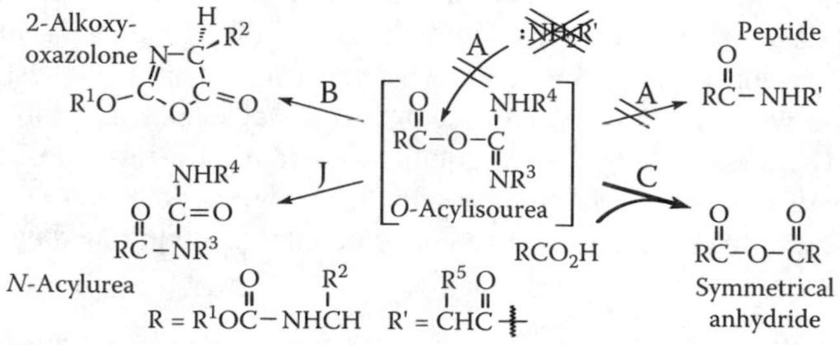

FIGURE 2.3 Factors affecting the course of events in carbodiimide-mediated reactions of *N*-alkoxycarbonylamino acids.

a solution of the acid and carbodiimide in the presence of tertiary amine. Highest yields of 2-alkoxy-5(4*H*)-oxazolone can be obtained by adding the *N*-alkoxycarbonylamino acid slowly to a solution of carbodiimide in an apolar solvent (see Section 1.18).[3–6]

3. R Rebek, D Feitler. Mechanism of the carbodiimide reaction. II. Peptide synthesis on the solid support. *J Am Chem Soc* 96, 1606, 1974.
4. DH Rich, J Singh. The carbodiimide method, in E Gross, J Meienhofer, eds. *The Peptides: Analysis, Synthesis, Biology*, Academic, New York, 1979, Vol 1, pp 241-261.
5. NL Benoiton. Quantitation and the sequence dependence of racemization in peptide synthesis, in E Gross, J Meienhofer, eds. *The Peptides: Analysis, Synthesis, Biology*, Academic, New York, 1981, Vol 5, pp 341-361.
6. M Slebioda, Z Wodecki, AM Kolodziejczyk. Formation of optically pure *N*-acyl-*N,N'*-dicyclohexylurea in *N,N'*-dicyclohexylcarbodiimide-mediated peptide synthesis. *Int J Pept Prot Res* 35, 539, 1990.

2.4 INTERMEDIATES AND THEIR FATE IN CARBODIIMIDE-MEDIATED REACTIONS OF *N*-ALKOXYCARBONYLAMINO ACIDS

2-Alkoxy-5(4*H*)-oxazolone is an intermediate in carbodiimide-mediated reactions of *N*-alkoxycarbonylamino acids, but it was omitted from Figure 2.3 to simplify the discussion. Figure 2.4 shows a simplified but all-inclusive scheme of the pertinent intermediates and reactions. The *O*-acylisourea is the first intermediate. The major source of peptide is aminolysis of the *O*-acylisourea (path A). A second major source of peptide is aminolysis of the symmetrical anhydride (path F) that originates from reaction of the *O*-acylisourea with the parent acid (path C). A third and minor source of peptide is the 2-alkoxy-5(4*H*)-oxazolone, formed by cyclization of the *O*-acylisourea (path B), that generates peptide directly (path E) or indirectly by first reacting with parent acid to give the symmetrical anhydride (path K). So there are three

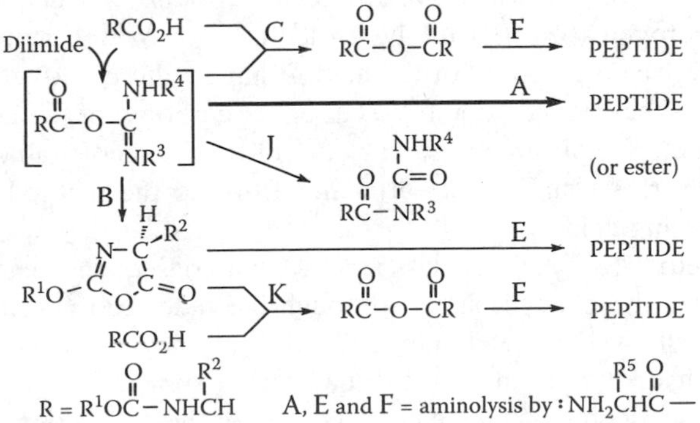

Diimide, RCO_2H → $RC(=O)-O-C(=NR^3)NHR^4$
C: + RCO_2H → $RC(=O)-O-C(=O)R$ —F→ PEPTIDE
A → PEPTIDE (or ester)
J → $RC(=O)-NR^3-C(=O)NHR^4$
B → 2-alkoxy-5(4*H*)-oxazolone (R^1O, R^2, H)
E → PEPTIDE
K: + RCO_2H → $RC(=O)-O-C(=O)R$ —F→ PEPTIDE

$R = R^1OC(=O)-NHCH(R^2)$ A, E and F = aminolysis by $:NH_2CH(R^5)C(=O)-$

FIGURE 2.4 Intermediates and their fate in carbodiimide-mediated reactions of *N*-alkoxycarbonylamino acids.

intermediates that are precursors of peptide. In addition, some of the *O*-acylisourea may react in another way to give *N*-acylurea (path J; see Section 2.3) that is not a precursor of peptide.

That being said, it must be recognized that the evidence that the *O*-acylisourea is the precursor of the 2-alkoxy-5(4*H*)-oxazolone is only circumstantial because experiments starting from the former have yet to be achieved. The oxazolone could theoretically come from the symmetrical anhydride. The latter generates 2-alkoxy-5(4*H*)-oxazolone in the presence of tertiary amines (see Section 4.16); even dicyclohexylcarbodiimide (DCC) was basic enough to generate 2-*tert*-butoxy-5(4*H*)-oxazolone from Boc-valine anhydride. However the weight of evidence points to *O*-acylisourea as the precursor of the 2-alkoxy-5(4*H*)-oxazolone. In the absence of *N*-nucleophile, such as in the preparation of esters, the major precursor of product is the symmetrical anhydride.[7,8]

7. NL Benoiton, FMF Chen. 2-Alkoxy-5(4*H*)-oxazolones from *N*-alkoxycarbonylamino acids and their implication in carbodiimide-mediated reactions in peptide synthesis. *Can J Chem* 59, 384, 1981.
8. NL Benoiton, FMF Chen. Reaction of *N*-t-butoxycarbonylamino acid anhydrides with tertiary amines and carbodiimides. New precursors for 2-*t*-butoxyoxazol-5(4*H*)-one and *N*-acylureas. *J Chem Soc Chem Commun* 1225, 1981.

2.5 PEPTIDE-BOND FORMATION FROM PREFORMED SYMMETRICAL ANHYDRIDES OF N-ALKOXYCARBONYLAMINO ACIDS

An alternative to the classical method of synthesis using carbodiimides is a variant in which the carbodiimide and acid are first allowed to react together in the absence of *N*-nucleophile (see Section 1.15). One-half of an equivalent of carbodiimide is employed. This generates half an equivalent of symmetrical anhydride, the formation of which (Figure 2.5, path C) can be rationalized in the same way as the reaction of acid with carbodiimide is rationalized; namely, protonation at the basic nitrogen ($C=NR^3$) of the *O*-acylisourea by the acid, followed by attack at the activated carbonyl of the acyl group by the carboxylate anion. The *N*-nucleophile is then added; aminolysis at either carbonyl of the anhydride (path F) gives peptide and half an equivalent of acid that is recoverable. Recovery of the acid, however, is usually not worth the effort, so the method is wasteful of 50% of the starting acid.

The symmetrical anhydride is less reactive and consequently more selective in its reactions than the *O*-acylisourea. Although the latter can acylate both *N*- and *O*-nucleophiles, the symmetrical anhydride will only acylate *N*-nucleophiles. This means that the hydroxyl groups of the side chains of serine, threonine, and tyrosine that have not been deprotonated are not acceptors of the acyl group of the symmetrical anhydride. An additional feature of this approach to carbodiimide-mediated reactions is that it avoids a possible side reaction between the carbodiimide and the *N*-nucleophile, which gives a trisubstituted guanidine $[(C_6H_{11}N)_2C=N–CHR^5CO–$

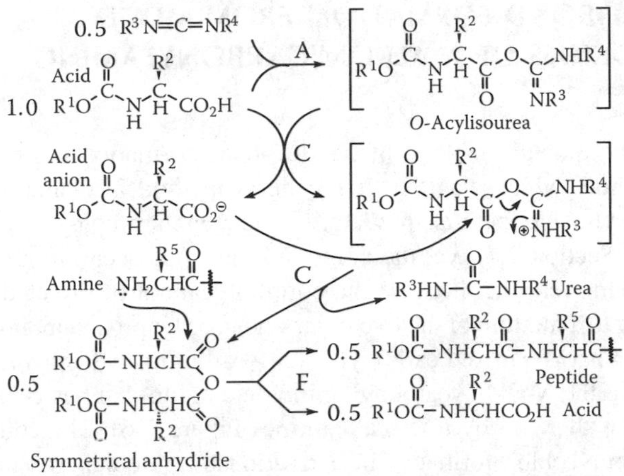

FIGURE 2.5 Peptide-bond formation from preformed symmetrical anhydrides of *N*-alkoxycarbonylamino acids (see Sections 1.15 and 1.16 for references). The reaction is effected in dichloromethane and the *N,N′*-dicyclohexylurea is removed by filtration. The dichloromethane is sometimes replaced by dimethylformamide.

from DCC]. This approach issued from the realization that the electrophile that undergoes aminolysis during synthesis on a solid support is the symmetrical anhydride.

When the reagent is dicyclohexylcarbodiimide, the reaction is carried out in dichloromethane, the *N,N′*-dicyclohexylurea is removed by filtration after 15–30 minutes, the solvent is sometimes replaced by dimethylformamide, and the solution is then added to the *N*-nucleophile. The *N,N′*-dicyclohexylurea is removed to help drive the coupling reaction to completion. The symmetrical anhydride is not prepared directly in the polar solvent because the latter suppresses its formation.

Symmetrical anhydrides are stable enough to be isolated but not stable enough to be stored for future use. They can be purified by repeated crystallization or by washing a solution of the anhydride that has been obtained using a soluble carbodiimide with aqueous solutions (see Section 1.16). The use of symmetrical anhydrides allows for synthesis of a peptide with less possibility of generation of side products. The anhydrides are particularly effective for acylating secondary amines (see Section 8.15).[9–11]

9. H Schüssler, H Zahn. Contribution on the course of reaction of carbobenzoxyamino acids with dicyclohexylcarbodiimide. *Chem Ber* 95, 1076, 1962.
10. F Weygand, P Huber, K Weiss. The synthesis of peptides with symmetrical anhydrides I. *Z Naturforsch* 22B, 1084, 1967.
11. H Hagenmeier, H Frank. Increased coupling yields in solid phase peptide synthesis with a modified carbodiimide coupling procedure. *Hoppe-Seyler's Z Physiol Chem* 353, 1973, 1972.

2.6 PEPTIDE-BOND FORMATION FROM MIXED ANHYDRIDES OF *N*-ALKOXYCARBONYLAMINO ACIDS

The second most popular method of peptide-bond formation has been the mixed-anhydride method, which was the first general method available. Only the acyl-chloride and acyl-azide methods predate it. The reagent employed is an alkyl chloroformate (see Section 2.7), yet the compound has not been blessed with the designation "coupling reagent" because the coupling cannot be effected by adding the reagent directly to a mixture of the two components. The procedure involves separate preparation of what is a mixed carboxylic acid–carbonic acid anhydride by addition of the reagent to the *N*-alkoxycarbonylamino acid anion that has been generated by deprotonation of the acid by a tertiary amine (Figure 2.6). The common reagents are ethyl or isobutyl chloroformate. The activation is very quick; reactants are usually left together for 1–2 minutes. Aminolysis is usually complete within an hour. The activation cannot be carried out in the presence of the *N*-nucleophile because the latter also reacts with the chloroformate. All stages of the reaction are carried out at low temperature to avoid side reactions (see Section 2.8). There is evidence that the tertiary amine used in the reaction is not merely a hydrogen chloride acceptor but an active participant in the reaction. A proposal is that the acylmorpholinium cation is first formed (Figure 2.6, boxed structures) and that it is the acceptor of the acid anion. The fact that use of diisopropylethylamine as base in a reaction failed to generate any mixed anhydride militates strongly in favor of the more elaborate mechanism. The reactivity of mixed anhydrides is similar to that of symmetrical anhydrides (see Section 2.5). A feature of symmetrical- and mixed-anhydride reactions that is different from that of carbodiimides and other coupling reagents is that the anhydrides can be used to acylate amino acid or peptide anions in partially aqueous solvent mixtures (see Section 7.21).

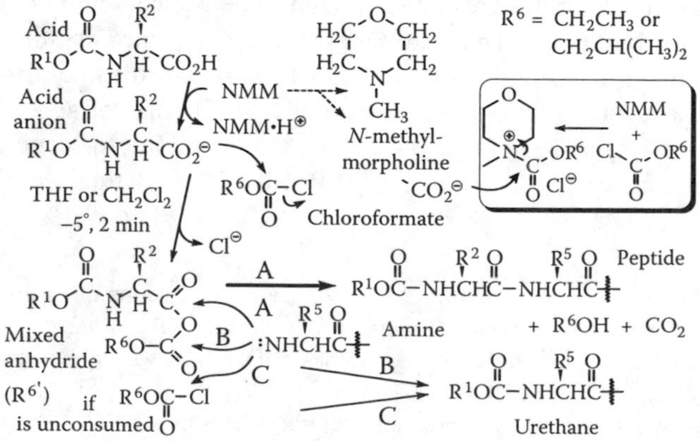

FIGURE 2.6 Peptide-bond formation from mixed anhydrides of *N*-alkoxycarbonylamino acids.[13,14,16]

The principal side reaction associated with the mixed-anhydride method is aminolysis at the carbonyl of the carbonate moiety (path B), giving a urethane. Because of the nature of alkyl group R^6, the side product is stable to all procedures of deprotection. In most cases the reaction is not of much significance, but it can reduce the yield of peptide by up to 10% for the hindered residues, where R^2 is a chain with a β-methyl group. The original mixed anhydrides were mixed carboxylic acid anhydrides (R^6 instead of R^6O) made from benzoyl chloride (C_6H_5COCl instead of R^6OCOCl), but it was found that there was insufficient selectivity in aminolysis between the two carbonyls. An oxygen atom was inserted next to the carbonyl of the second acid moiety to reduce the electrophilicity of the carbonyl carbon atom. Isobutyl chloroformate was introduced instead of ethyl chloroformate, with the same objective in mind: to drive the aminolysis to the acyl carbonyl by increasing the electron density at the carbonyl of the carbonate moiety. Two examples of mixed carboxylic acid anhydrides remain of interest to peptide chemists; namely, those formed from pivalic (see Section 8.15) and dichlorobenzoic (see Section 5.20) acids.

An additional minor source of urethane can be the reaction of unconsumed reagent with *N*-nucleophile (path C). Aminolysis of chloroformate occurs if there is an excess of reagent or if the anhydride-forming reaction is incomplete. The latter is more likely when the residues activated are hindered. This side reaction can be avoided by limiting the amount of reagent and extending the time of activation. A third side reaction that is of little consequence is disproportionation of the mixed anhydride to the symmetrical anhydride and dialkyl pyrocarbonate (see Section 7.5).

The success of mixed-anhydride reactions has been considered to be extremely dependent on the choice of conditions employed. Much attention was directed to defining conditions that would minimize the epimerization that occurs during the coupling of peptides (see Section 1.9). It was concluded that superior results are achieved by carrying out the activation for only 1–2 minutes at –5° to –15°, using *N*-methylmorpholine as the base in anhydrous solvents other than chloroform and dichloromethane.[12] Triethylamine or tri-*n*-butylamine had been used initially as base, but it was found that the weaker and less hindered cyclic amine leads to less isomerization. With the passage of time, the stringent attention to detail that was recommended for coupling peptides emerged as conventional wisdom for achieving best results in the coupling of *N*-alkoxycarbonylamino acids. In fact, not all the precautions required for efficient coupling of peptides are essential for efficient coupling of *N*-alkoxycarbonylamino acids. Once more was known about the properties of mixed anhydrides, it became apparent that not the low temperature, nor the absence of moisture, nor the need to avoid halogen-containing solvents is critical for achieving efficient preparation and coupling of mixed anhydrides of *N*-alkoxycarbonylamino acids (see Section 2.8).[12–17]

12. GW Anderson, JE Zimmerman, FM Callahan. A reinvestigation of the mixed carbonic anhydride method of peptide synthesis. *J Am Chem Soc* 89, 5012, 1967.
13. RA Boissonnas. A new method of peptide synthesis. *Helv Chim Acta* 34, 874, 1951.
14. JR Vaughan. Acylalkylcarbonates as acylating agents for the synthesis of peptides. *J Am Chem Soc* 73, 3547, 1951.

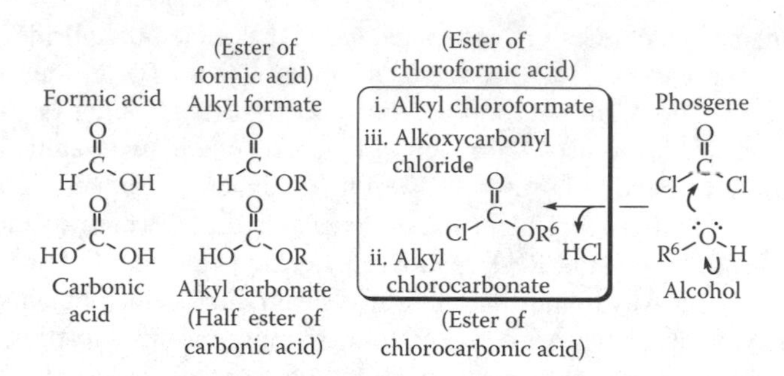

FIGURE 2.7 Alkyl chloroformates and their nomenclature.

15. JR Vaughan, RL Osato. The preparation of peptides using mixed carbonic-carboxylic acid anhydrides. *J Am Chem Soc* 74, 676, 1952.
16. T Wieland, H Bernhard. On the synthesis of peptides. Part 3. The use of anhydrides of N-acylated amino acids and derivatives of inorganic acids. *Ann Chem* 572, 190, 1951.
17. J Meienhofer. The mixed carbonic anhydride method of peptide synthesis, in E Gross, J Meienhofer, eds. *The Peptides: Analysis, Synthesis, Biology*, Academic, New York, 1979, Vol 1, pp 263-314.

2.7 ALKYL CHLOROFORMATES AND THEIR NOMENCLATURE

The reagents used to generate mixed anhydrides are named in three different ways because they can be considered to be derivatives of formic acid, carbonic acid, or hydrogen chloride (Figure 2.7). They are alkyl chloroformates that are esters of chlorinated formic acid, alkyl chlorocarbonates that are monoesters of carbonic acid with the second hydroxyl group replaced by chloro, and alkoxycarbonyl chlorides that are acyl-substituted hydrogen chlorides. The compounds are obtained by reaction of phosgene with the pertinent alcohol (see Section 3.3). In addition to their role as reagents for mixed-anhydride couplings, with one notable exception they are used for preparing the *N*-alkoxycarbonylamino acids that are employed for synthesis (see Section 3.3). Most chloroformates are readily prepared and inexpensive, but there are two whose properties create difficulties. Isopropenyl chloroformate cannot be obtained directly from the alcohol or acetone, and *tert*-butyl chloroformate is not stable enough for routine use because its boiling point is only slightly above ambient temperature.

2.8 PURIFIED MIXED ANHYDRIDES OF *N*-ALKOXYCARBONYLAMINO ACIDS AND THEIR DECOMPOSITION TO 2-ALKOXY-5(4*H*)-OXAZOLONES

Up until the late 1980s, it was conventional wisdom that mixed-anhydride reactions should be carried out at low temperature under anhydrous conditions and not in halogen-containing solvents. It was believed that the anhydrides were not very stable and that they decomposed by disproportionation to give the symmetrical anhydride and the dialkylpyrocarbonate (see Section 7.5). Knowledge of the comportment of mixed anhydrides was based on observations made of solutions of anhydrides containing tertiary-amine salts because that was the only way of getting the anhydrides. There was no method available to purify them. As a result of research by Chen and Benoiton, it is now apparent that mixed anhydrides are not so sensitive to water that they cannot be purified by washing them with aqueous solutions. In addition, it appears that halogen-containing solvents are good solvents for mixed-anhydride reactions if the tertiary amine used is not triethylamine. It transpires that mixed-anhydride-forming reactions are slowed down considerably by the combination of halogen-containing solvent and triethylamine. These findings issued from studies on urethane formation and failed attempts to distinguish between urethane generated by aminolysis at the carbonate moiety of the anhydride from that generated by aminolysis of unconsumed reagent. The impasse prompted examination of the possibility of purifying the anhydrides by washing them with water, as had been done for purifying symmetrical anhydrides (see Section 1.15) and 2-alkyl-5(4*H*)-oxazolones (see Section 1.17). It was found indeed that mixed anhydrides can be purified by washing a solution of the anhydride and tertiary-amine salt in dichloromethane with aqueous acid. They are also obtainable from reactions carried out at room temperature. Access to purified mixed anhydrides has allowed accurate scrutiny of their properties. Mixed anhydrides of *N*-alkoxycarbonylamino acids can be obtained in a pure state by reacting the acid with alkyl chloroformate in dichloromethane at room temperature in the presence of *N*-methylmorpholine or *N*-methylpiperidine for 2 minutes, followed by washing the solution with aqueous acid (Figure 2.8). The products are reasonably stable, slowly decomposing by attack at the activated carbonyl of the anhydride by the carbonyl oxygen of the urethane to give the 2-alkoxy-5(4*H*)-oxazolone, the alcohol and carbon dioxide. The ease of cyclization depends on the nature of the alkyl group of the carbonate moiety, with the anhydride from a secondary alkyl (isopropyl) chloroformate being more stable than those from primary alkyl chloroformates. The anhydride from isopropenyl chloroformate is the

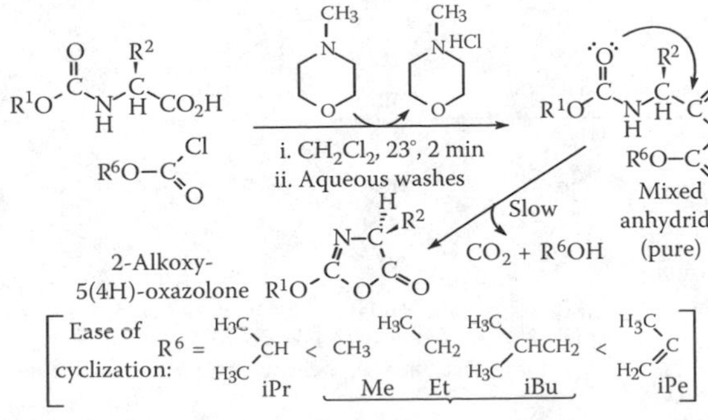

FIGURE 2.8 Purified mixed anhydrides of *N*-alkoxycarbonylamino acids and their decomposition to 2-alkoxy-5(4*H*)-oxazolones[18]; iPr = isopropyl, iPe = isopropenyl.

least stable anhydride. An important corrollary issues from these observations: Couplings of peptides by the mixed-anhydride method using isopropyl chloroformate should lead to less epimerization than couplings using ethyl or isobutyl chloroformate. This was shown to be the case (see Section 7.3). An additional practical use of the observations has been development of a general method of preparation of 2-alkoxy-5(4*H*)-oxazolones from mixed anhydrides generated using isopropenyl chloroformate.

There is no obvious advantage of purifying mixed anhydrides of *N*-alkoxycarbonylamino acids before reacting them with an *N*-nucleophile to form a peptide bond. However, there are a few reports of success with the reaction of a purified mixed anhydride after the reaction with the anhydride in the presence of tertiary-amine salt had failed. It would seem that the prudent course of action is to try the reaction using both the crude and purified anhydrides in case one approach is better.[18,19]

18. FMF Chen, NL Benoiton. The preparation and reactions of mixed anhydrides of *N*-alkoxycarbonylamino acids. *Can J Chem* 65, 619, 1987.
19. NL Benoiton, FMF Chen. Preparation of 2-alkoxy-5(4*H*)-oxazolones from mixed anhydrides of *N*-alkoxycarbonylamino acids. *Int J Pept Prot Res* 42, 455, 1993.

2.9 PEPTIDE-BOND FORMATION FROM ACTIVATED ESTERS OF *N*-ALKOXYCARBONYLAMINO ACIDS

A unique approach to the synthesis of peptides is the preparation of a derivative of the *N*-alkoxycarbonylamino acid that is stable enough to be stored and yet reactive enough to combine with an amino group when the two are mixed together. Such compounds are created by converting the acid into what is referred to as an activated ester by reacting it with either a substituted phenol or a substituted hydroxylamine HOR^7 (Figure 2.9). The substituents R^7 are designed to render the carbonyl of the

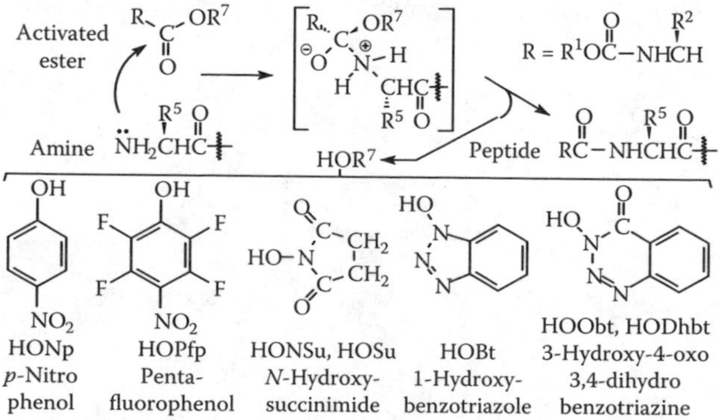

FIGURE 2.9 Peptide-bond formation from activated esters of *N*-alkoxycarbonylamino acids [Wieland, 1951; Schwyzer, 1955; Bodanszky 1955]. Some hydroxy compounds have two abbreviations. HONSu conveys the notion of bonding through a nitrogen atom and is consistent with HONPht = *N*-hydroxyphthalimide. Su can be interpreted as succinyl/oyl.

acyl moiety susceptible to nucleophilic attack by an amine at room temperature. The esters undergo aminolysis by a two-step reaction involving a tetrahedral intermediate whose rearrangement to the peptide and starting hydroxy compound is the rate-limiting step. They are prepared most often by reaction of the acid and the phenol or hydroxylamine, using dicyclohexylcarbodiimide. They are really mixed anhydrides formed from a carboxylic acid and a phenolic or hydroxamic acid. Many different types of activated esters are available; the few in common use today appear in Figure 2.9.

Development of the activated-ester method of coupling issued from studies on reactions of vinyl, thiophenyl, and cyanomethyl esters. Literally dozens of activated esters of the type -C(=O)OCR^7 were designed; the *p*-nitrophenyl and then the more reactive pentafluorophenyl esters emerged as the most popular. These compounds are activated by virtue of the electron-withdrawing nature of the substituted phenyl ring. The phenols liberated by aminolysis are not soluble in water, and they consequently present an obstacle to the purification of desired products. A different type of activated ester derived from hydroxamic acids instead of phenols then surfaced. The first were the *o*-phthalimido [$R^7 = C_6H_4(CO)_2$=N] esters that liberate a water-insoluble side product, but these were soon replaced by the more versatile succinimido esters [$R^7 = C_2H_4(CO)_2$=N; Figure 2.9]. The latter generate water-soluble *N*-hydroxysuccinimide that is easy to remove from target peptides. Additional examples of esters derived from hydroxamic acids are benzotriazolyl and 4-oxo-3,4-dihydrobenzotriazinyl esters (Figure 2.9). These are activated not so much because of the electron-withdrawing effects of the ring moieties but because of the nature and juxtaposition of the atoms of the heterocyclic rings (see Section 2.11).

The more activated the ester, the less stable is the compound. All the esters mentioned above can be used as shelf-stable reagents except benzotriazolyl esters, which decompose too readily. In addition to their use as activated forms of the *N*-alkoxycarbonylamino acids, the esters derived from hydroxamic acids are implicated as intermediates in coupling reactions in which the *N*-hydroxy compounds have been added to promote efficient coupling between an acid and a primary or secondary amine (see Section 2.10). It is pertinent to mention that the *O*-acylisourea generated from carbodiimides (see Section 2.02) is an activated ester but one of nature different than those alluded to above.

The aminolysis of activated esters generally occurs more readily in polar solvents and is catalyzed by mild acid (see Section 7.6) or 1-hydroxybenzotriazole. Trans-esterification and through mixed anhydrides are other methods by which activated esters can be obtained (see Section 7.8).[20–27]

20. T Wieland, W Schäfer, E Bokelmann. Peptide syntheses. V. A convenient method for the preparation of acylthiophenols and their application in the syntheses of amides and peptides. *Ann Chem* 573, 99, 1951.
21. R Schwyzer, M. Feuer, B Iselin. Activated esters. III. Reactions of activated esters of amino acid and peptide derivatives with amines and amino acid esters. *Helv Chim Acta* 38, 83, 1955.
22. M Bodanszky. Synthesis of peptides by aminolysis of nitrophenyl esters. *Nature (London)* 75, 685, 1955.

23. GHL Nefkens, GI Tesser. A novel activated ester in peptide synthesis (phthalimido esters). *J Am Chem Soc* 26, 1263, 1961.
24. JW Anderson, JE Zimmerman, FM Callahan. The use of esters of N-hydroxysuccinimide in peptide synthesis. *J Am Chem Soc* 86, 1839, 1964.
25. J Kovacs, R Gianotti, A Kapoor. Polypeptides with known repeating sequence of amino acids. Synthesis of poly-L-glutamyl-L-alanyl-L-glutamic acid and polyglycyl-L-phenylalanine through pentachlorophenyl active ester. *J Am Chem Soc* 88, 2282, 1966.
26. L Kisfaludy, MQ Ceprini, B Rakoczy, J Kovacs. Pentachlorophenyl and pentafluorophenyl esters of peptides and the problem of racemization II, in HC Beyerman, A van de Linde, W Massen van den Brink, eds. *Peptides, Proceedings of the 8th European Peptide Symposium*, North-Holland, Amsterdam, 1967, pp 25-27.
27. M Bodanszky. Active esters in peptide synthesis, in E Gross, J Meienhofer, eds. *The Peptides: Analysis, Synthesis, Biology*, Academic, New York, 1979, pp 105-196.

2.10 ANCHIMERIC ASSISTANCE IN THE AMINOLYSIS OF ACTIVATED ESTERS

Activated esters undergo aminolysis because of the electon-withdrawing property of the ester moiety. However, the esters formed from substituted hydroxamic acids are so highly activated that their reactivity cannot be explained on the basis of this property alone. An additional phenomenon is operative; neighboring atoms assist in the union of the two reactants. Inspection of the structures in Figure 2.10 reveals that in these compounds, the ester oxygen is attached to a nitrogen atom that is linked to a double-bonded atom that is either the carbon atom of a carbonyl or the nitrogen atom of a triaza sequence of atoms. Each carbonyl or triaza group bears a pair of unshared electrons. It is the presence of these electrons in the ester moiety in proximity to the carbonyl that is responsible for the high activation. The electron-dense atoms promote attack at the carbonyl by the nucleophile by forming a hydrogen bond with a hydrogen atom of the amino group (Figure 2.10). This neighboring-group participation in the formation of a new chemical bond is referred to as anchimeric assistance. The reaction is, in fact, an intramolecular general base-catalyzed reaction. The effect of the neighboring groups on the reactivity of the

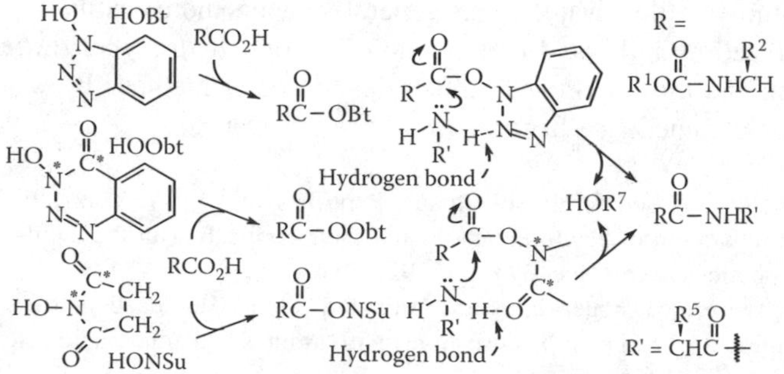

FIGURE 2.10 Anchimeric assistance in the aminolysis of activated esters.

electrophile is referred to as an *ortho* effect. Recognition of the beneficial effects of a favorable juxtaposition of potentially reactive groups issued from observations on catalysis of the hydrolysis of esters; it was realized that *o*-hydroxyphenylamine is a much better catalyst for hydrolysis than a mixture of phenol and phenylamine (aniline). This rationalization is the same as that used to explain the efficiency of enzymes as catalysts and the supernucleophilicity of hydrazine (H_2NNH_2). Other activated esters exist; for example, with R^7 (Figure 2.9) = pyridine linked at C-2, where the juxtaposition of pertinent atoms is different from that described in Figure 2.10.[27–29]

27. M Bodanszky. Active esters in peptide synthesis, in E Gross, J Meienhofer, eds. *The Peptides: Analysis, Synthesis, Biology*, Academic, New York, 1979, pp 105-196.
28. JH Jones, GT Young. Anchimeric acceleration of aminolysis of esters and its application to peptide synthesis. *Chem Commun* 35, 1967.
29. W König, R Geiger. New catalysts in peptide synthesis, in J Meienhofer, ed. *Chemistry and Biology of Peptides. Proceedings of the 3rd American Peptide Symposium*, Ann Arbor Science, Ann Arbor, MI, 1972, pp 343-350.

2.11 ON THE ROLE OF ADDITIVES AS AUXILIARY NUCLEOPHILES: GENERATION OF ACTIVATED ESTERS

The substituted hydroxamic acids commonly found in the ester moiety of activated esters also play a prominent role as "additives" for carbodiimide-mediated reactions. It has been found that the presence of a compound of this type in a mixture containing reactants and carbodiimide significantly improves the efficiency of coupling. The additive of choice has traditionally been 1-hydroxybenzotriazole. A modified version, the 7-aza analogue, has recently been introduced. *N*-Hydroxysuccinimide and HOObt, which is an analogue of HOBt with a carbonyl group inserted into the heterocyclic ring, are the other two common additives. The structures and names of the compounds appear in Figure 2.11. In the absence of the amine nucleophile, addition of carbodiimide to a mixture of carboxylic acid and hydroxamic acid gives the activated ester. In the presence of amine nucleophile, the additive competes with the *N*-nucleophile for the *O*-acylisourea, reacting with the latter to generate an activated ester before the *O*-acylisourea has time to undergo secondary reactions. It thus acts as an auxiliary nucleophile. If some of the *O*-acylisourea has cyclized to the 5(4*H*)-oxazolone, the additive reacts with it also before it has time to enolize. In this way, both *N*-acylurea formation (see Section 2.12) and oxazolone isomerization (see Section 2.12) are suppressed. The activated ester, instead of the *O*-acylisourea, becomes the precursor of the peptide. Additives are also employed as auxiliary nucleophiles occasionally, with the newer onium salt-based reagents (see Section 2.16).[30–34]

30. E Wünsch, F Drees. On the synthesis of glucagon. X. Preparation of sequence 22-29. (*N*-hydroxysuccinimide) *Chem Ber* 99, 110, 1966.

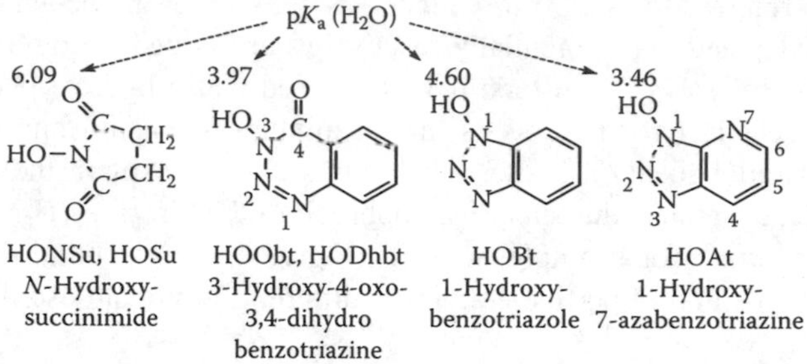

pK$_a$ (H$_2$O)

6.09 — HONSu, HOSu, *N*-Hydroxysuccinimide

3.97 — HOObt, HODhbt, 3-Hydroxy-4-oxo-3,4-dihydro benzotriazine

4.60 — HOBt, 1-Hydroxybenzotriazole

3.46 — HOAt, 1-Hydroxy-7-azabenzotriazine

FIGURE 2.11 Structures and nomenclature of compounds that serve as auxiliary nucleophiles. Generation of activated esters. Substituted hydroxamic acids are sometimes added to carbodiimides or other reactions to improve the efficiency of couplings. The "additive" suppresses side reactions by converting activated species into activated esters (see Section 2.10) before they have time to undergo secondary reactions. pK_a (Me$_2$SO): HOBt 9.30, HOAt 8.70.

31. JE Zimmerman, GW Callahan. The effect of active ester components on racemization in the synthesis of peptides by the dicyclohexylcarbodiimide method. *J Am Chem Soc* 89, 7151, 1967.
32. W König, R Geiger. A new method for the synthesis of peptides: activation of the carboxyl group with dicyclohexylcarbodiimide and 1-hydroxybenzotriazoles. *Chem Ber* 103, 788, 1970.
33. W König, R Geiger. A new method for the synthesis of peptides: activation of the carboxyl group with dicyclohexylcarbodiimide and 3-hydroxy-4-oxo-3,4-dihydro-1.2.3-benzotriazine. *Chem Ber* 103, 2034, 1970.
34. LA Carpino. 1-Hydroxy-7-azabenzotriazole. An efficient peptide coupling additive. *J Am Chem Soc* 115, 4397, 1993.

2.12 1-HYDROXYBENZOTRIAZOLE AS AN ADDITIVE THAT SUPPRESSES *N*-ACYLUREA FORMATION BY PROTONATION OF THE *O*-ACYLISOUREA

In carbodiimide-mediated reactions (see Section 2.2), peptide is formed by aminolysis of the *O*-acylisourea at the activated carbonyl (Figure 2.12, path A). A competing

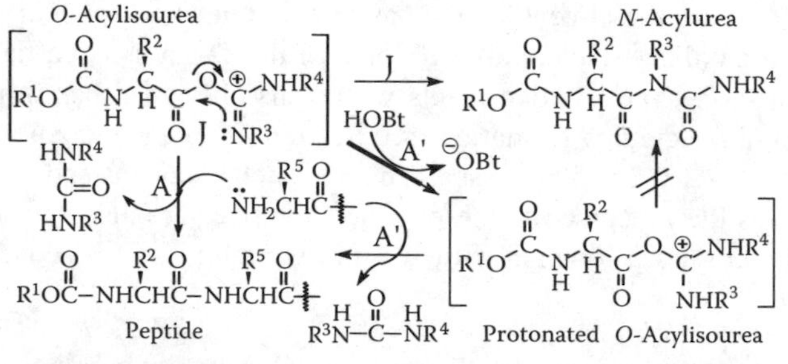

O-Acylisourea — *N*-Acylurea — HOBt — Peptide — Protonated *O*-Acylisourea

FIGURE 2.12 1-Hydroxybenzotriazole as an additive that suppresses *N*-acylurea formation by protonation of the *O*-acylisourea (see review by Rich and Singh[4]).

intramolecular reaction can occur whereby the basic imine nitrogen of the *O*-acylisourea attacks the same carbonyl generating *N*-acylurea (path J) that does not give rise to peptide. If this basic nitrogen atom is protonated (path A′), the nucleophilicity is eliminated and the *O*-to-*N* shift of the acyl group (path J) cannot occur. 1-Hydroxybenzotriazole increases the efficiency of carbodiimide-mediated reactions. One of the ways by which this is achieved is the elimination of *N*-acylurea formation. HOBt is weakly acidic (see Section 2.11). The beneficial effect of HOBt in suppressing *N*-acylurea formation is attributed to its role as an acid that protonates the *O*-acylisourea, thus preventing the intramolecular reaction from occurring. In addition, inspection of the structure of the protonated *O*-acylisourea reveals that it is a better electrophile than the unprotonated form. Thus, protonation also favors consumption of the *O*-acylisourea by enhancing its electrophilicity. At the same time, protonation by HOBt generates the benzotriazolyloxy anion (path A′), which is a very good acceptor of electrophiles. Thus, protonation by HOBt simultaneously provides an additional and stronger nucleophile for consumption of the *O*-acylisourea. The corollary to the above is that deprotonation of the *O*-acylisourea favors the intramolecular reaction. Thus, the presence of tertiary amines is deleterious to carbodiimide-mediated reactions because they promote *N*-acylurea formation. A more polar solvent as well as higher temperature also promote *N*-acylurea formation. The latter is why carbodiimide reactions are carried out at temperatures lower than ambient temperature. The *O*-to-*N* shift (see Section 6.6) of the acyl group is also favored by any delay in consumption of the *O*-acylisourea. *N*-Acylurea formation is thus more prevalent when side-chain R^2 interferes with the approach of the amine nucleophile, which is the case for β-methylamino acids [Val, Ile, Thr(R)].

Just as protonation of the *O*-acylisourea enhances its electrophilicity, and consequently its consumption, it can be postulated that the reactivity with nucleophiles of any 2-alkoxy-5(4*H*)-oxazolone that is formed is enhanced by its protonation. Thus, it is reasonable to assert that a beneficial effect of HOBt in improving efficiency in couplings of *N*-alkoxycarbonylamino acids is protonation of the oxazolone (see Section 2.25) that facilitates its consumption.[4,32]

4. DH Rich, J Singh. The carbodiimide method, in E Gross, J Meienhofer, eds. *The Peptides: Analysis, Synthesis, Biology*, Academic, New York, 1979, Vol 1, pp 241-261.
32. W König, R Geiger. A new method for the synthesis of peptides: activation of the carboxyl group with dicyclohexylcarbodiimide and 1-hydroxybenzotriazoles. *Chem Ber* 103, 788, 1970.

2.13 PEPTIDE-BOND FORMATION FROM AZIDES OF *N*-ALKOXYCARBONYLAMINO ACIDS

The acyl-azide method of coupling (Figure 2.13) has been available for about a century, but it is not attractive for routine use because it involves four distinct steps that include two stable intermediates that require purification. In addition, aminolysis of the azide is slow. The first step involves preparation of the ester (see Section 3.17), which can be methyl, ethyl, or benzyl. The ester is converted to the hydrazide by reaction in alcohol with excess hydrazine at ambient or higher

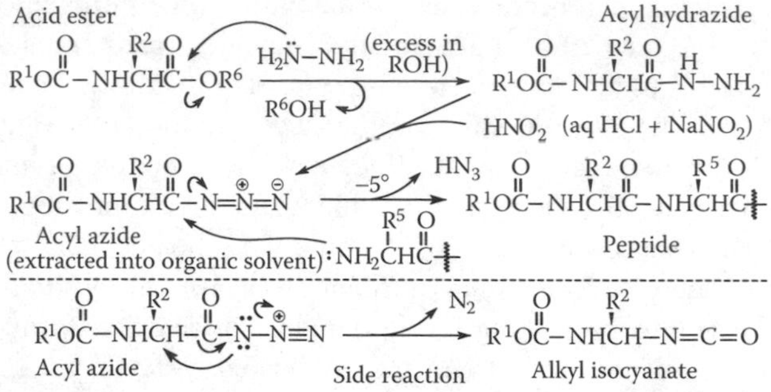

Acid ester
$R^1OC(=O)-NHCH(R^2)C(=O)-OR^6 \xrightarrow{H_2N-NH_2 \text{ (excess in ROH)}} R^1OC(=O)-NHCH(R^2)C(=O)-NH-NH_2$ + R^6OH
Acyl hydrazide
HNO_2 (aq HCl + $NaNO_2$)
$R^1OC(=O)-NHCH(R^2)C(=O)-N{=}\overset{\oplus}{N}{=}\overset{\ominus}{N} \xrightarrow{-5^\circ} R^1OC(=O)-NHCH(R^2)C(=O)-NHCH(R^5)C(=O)-$ + HN_3
Acyl azide
(extracted into organic solvent) : $NH_2CH(R^5)C(=O)-$
Peptide

$R^1OC(=O)-NHCH(R^2)-C(=O)-\ddot{N}-\overset{\oplus}{N}{\equiv}N \longrightarrow R^1OC(=O)-NHCH(R^2)-N{=}C{=}O$ + N_2
Acyl azide
Side reaction
Alkyl isocyanate

FIGURE 2.13 Peptide-bond formation from azides of *N*-alkoxycarbonylamino acids (see review by Meienhofer).[35,36]

temperatures. The excess of hydrazine is required to ensure that no diacylated hydrazide is produced. The hydrazide often crystallizes out of solution or after removal of solvent. The purified hydrazide is transformed into the azide by the action of nitrous acid; this is usually achieved by the addition of sodium nitrite to a cold solution of the hydrazide in a mixture of acetic and hydrochloric acids. The azide is generated at low temperature because it readily decomposes with the release of nitrogen at ambient temperature. The azide is extracted into an organic solvent, and the peptide is obtained by leaving the dried solution in the presence of the amine-nucleophile in the cold for several hours. An additional side reaction that occurs at higher temperature is rearrangement of the acyl azide to the alkyl isocyanate (Figure 2.13), which can react with the nucleophile to yield a peptide urea that is difficult to remove from the product. The side product is neutral and not easy to remove from the peptide. Because of the time and effort required, the acyl-azide method is not suitable for repetitive syntheses, but it has two characteristic features that make it a popular option for coupling in selected cases. The first relates to the strategy of minimum protection (see Section 7.16). This method can be used for the activation of serine, threonine (see Section 6.5), and histidine (see Section 6.11) derivatives with unprotected side chains—the latter being unaffected by the reactions employed. The second relates to the coupling of segments. It is the only method that just about guarantees the preservation of chiral integrity during peptide-bond formation between segments (see Sections 2.24 and 7.16). The latter is possible because the acyl azide does not generate oxazolone (see Section 2.23). It is the only activated form of an *N*-acylamino acid or peptide that can be isolated for which cyclization to the oxazolone has not been demonstrated.[35,36]

35. T Curtius. Synthetic studies on hippuramide. *Ber Deutsch Chem Ges* 35, 3226, 1902.
36. J Meienhofer. The azide method in peptide synthesis, in E Gross, J Meienhofer, eds. *The Peptides: Analysis, Synthesis, Biology*, Academic, New York, 1979, pp 197-239.

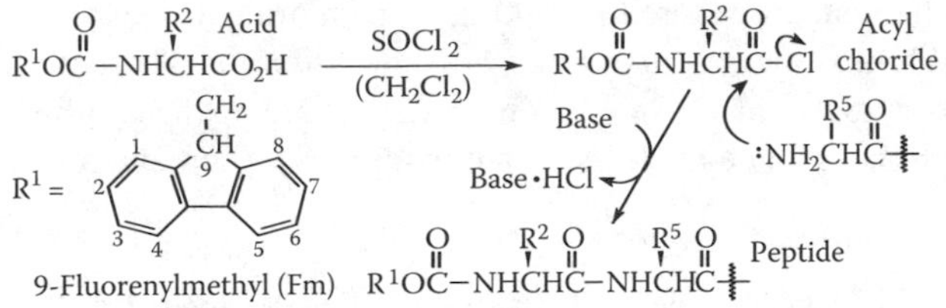

FIGURE 2.14 Peptide-bond formation from chlorides of *N*-alkoxycarbonylamino acids. *N*-9-Fluorenylmethoxycarbonylamino-acid chlorides.[41] The base is $NaHCO_3$, Na_2CO_3, or a tertiary amine. The reaction is carried out in a one- or two-phase system. The latter is used to try to suppress formation of the 2-alkoxy-5(4*H*)-oxazolone that is generated by the action of the base on the acid chloride. The method is applicable primarily to Fmoc-amino-acid derivatives that do not have acid-sensitive protecting groups on their side chains.

2.14 PEPTIDE-BOND FORMATION FROM CHLORIDES OF *N*-ALKOXYCARBONYLAMINO ACIDS: *N*-9-FLUORENYLMETHOXYCARBONYLAMINO-ACID CHLORIDES

Acid chlorides have been available since the earliest times of peptide synthesis. Glycyl-peptides were originally obtained by acylation of an amino acid with chloroacetyl chloride, followed by ammonolysis, but their use was of limited applicability until the discovery in 1932 of the cleavable benzyloxycarbonyl group, which permitted peptide-bond formation using *N*-benzyloxycarbonylamino-acid chlorides (Figure 2.14; $R^1 = C_6H_5CH_2$) — the latter obtained by reaction of the parent acid with phosphorus pentachloride. Their reign was short-lived, however. The emergence of simpler coupling methods in the 1950s, combined with the incompatibility of the acid chloride method of the time with derivatives bearing acid-sensitive tertiary-butyl based protectors, just about eliminated their use. Things changed dramatically, however, with the introduction of N^α-protection by the acid-stable 9-fluorenylmethoxycarbonyl group (Figure 2.14). Fmoc-amino-acid chlorides are generated by reaction of the parent acid with thionyl chloride in hot dichloromethane. Though sensitive to water, they are stable enough to be purified by recrystallization. They acylate amino groups readily in the presence of a base that is required to neutralize the hydrogen chloride that is liberated. The base is necessary, but its presence complicates the issue, converting the acid chloride to the 2-alkoxy-5(4*H*)-oxazolone, which is aminolyzed at a slower rate. In one variant, the aminolysis is carried out in a two-phase system of chloroform-aqueous carbonate to minimize contact of acid chloride with the base. Another option allowing efficient coupling is the use of the potassium salt of 1-hydroxybenzotriazole instead of tertiary amine for neutralizing the acid. Optimum conditions for assembly of a peptide chain using Fmoc-amino acid chlorides have been elucidated, but the method has not been adopted for general use because of the attendant obstacles. Fmoc-amino-acid chlorides have nevertheless proven efficient in solid-phase synthesis for attaching the first residue to the hydroxymethyl group of a linker-resin and for coupling hindered residues. They also can be obtained by a general procedure for making acid chlorides using oxalyl chloride and by reaction of a mixed anhydride with hydrogen chloride. The chlorides of derivatives with tertiary-butyl-based side-chain protectors are not accessible by the general procedures, but they can be made using a phosgene replacement, and probably using oxalyl chloride (see Section 7.11).[37–45]

37. E Fischer, E Otto. Synthesis of derivatives of some dipeptides. *Ber Deutsch Chem Ges* 36, 2106, 1903.
38. M Bergmann, L Zervas. On a general method of peptide synthesis. *Ber Deutsch Chem Ges* 65, 1192, 1932.
39. S Pass, B Amit, A Parchornik. Racemization-free photochemical coupling of peptide segments. (Fmoc-amino-acid chlorides) *J Am Chem Soc* 103, 7674, 1981.
40. H Kunz, H-H Bechtolsheimer. Synthesis of sterically hindered peptides and depsipeptides by an acid chloride method with 2-phosphonioethoxycarbonyl-(Peoc)-amino acids and hydroxy acids. *Liebigs Ann Chem* 2068, 1982.
41. LA Carpino, BJ Cohen, KE Stephens, SY Sadat-Aalaee, J-H Tien, DC Lakgridge. (9-Fluorenylmethyl)oxycarbonyl (Fmoc) amino acid chlorides. Synthesis, characterization, and application to the rapid synthesis of short peptide segments. *J Org Chem* 51, 3732, 1986
42. LA Carpino, HG Ghao, M Beyermann, M Bienert. (9-Fluorenylmethyl)oxycarbonylamino acid chlorides in solid-phase peptide synthesis. *J Org Chem* 56, 2635, 1991.
43. FMF Chen, YC Lee, NL Benoiton. Preparation of *N*-9-fluorenylmethoxycarbonylamino acid chlorides from mixed anhydrides by the action of hydrogen chloride. *Int J Pept Prot Res* 38, 97, 1991.
44. KM Sivanandaiah, VV Suresh Babu, SC Shankaramma. Synthesis of peptides mediated by KOBt. *Int J Pept Prot Res* 44, 24, 1994.
45. LA Carpino, M Beyermann, H Wenschuh, M Bienert. Peptide Synthesis via amino acid halides. *Acc Chem Res* 29, 268, 1997.

2.15 PEPTIDE-BOND FORMATION FROM 1-ETHOXYCARBONYL-2-ETHOXY-1,2-DIHYDROQUINOLINE-MEDIATED REACTIONS OF *N*-ALKOXYCARBONYLAMINO ACIDS

One of very few reagents that resemble carbodiimides in that they effect couplings between two reactants without the addition of a fourth compound is 1-ethoxycarbonyl-2-ethoxy-1,2-dihydroquinoline (Figure 2.15), known by its abbreviation, EEDQ. EEDQ emerged as a peptide-bond forming reagent from studies on the inhibition of choline esterase when it was realized that the inhibition involved reaction of the reagent with a carboxyl group of the enzyme. The carboxyl group of an *N*-alkoxycarbonylamino acid displaces the 2-ethoxy group of the reagent, probably by attack of the anion on the protonated reagent. A spontaneous rearrangement with concomitant expulsion of the weakly basic quinoline follows, with the acyloxy and alkoxycarbonyl groups reacting with each other to form the mixed anhydride. Peptide is produced by aminolysis of the mixed anhydride, that takes place immediately. The sequence of reactions is relatively slow, requiring more than 1 hour. There is no

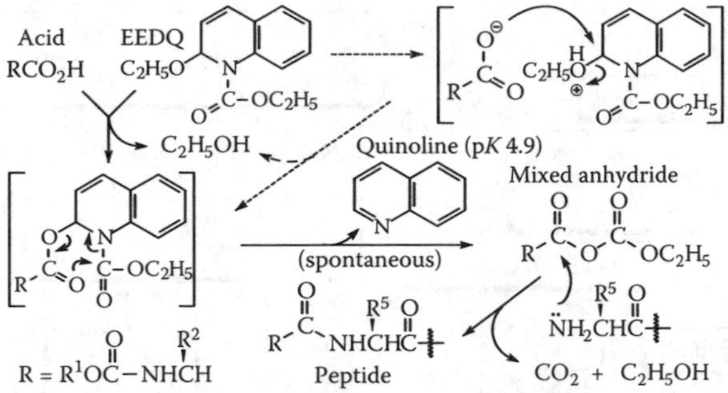

FIGURE 2.15 Peptide-bond formation from 1-ethoxycarbonyl-2-ethoxy-1,2-dihydroquinoline-mediated reactions of *N*-alkoxycarbonylamino acids.[46] The intermediate is the mixed anhydride that is slowly generated in the presence of the attacking nucleophile without a tertiary amine having been added.

build-up of anhydride, and hence there are no side reactions resulting from its decomposition. However, the side reaction of aminolysis at the carbonyl of the ethyl carbonate moiety of the anhydride that is associated with the mixed-anhydride reaction (see Section 2.6) also occurs when EEDQ is used. For these reasons, EEDQ is not employed for solid-phase synthesis, but it is sometimes used routinely instead of DCC for synthesis in solution by operators who have developed skin sensitivity (see Section 7.1) to the carbodiimide. The isobutyl equivalent of EEDQ has been developed to try to minimize the side reaction.[46,47]

46. B Belleau and G Malek. A new convenient reagent for peptide syntheses. *J Am Chem Soc* 90, 1651, 1968.
47. Y Kiso, H Yajima. 2-Isobutoxy-1-isobutoxycarbonyl-1,2-dihydro-quinoline as a coupling reagent in peptide synthesis. *J Chem Soc Chem Commun* 942, 1972.

2.16 COUPLING REAGENTS COMPOSED OF AN ADDITIVE LINKED TO A CHARGED ATOM BEARING DIALKYLAMINO SUBSTITUENTS AND A NONNUCLEOPHILIC COUNTER-ION

The late 1970s saw the beginning of a new era in coupling technologies, with the introduction of reagents that incorporate 1-hydroxybenzotriazole in the molecule. These reagents effect couplings efficiently and at high speeds and consequently are very attractive for use in automated instruments. They comprise oxybenzotriazole that is linked through the oxygen atom to an atom bearing dialkylamino substituents insufficient in number, such that the linking atom is positively charged (Figure 2.16). The charged atom is either phosphorus or carbon; the substituents are either dimethylamino or the five-membered pyrrolidine ring formed from a tetramethylene chain joined at both ends to the nitrogen atom. The positive charge is neutralized by a nonnucleophilic counter-ion, either hexafluorophosphate or tetrafluoroborate. The *tris*-nitrogen-substituted phosphorus-containing reagents are phosphonium salts and

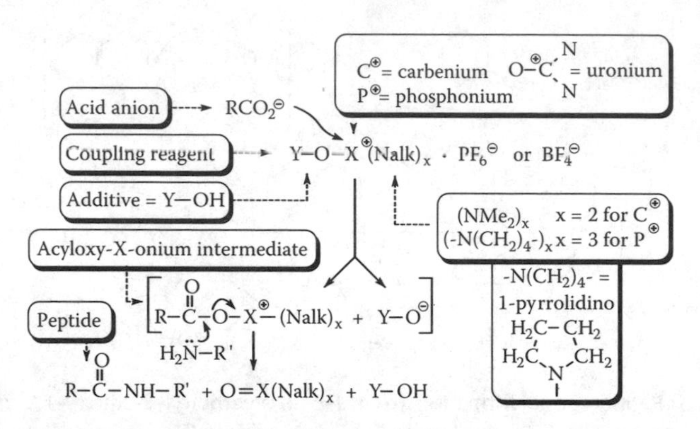

FIGURE 2.16 Coupling reagents made up of an additive linked to a charged atom bearing dialkylamino substituents and a non-nucleophilic counter ion.

are named as such. The *bis*-nitrogen-substituted carbon-containing reagents are carbenium salts, but they are named differently, as uronium salts on the basis of the trivial name of the fully substituted parent compound urea. In terms of their reactivity, the reagents are unique in that they do not react with carboxyl groups but only with carboxylate anions, and they do not react with amino groups if the anion is present. The coupling reactions are, therefore, initiated by converting the carboxyl groups to carboxylate anions by the addition of a tertiary amine. The anions react immediately with the charged atoms of the reagents because the benzotriazolyloxy anion is a good leaving group. The acyloxy-onium intermediate that is formed has a highly activated carbonyl that is quickly attacked by the nitrogen nucleophile to give the peptide or by the benzotriazolyloxy anion to give the benzotriazolyl ester that can also be the precursor of the peptide. Both activation and aminolysis take place in the same solvent, and there is no known general side reaction associated with these reagents that diminishes the yield of peptide by consuming the activated species. When undesirable developments arise, as they occasionally do, they are usually a result of the tertiary amine that is necessary to propel the reaction. Newer reagents that contain other additives such as *N*-hydroxysuccinimide and 1-hydroxy-7-azabenzotriazole have also been developed.

2.17 PEPTIDE-BOND FORMATION FROM BENZOTRIAZOL-1-YL-OXY-*tris*(DIMETHYLAMINO)PHOSPHONIUM HEXAFLUOROPHOSPHATE-MEDIATED REACTIONS OF *N*-ALKOXYCARBONYLAMINO ACIDS

The first of a new generation of coupling reagents adopted by the peptide community that was created from an additive that is linked to a positively charged atom is the title compound $BtOP^{+}(NMe_2)_3 \cdot PF_6^{-}$ (Figure 2.17), known by its abbreviation BOP. It emerged from studies in different laboratories on benzotriazolyl sulfonates and

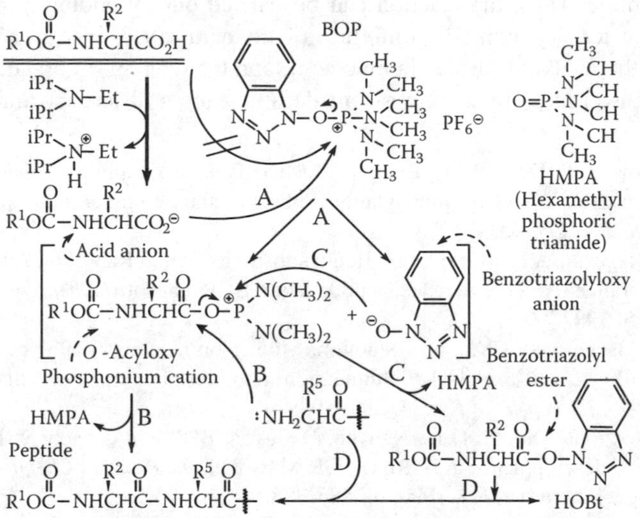

FIGURE 2.17 Peptide-bond formation from benzotriazol-1-yl-oxy-*tris*(dimethylamino)phosphonium hexafluorophosphate-mediated reactions of *N*-alkoxycarbonylamino acids.[48] The peptide can originate by aminolysis of either of two precursors: the acyloxyphosphonium cation and the benzotriazolyl ester.

diphosphonium compounds, and in particular on the effects of HOBt on couplings mediated by phosphonium halides. As for other reagents of this type, BOP does not react with carboxylic acids; the acid must first be converted to its anion by addition of a tertiary amine. Though introduced with triethylamine as the base, it soon became clear that the more basic diisopropylethylamine is the reagent of choice for the deprotonation. The base is added to a mixture of the acid, the amine nucleophile, and BOP. The carboxy anion attacks the positively charged phosphorus atom, displacing the oxybenzotriazolyl moiety as the anion and generating the acyloxyphosphonium intermediate still bearing the positive charge (path A). Peptide is then produced by one of two possible routes, aminolysis at the acyl carbonyl of the intermediate (path B) or attack by the oxybenzotriazolyl anion at the acyl carbonyl to give the benzotriazolyl ester (path C), which then undergoes aminolysis (path D) (see Section 2.20). The reaction goes to completion only if the HOBt that is liberated is neutralized. Thus, two equivalents of tertiary amine are added at the beginning. The reaction occurs quickly, usually within 15 minutes. The liberated hexamethylphosphoric triamide, commonly known as hexamethylphosphoramide, and HOBt can be separated from the product by extraction into water or mild alkali. The reaction can be applied to derivatives of serine and threonine whose hydroxyl groups are not protected. Though applicable for general use, BOP and similarly reacting reagents are particularly suited for the coupling of Boc-amino acids for two reasons. First, the basic milieu prevents the decomposition of activated intermediates that are sensitive to acid (see Section 7.5). Second, manipulation is minimized because neutralization of the acid that binds to the amino group after it is deprotected can be achieved immediately preceding the next coupling by adding an additional equivalent of tertiary amine. Thus, the reaction can be carried out by adding a total of three equivalents of tertiary amine to a mixture of the acid salt of the amino-containing component, the *N*-alkoxycarbonylamino acid, and the reagent. In practice, optimum performance is ensured by adding a considerable excess of tertiary amine.[48–51]

48. B Castro, JR Dormoy, G Evin, B Castro. Peptide coupling reagents IV (1) — benzotriazole *N*-oxytrisdimethylamino phosphonium hexafluorophosphate (B.O.P.) *Tetrahedron Lett* 1219, 1975.
49. D Le-Nguyen, A Heitz, B Castro. Renin substrates. Part 2. Rapid solid phase synthesis of the ratine sequence tetradecapeptide using BOP reagent. *J Chem Soc Perkin Trans 1* 1915, 1987.
50. R Steinauer, FMF Chen, NL Benoiton. Studies on racemization associated with the use of benzotriazol-1-yl-*tris*(dimethylamino)phosphonium hexafluorophosphate (BOP). *Int J Pept Prot Res* 34, 295, 1989.
51. J Coste, M-N Dufour, D Le-Nguyen, B Castro. BOP and congeners: Present status and new developments, in JE Rivier, GR Marshall, eds. *Peptides Chemistry, Structure, Biology.* Escom, Leiden, 1990, pp 885-888.

2.18 PEPTIDE-BOND FORMATION FROM *O*-BENZOTRIAZOL-1-YL-*N,N,N′,N′*-TETRAMETHYLURONIUM HEXAFLUOROPHOSPHATE- AND TETRAFLUOROBORATE-MEDIATED REACTIONS OF *N*-ALKOXYCARBONYLAMINO ACIDS

A different type of coupling reagent that incorporates 1-hydroxybenzotriazole in the molecule was developed very shortly after BOP, but it elicited little attention until more than a decade later, when its successful use was reported by other researchers. In this reagent, the dimethylamino groups are linked to a carbon instead of a phosphorus atom, giving $BtOC^+(NMe_2)_2$ (Figure 2.18). The counter-ion of the original compound known as HBTU (*O*-benzotriazol-1-yl-*N,N,N′,N′*-tetramethyluronium hexafluorophosphate), prepared from tetramethylurea using phosgene and then HOBt, is hexafluorophosphate. A second form known as TBTU (*O*-benzotriazolyltetramethyluronium tetrafluoroborate), with tetrafluoroborate as the counter-ion, became available once the carbenium moiety was prepared using oxalyl chloride instead of phosgene. Two points relating to the nomenclature of these reagents warrant mention. First, the compounds are not named as substituted carbenium (C^+) derivatives but as substituted uronium (OC^+N_2) derivatives. Second, the abbreviations incorporate a designation of the counter-ions (the first letter) and are based on the names of the compounds in French, the language of their developers, in which the anion of a salt is mentioned first (NaCl is chlorure de sodium in French). The mode of action of the reagents is the same as that for BOP except that the atom that is acylated by the carboxylate anion is carbon instead of phosphorus. They do not react with carboxylic acids, only with carboxylate anions. The original reports did not address the issue of the nature of the tertiary amine. This came later, and again

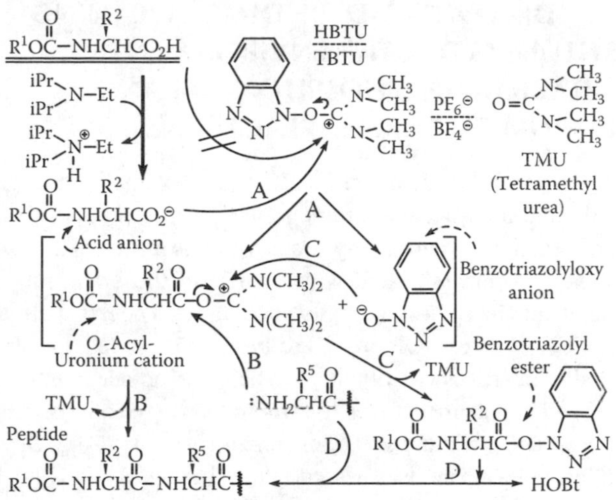

FIGURE 2.18 Peptide-bond formation from *O*-benzotriazol-1-yl-*N,N,N′,N′*-tetramethyluronium hexafluorophosphate– and tetrafluoroborate-mediated reactions of *N*-alkoxycarbonylamino acids.[52,54] The peptide can originate by aminolysis of either of two precursors: the acyloxycarbocation and the benzotriazolyl ester.

diisopropylethylamine emerged as the superior base. Once generated, the carboxylate attacks the reagent, giving the acyloxycarbenium intermediate that is either aminolyzed or converted to the benzotriazolyl ester (see Section 2.20). Peptide-bond formation occurs as quickly as with BOP, and the tetramethylurea liberated is miscible with water, so it is even easier to get rid of than hexamethylphosphoramide. The uronium compounds are slightly less reactive than BOP, but this makes them preferred for solid-phase synthesis because they are more stable during storage. There is little difference between HBTU and TBTU except for their classification for transportation purposes. TBTU is classified as flammable, albeit in the least dangerous of the four categories of such compounds; no transit hazard is associated with HBTU (see Section 7.17 for the revised structures of these compounds).[52–55]

52. V Dourtoglou, J-C Ziegler, B Gross. L'hexafluorophosphate de *O*-benzotriazolyl-*N,N*-tetramethyluronium hexafluorophosphate: a new and efficient peptide coupling reagent. Tetrahedron Lett 1269, 1978.
53. V Dourtoglou, B Gross, V Lambropoulou, C Zioudrou. *O*-Benzotriazolyl-*N,N,N′,N′*-tetramethyluronium hexafluorophosphate as coupling reagent for the synthesis of peptides of biological interest. *Synthesis* 572, 1984.
54. R Knorr, A Trzeciak, W Bannwarth, D Gillesen. New coupling reagents in peptide chemistry. *Tetrahedron Lett* 30, 1927, 1989.
55. GE Reid, RJ Simpson. Automated solid-phase peptide synthesis: use of 2-(1*H*-benzotriazol-1-yl)-1,1,3,3-tetramethyluronium tetrafluoroborate for coupling of *tert*-butyloxycarbonyl amino acids. *Anal Biochem* 200, 301, 1992.

2.19 PYRROLIDINO INSTEAD OF DIMETHYLAMINO SUBSTITUENTS FOR THE ENVIRONMENTAL ACCEPTABILITY OF PHOSPHONIUM AND CARBENIUM SALT-BASED REAGENTS

The side product from BOP-mediated reactions is hexamethylphosphoric triamide, which is volatile and a suspected carcinogen. The side product from HBTU- and TBTU-mediated reactions is tetramethylurea, which is more volatile and is cytotoxic. Realization of the hazards associated with the use of these reagents led to a search for a variant that would be environmentally acceptable. The *tris*(diethylamino) equivalent [$BtOP^+(NEt_2)_3 \cdot PF_6^-$] of BOP proved to be much less reactive, but a compound with the desired properties was obtained when the adjacent ethyl groups of the former were joined together to form five-membered pyrrolidine rings. This compound, $BtOP^+(Pyr)_3 \cdot PF_6^-$ (Figure 2.19), or benzotriazol-1-yl-oxytripyrrolidinophosphonium hexafluorophosphate, was assigned the trade name PyBOP. PyBOP has all the attractive properties of BOP, and the tripyrrolidinophosphoric oxide that is liberated by its reactions is innocuous though it may require chromatography on silica gel to remove it from a product. The corresponding equivalent of HBTU, describable as $BtOC^+(Pyr)_2 \cdot PF_6^-$ or $BtOC^+(\text{-NBu-})_2 \cdot PF_6^-$ and known as HBPyU or BCC became available shortly after. It must be pointed out that care is in order when naming uronium compounds containing pyrrolidino substituents. The rings cannot be named as such because the nitrogen atom is already included in the term uronium, and the current abbreviations add to the confusion. Precise names for the compound represented by HBPyU are *O*-(benzotriazol-1-yl)-*N,N,N′,N′*-*bis*(tetramethylene)uronium, 2-(benzotriazol-1-yl)-1,1,3,3-*bis*(tetramethylene)uronium, and benzotriazol-1-yl-oxy-*bis*(pyrrolidino)carbenium hexafluorophosphate (see Section 7.17 for a revision of the structures). A pentamethylene equivalent also exists. It appears that the pyrrolidino-substituted reagents generally perform as well or better than their dimethylamino counterparts.[56,57]

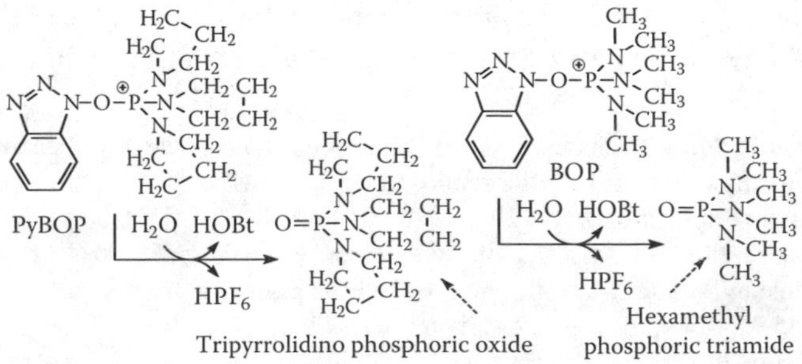

FIGURE 2.19 Pyrrolidino instead of dimethylamino substitutents for the environmental acceptability of phosphonium and carbenium salt-based reagents.[56] Tetramethylurea from *O*-benzotriazol-1-yl-*N,N,N,N′*-tetramethyluronium hexafluorophosphate and tetrafluoroborate is more volatile and is cytotoxic. The product released from PyBOP is not environmentally objectionable. PyBOP = benzotriazol-1-yl-oxytripyrrolidinophosphonium hexafluorophosphate.

56. J Coste, D Le-Nguyen, B Castro. PyBOP: A new peptide coupling reagent devoid of toxic by-product. *Tetrahedron Lett* 31, 205, 1990.
57. S Chen, J Xu. A new coupling reagent for peptide synthesis. Benzotriazol-yl-bis(pyrrolidino)-carbonium hexafluorophosphate (BCC). *Tetrahedron Lett* 33, 647, 1992.

2.20 INTERMEDIATES AND THEIR FATE IN BENZOTRIAZOL-1-YL-OXYPHOSPHONIUM AND CARBENIUM SALT-MEDIATED REACTIONS

The first step in BOP-mediated coupling reactions is displacement of the benzotriazolyloxy moiety of the reagent by the anion of the starting acid to give the acyloxyphosphonium cation and the benzotriazolyloxy anion (Figure 2.20, path A). Two alternative courses of action are then possible. The phosphonium intermediate can undergo aminolysis to give the peptide (path B), or it can undergo attack by the benzotriazolyloxy anion to give the benzotriazolyl ester (path C), which is then aminolyzed to give the peptide (path D). The acyloxyphosphonium ion is an intermediate that has been postulated; determined efforts to try to detect it have proven unsuccessful, except in the case of the highly hindered 2,4,6-trimethylphenylacetic acid. When BOP was introduced, it was suggested that the precursor of the peptide was the benzotriazolyl ester. However, there are a few observations that are difficult to reconcile with this theory. The first is that peptide-bond formation is so quick as to be inconsistent with the rate of aminolysis of an activated ester. A second observation indicating that the benzotriazolyl ester is not the only immediate precursor of product is the fact that *N*-alkoxycarbonyl-*N*-methylamino acids can be coupled to the amino group of a peptide chain using BOP, yet the benzotriazolyl ester of the

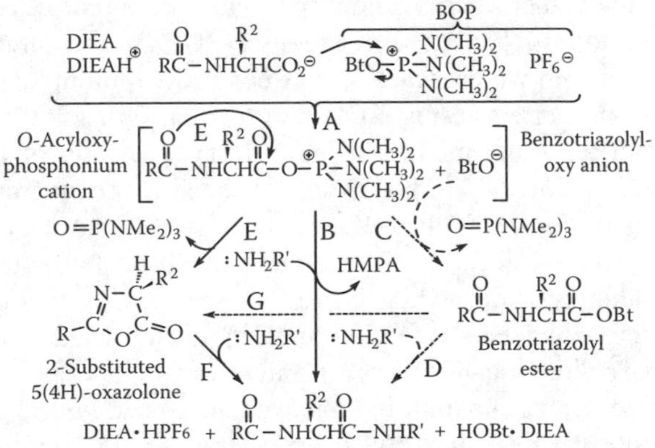

FIGURE 2.20 Intermediates and their fate in benzotriazol-1-yl-oxyphosphonium salt–mediated reactions. Indirect evidence (Figure 2.21) is not compatible with the tenet that the precursor of the peptide is the benzotriazolyl ester (path C). The evidence indicates that the peptide originates from the acyloxyphosphonium cation (path B). Conversion of this intermediate into the oxazolone (path E) can account for the epimerization that occurs during segment couplings.

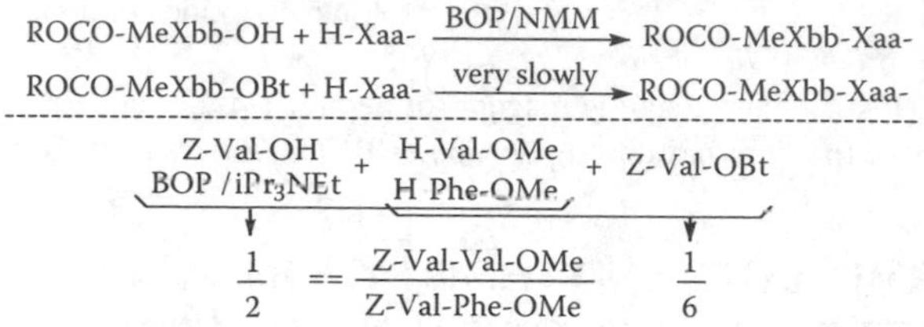

ROCO-MeXbb-OH + H-Xaa- $\xrightarrow{\text{BOP/NMM}}$ ROCO-MeXbb-Xaa-

ROCO-MeXbb-OBt + H-Xaa- $\xrightarrow{\text{very slowly}}$ ROCO-MeXbb-Xaa-

Z-Val-OH + BOP / iPr$_3$NEt + H-Val-OMe / H Phe-OMe + Z-Val-OBt

$\frac{1}{2}$ == $\frac{\text{Z-Val-Val-OMe}}{\text{Z-Val-Phe-OMe}}$ $\frac{1}{6}$

FIGURE 2.21 Experimental evidence indicates that products from BOP-mediated reactions do not originate from the benzotriazolyl ester. The use of BOP allows successful coupling of *N*-alkoxycarbonyl-*N*-methylamino acids, whereas the benzotriazolyl esters of these acids undergo aminolysis only with great difficulty. The higher ratio of products obtained from the BOP-mediated reaction in the competing reactions described implies a compound other than the benzotriazolyl ester as the precursor of the peptides.

same *N*-alkoxycarbonyl-*N*-methylamino acids react exceptionally slowly with amino groups (Figure 2.21). Some intermediate other than the benzotriazolyl ester must be invoked to account for success in BOP-mediated couplings of *N*-methylamino acids. A third argument resides in the fact that BOP-mediated couplings of peptide segments are not free from epimerization at the activated residue. The isomerization can attributed to formation (Figure 2.20, path E) and enolization of the 2-alkyl-5(4*H*)-oxazolone, which is in the presence of the base (see Section 1.9). The developers of BOP suggested that the oxazolone arose from the benzotriazolyl ester (Figure 2.20, path G). However, it is just as reasonable to hold that the precursor of the oxazolone is the phosphonium intermediate (path E) than to hold that it is the benzotriazolyl ester (path G).

This argument was put forth in a lecture by N. L. Benoiton in 1988 at a symposium in Montpellier, France, and it induced the group of B. Castro to refocus their attention on the mechanism of reaction of BOP. The researchers carried out competitive reactions in which Z-valine was coupled with a mixture of valine and phenylalanine methyl esters, using BOP, and the benzotriazolyl ester of Z-valine was coupled with the same mixture (Figure 2.21). The ratio of the two peptide products, Z-Val-Val-OMe/Z-Val-Phe-OMe, with the former containing two hindered residues, was higher for the BOP-mediated reaction. As a consequence, they concluded that the products in the BOP-mediated reaction could not have originated exclusively by aminolysis of the benzotriazolyl ester. In fact, it is not necessary to invoke the benzotriazolyl ester as an intermediate in BOP-mediated reactions. The results can be explained on the basis of aminolysis of the acyloxyphosphonium cation. However, the phosphonium intermediate does react with the benzotriazolyloxy anion, giving the benzotriazolyl ester when it is not consumed by another nucleophile. In such cases, the benzotriazolyl ester is the equivalent of the symmetrical anhydride in carbodiimide-mediated reactions (see Section 2.4)—each is generated when the initial activated species, the acyloxyphosphonium ion and the *O*-acylisourea, respectively, are not consumed by another nucleophile. The uronium-based reagents are believed to effect reactions by the same mechanism. That being said, most scientists take it for granted and report results on the basis that the peptide is formed by aminolysis of the benzotriazolyl ester. In reality, the question has little practical significance because it has no effect on the reactions if they are carried out according to the original protocols (see Section 7.20).[51,58,59]

51. J Coste, M-N Dufour, D Le-Nguyen, B Castro. BOP and congeners: Present status and new developments, in JE Rivier, GR Marshall, eds. *Peptides Chemistry, Structure, Biology.* Escom, Leiden, 1990, pp 885-888.
58. B Castro, J-R Dormoy, G Evin, C Selve. Peptide coupling reagents. Part VII. Mechanism of the formation of active esters of hydroxybenzotriazole in the reaction of carboxylate ions on the BOP reagent for peptide coupling. A comparison with Itoh's reagent. *J Chem Res (S)* 82, 1977.
59. J Coste, E Frérot, P Jouin, B Castro. Oxybenzotriazole free peptide coupling reagents for N-methylated amino acids. *Tetrahedron Lett* 32, 1967, 1991.

2.21 1-HYDROXYBENZOTRIAZOLE AS ADDITIVE IN COUPLINGS OF *N*-ALKOXYCARBONYLAMINO ACIDS EFFECTED BY PHOSPHONIUM AND URONIUM SALT-BASED REAGENTS

Phosphonium and uronium salt-based reagents effect couplings by first reacting with the anion of the starting acid (Figure 2.22, path A). The benzotriazolyl ester is then one of the two possible precursors of the peptide. Operating in a research climate in which HOBt was commonly used as additive, Hudson rationalized that if the ester is the precursor of the peptide, additional HOBt in the form of the anion would be beneficial because it would by mass action promote formation of the ester. A favorable effect from adding HOBt was reported, so one variant of the use of

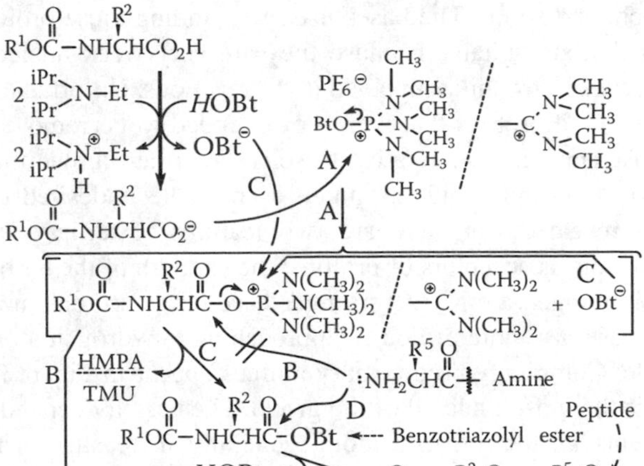

FIGURE 2.22 Couplings using phosphonium and uronium salt-based reagents with 1-hydroxybenzotriazole as additive.[60] The additional HOBt promotes formation of the benzotriazolyl ester, which is the precursor of the peptide.

phosphonium and uronium salt-based reagents is to carry out the reaction in the presence of HOBt and the additional tertiary amine required to transform it into the anion. The practice became common enough, as it was recommended as the best protocol by an instrument manufacturer. However, subsequent experience and reinvestigation have changed the views of the peptide community on this issue. Cases have been identified in which the addition of HOBt to couplings of valine and isoleucine derivatives resulting from onium salt-based reagents has been detrimental, and the instrument manufacturer has withdrawn its recommendation. The deleterious effect of the additive can be explained on the basis that the esters generated from hindered residues are less readily aminolyzed, as has been shown for the esters of *N*-methylamino acids (see Section 2.20). The latest developments seem to indicate that the routine supplementation of onium salt-mediated reactions of *N*-alkoxycarbonylamino acids with HOBt is open to question.[59,60]

59. J Coste, E Frérot, P Jouin, B Castro. Oxybenzotriazole free peptide coupling reagents for N-methylated amino acids. *Tetrahedron Lett* 32, 1967, 1991.
60. D Hudson. Methodological implications of simultaneous solid-phase peptide synthesis. 1. Comparison of different coupling procedures. *J Org Chem* 53, 617, 1988.

2.22 SOME TERTIARY AMINES USED AS BASES IN PEPTIDE SYNTHESIS

There are many cases in peptide synthesis where reactions involve the removal of protons. Amino groups are deprotonated so they can act as nucleophiles in aminolysis reactions, carboxyl groups are deprotonated so that they will react with coupling reagents, acids liberated during aminolysis are neutralized so that the reactions will go to completion, and so on. The bases used for binding these protons are usually tertiary amines that are suitable because they are very weak nucleophiles that do not compete for the electrophiles involved in the reactions. Unfortunately, the tertiary amines may also bind protons that were not intended to be removed, which leads to side reactions. Thus, it is imperative to select a tertiary amine that achieves the desired deprotonation and avoids as much as possible undesired deprotonations. Experience has revealed that there are two features of tertiary amines that are pertinent to their role as acceptors of protons: the strength of the base and hindrance around the basic nitogen atom. The stronger base binds protons more tightly. The more hindered base has more difficulty approaching a hydrogen atom to remove it from a molecule. Commonly used tertiary amines appear in Figure 2.23. Diisopropylethylamine is the most hindered of the group. The basicity depends on the nature of the substituents on the nitrogen atom, generally increasing as the number of electron-donating methyl and methylene groups increases, though diisopropylethylamine is not more basic than triethylamine. Pyridine is the least basic because of the aromatic ring that is electron deficient. It is not basic enough for most purposes but is useful as a solvent for preparing activated esters (see Section 7.13). The base traditionally used by peptide chemists has been *N*-methylmorpholine, a practice that emerged from in-depth studies on the mixed-anhydride method of coupling. It was shown that this weaker and less-hindered base led to less isomerization during

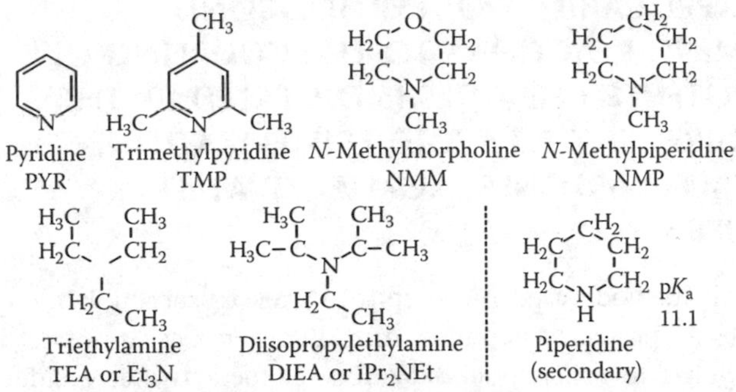

Pyridine PYR | Trimethylpyridine TMP | *N*-Methylmorpholine NMM | *N*-Methylpiperidine NMP

Triethylamine TEA or Et_3N | Diisopropylethylamine DIEA or iPr_2NEt | Piperidine (secondary) pK_a 11.1

Hindrance: PYR < NMM = NMP < TMP < TEA < DIEA

Aqueous pKa: 5.2 7.38 10? 7.42 10.7 10.4

FIGURE 2.23 Some tertiary amines used as bases in peptide synthesis. The less hindered amines are able to abstract protons more readily, but the more basic amines bind the protons more tightly.

segment couplings than triethylamine and tri-*n*-butylamine, which had been popular at the time. The apparent anomaly may be rationalized on the basis of the fact that the tertiary amine participates in mixed-anhydride reactions not only as a proton acceptor but also as an acyl carrier (see Section 2.6). Regardless, *N*-methylmorpholine became the base of choice for all purposes for more than two decades. This statement is substantiated by the fact that the earlier experiments carried out by its developers and others with BOP (see Section 2.17), the first of the onium salt–type reagents, were carried out with *N*-methylmorpholine as the base. It was soon realized, however, that maximum efficiency both for phosphonium and uronium salt-mediated reactions was achieved using diisopropylethylamine. The stronger base is required to drive the reaction that requires removal of the proton from the carboxyl group of the acid-containing moiety. However, this stronger base does not give any mixed anhydride when added to a mixture of acid and chlorofomate. This can only be explained by the fact that it is too hindered to initiate the reaction. *N*-Methylpiperidine has proven to be superior to *N*-methylmorpholine for the mixed-anhydride reactions. In terms of minimizing isomerization during segment couplings, a recent report indicates that trimethylpyridine might be superior to other tertiary amines.[13,51,61–63]

13. RA Boissonnas. A new method of peptide synthesis. *Helv Chim Acta* 34, 874, 1951.
51. J Coste, M-N Dufour, D Le-Nguyen, B Castro. BOP and congeners: Present status and new developments, in JE Rivier, GR Marshall, eds. *Peptides Chemistry, Structure, Biology.* Escom, Leiden, 1990, pp 885-888.
61. FMF Chen, Y Lee, R Steinauer, NL Benoiton. Mixed anhydrides in peptide synthesis. Reduction of urethane formation and racemization using *N*-methylpiperidine as tertiary amine base. *J Org Chem* 48, 2939, 1983.
62. LA Carpino, D Ionescu, A El-Faham. Peptide segment coupling in the presence of highly hindered tertiary amines. *J Org Chem* 61, 2460, 1996.
63. D Perrin. *Dissociation Constants of Organic Bases in Aqueous Solution*, Butterworths, London, 1965, Supplement, 1972.

2.23 THE APPLICABILITY OF PEPTIDE-BOND FORMING REACTIONS TO THE COUPLING OF *N*-PROTECTED PEPTIDES IS DICTATED BY THE REQUIREMENT TO AVOID EPIMERIZATION: 5(4*H*)-OXAZOLONES FROM ACTIVATED PEPTIDES

In principle, all methods used for coupling *N*-alkoxycarbonylamino acids can be used to couple N^{α}-protected peptides. Activated peptides, however, have a strong tendency to cyclize to generate an oxazolone at the activated residue (see Section 1.9). The oxazolone produces the target peptide by aminolysis, but its formation usually results in an isomerized product when the activated residue is chiral (see Section 1.7). Hence, only those methods for which the cyclization reaction can be suppressed considerably or does not occur are used in practice. The ease with which activated peptides generate oxazolones is illustrated by the following observations: When an *N*-acylamino acid or N^{α}-protected peptide is mixed with a carbodiimide at 0°C in the absence of an amine nucleophile, the product is the oxazolone (Figure 2.24, path A) and not the symmetrical anhydride, as is the case for *N*-alkoxycarbonylamino acids (see Section 1.16). Similarly, when a chloroformate is added to an *N*-acylamino acid or N^{α}-protected dipeptide in the presence of *N*-methylmorpholine at 0°C and the salts are removed within 2 minutes by washing the solution with water, the product is the oxazolone (Figure 2.24, path B) and not the mixed anhydride, as is the case for *N*-alkoxycarbonylamino acids (see Section 2.8). The 2-alkyl-5(4*H*)-oxazolones isomerize readily (see Section 4.4); the longer they exist in solution, the greater the isomerization. As a consequence, the avoidance of oxazolone formation and its subsequent enolization are primordial to the applicability of peptide-bond-forming reactions in the coupling of peptides. Special precautions to

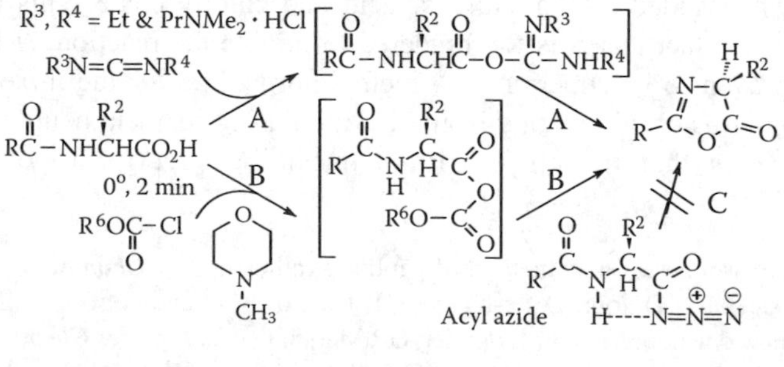

FIGURE 2.24 Oxazolones from activated dipeptides. Reaction of the acid with soluble carbodiimide in dichloromethane at 0°C (path A) followed by washing with water gave enantiomerically pure oxazolones in high yield.[65] (see section 1-17). Generation of the mixed anhydride of the acid (path B) followed by washing with water gave chemically pure oxazolones that were close to enantiomerically pure, 95% enantiomeric excess for Z-glycylleucine.[66] An acyl azide does not form an oxazolone (path C). Cyclization is probably prevented by the hydrogen bond formed between the azido nitrogen atom and the NH-proton.[67]

suppress oxazolone formation must be taken for all methods except the acyl-azide method. Acyl azides do not form oxazolones, most likely because cyclization (Figure 2.24, path C) is prevented by a hydrogen bond formed between the nitrogen atom of the azide group and the NH proton of the activated residue. In terms of the nature of the activated residue, the more hindered valine and isoleucine have a greater tendency to cyclize. This tendency to cyclize is so strong for the extremely hindered aminoisobutyric acid that it is unavoidable during activation of the latter. In fact, the 2-substituted-4,4-dimethyl-5(4*H*)-oxazolone that is produced is often isolated and used as the activated form of the segment in question. The hindered *N*-methylamino acids also form the equivalent of oxazolones more readily than their parent counterparts (see Section 8.14). In contrast, acyl-prolines form oxazolones so slowly that they are of no consequence. The reason for this is that formation of the product with two contiguous rings is energetically unfavored. Thus, activated peptides have a tendency to form oxazolones, except when the carboxy-terminal residue is proline or when the activating group is an azide group. Formation of the oxazolone is inconsequential when the carboxy-terminal residue is glycine or aminoisobutyric acid because they are not chiral residues (see Section 7.23).[64–68]

64. I Antonovics, AL Heard, J Hugo, MW Williams, GT Young. Current work on the racemization problem, in L Zervas, ed. *Peptides. Proceedings of the 6th European Peptide Symposium*. Pergamon, Oxford, 1966, pp 121-130.
65. FMF Chen, K Kuroda, NL Benoiton. A simple preparation of 5-oxo-4,5-dihydro-1,3-oxazoles (Oxazolones). *Synthesis* 230, 1979.
66. FMF Chen, M Slebioda, NL Benoiton. Mixed carboxylic-carbonic acid anhydrides of acylamino acids and peptides as a convenient source of 2,4-dialkyl-5(4*H*)-oxazolones. *Int J Pept Prot Res* 31, 339, 1988.
67. M Crisma, V Moretto, G Valle, F Formaggio, C Toniolo. First characterization at atomic resolution of the *C*-activating groups in a peptide synthesis acid chloride, acid azide and carboxylic-carboxylic mixed anhydride. *Int J Pept Prot Res* 42, 378, 1993.
68. P Wipf, H Heimgartner. Coupling of peptides with C-terminal α,α-disubstituted α-amino acids *via* oxazol-5(4*H*)-one. *Helv Chim Acta* 69, 1153, 1986.

2.24 METHODS FOR COUPLING *N*-PROTECTED PEPTIDES

The chemistry of the reactions involved in coupling peptides is the same as that for coupling *N*-alkoxycarbonylamino acids. However, the oxazolone that is formed by the activated peptide is chirally unstable, it is formed more readily, and there is an added impetus for it to form because the rate of bond formation between segments is lower. In addition, segments usually have to be coupled in polar solvents because they are insoluble in nonpolar solvents, and polar solvents promote the undesirable side reaction. The result is that the number of procedures actually used for coupling peptides is rather small. The methods in question are addressed below.

Symmetrical anhydrides (see Section 2.05): There is no evidence that anhydrides of peptides exist. Methods that might be used to produce them generate oxazolones.

Acyl halides (see Section 2.14): There have been no reports of their use. The tendency for them to form oxazolones is exceedingly high.

Carbodiimides (see Section 2.2): These are rarely used alone because of isomerization and *N*-acylurea formation that can be suppressed considerably by the use of additives.

Phosphonium and uronium salts (see Sections 2.17–2.19): These are rarely used alone because of isomerization that is promoted by the tertiary amine that is required to effect the reaction.

Activated esters (see Section 2.9): Activated esters of peptides are rarely used because there is no general method available for converting an N^{α}-protected peptide into the ester with a guarantee that it will be a single isomer. Attempts have been made to overcome this obstacle (see Section 7.8). However, solid phase synthesis allows the preparation of thioesters of segments (see Section 7.10). Once the ester is in hand, it can be aminolyzed without generation of a second isomer if suitable conditions are employed.

Acyl azides (see Section 2.13): The acyl-azide method of coupling is unique for two reasons. First, it is the only case in which the immediate precursor of the activated form of the peptide is not the parent acid. The starting material is the peptide ester that is obtained from the amino acid ester by usual chain assembly (Figure 2.25, path A). Second, it is the only method that just about guarantees production of a peptide that is enantiomerically pure, provided scrupulous attention is paid to details of procedure. There is no danger for loss of chirality during conversion of the ester to the hydrazide and then the azide, but care must be taken to avoid contact of

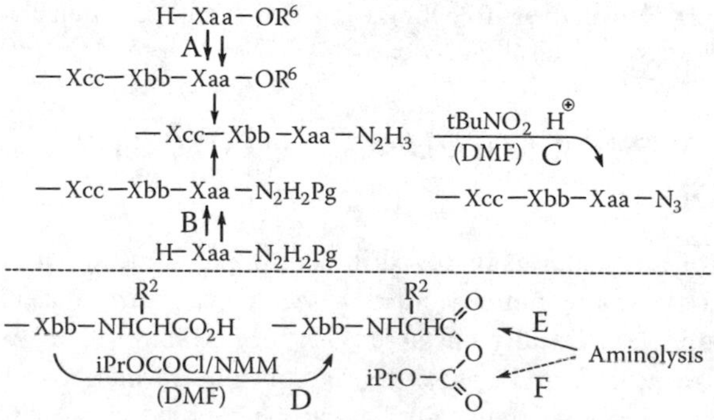

H–Xaa–OR^6
A
–Xcc–Xbb–Xaa–OR^6
–Xcc–Xbb–Xaa–N_2H_3 — $tBuNO_2$ $H^{\oplus}$ (DMF) C → –Xcc–Xbb–Xaa–N_3
–Xcc–Xbb–Xaa–N_2H_2Pg
B
H–Xaa–N_2H_2Pg

–Xbb–NHCH(R^2)CO_2H — iPrOCOCl/NMM (DMF) D → –Xbb–NHCH(R^2)C(=O)–O–C(=O)–OiPr ← E, F Aminolysis

FIGURE 2.25 Methods for coupling N^{α}-protected peptide segments. Different reagents than those used for coupling *N*-alkoxycarbonylamino acids are employed for generating acyl azides (path C) and mixed anhydrides (path D). To permit use of side-chain protectors that are sensitive to hydrazine, the acyl hydrazide can be obtained by a different approach (path B). Pg = protecting group.

the acyl azide with base, even aqueous sodium hydrogen carbonate, which has an adverse effect on its stereochemistry. Maximum efficiency in transforming the hydrazide to the azide is achieved at 0°C or less in an organic solvent using acid and *tert*-butyl nitrite instead of sodium nitrite (Figure 2.25, path C), and diisopropylethylamine for neutralizing the acid. Hydrazine-sensitive side-chain protectors are incompatible with the acyl-azide method, but this obstacle can be circumvented by preparing the hydrazide at the stage of the first residue in a protected form (Figure 2.25, path B), assembling the peptide in the usual manner, and then deprotecting the hydrazide to effect the coupling (see Section 7.16).

Mixed anhydrides (see Section 2.6): The mixed-anhydride method provides efficient coupling of peptides with minimal isomerization if the established protocol is strictly adhered to. This includes a short activation time at low temperature, isopropyl chloroformate as the reagent, and *N*-methylmorpholine or *N*-methylpiperidine as the tertiary amine (Figure 2.25, path D). In what is an apparent anomaly with respect to conventional wisdom, a polar solvent such as dimethylformamide seems to be preferable to apolar solvents for minimizing isomerization. Aminolysis at the wrong carbonyl of the anhydride of a peptide (path F) is less than that for the anhydride from the corresponding *N*-alkoxycarbonylamino acid.

EEDQ (see Section 2-15): EEDQ is the only coupling reagent that is used without an additive.

DCC or EDC with an additive: This is probably the most common method of coupling segments, with HOBt and HOObt competing as the most efficient additives. HOAt may be on par with the other two. The additive is essential to reduce isomerization to acceptable levels (see Sections 2.25 and 2.26). An important variant is supplementation of the reaction mixture with a cupric ion that minimizes or eliminates isomerization by preventing any oxazolone that is formed from enolizing (see Section 7.2).

BOP, PyBOP, HBTU, HATU, and so forth with an additive: It has been considered essential to use an additive with these reagents because the tertiary amine required to effect the coupling promotes isomerization. Diisopropylethylamine or possibly trimethylpyridine are the bases of choice to minimize the side reaction, but the additive may increase isomerization (see Section 7.18).

In general, products from the coupling of segments are easier to purify than products from the coupling of amino acid derivatives because the differences in physical properties between the reactants and products are greater.[12,17,69–77]

12. GW Anderson, JE Zimmerman, FM Callahan. A reinvestigation of the mixed carbonic anhydride method of peptide synthesis. *J Am Chem Soc* 89, 5012, 1967.
17. J Meienhofer. The mixed carbonic anhydride method of peptide synthesis, in E Gross, J Meienhofer, eds. *The Peptides: Analysis, Synthesis, Biology*, Academic, New York, 1979, Vol 1, pp 263-314.

69. NL Benoiton, Y Lee, FMF Chen. Isopropyl chloroformate as a superior reagent for mixed anhydride generation and couplings in peptide synthesis. *Int J Pept Prot Res* 31, 577, 1988.
70. J Honzl, J Rudinger. Amino acids and peptides. XXXIII. Nitrosyl chloride and butyl nitrite as reagents in peptide synthesis by the azide method; suppression of amide formation. *Coll Czech Chem Commun* 26, 2333, 1961.
71. H Romovacek, SR Dowd, K Kawasaki, N Nishi, K Hofmann. Studies on polypeptides. 54. The synthesis of a peptide corresponding to positions 24-104 of the peptide chain of ribonuclease T_1. *J Am Chem Soc* 101, 6081, 1979.
72. E Wünsch, H-G Heidrich, W Grassmann. Synthesis of $Lys^{1,9}$-bradykinin and $Lys^{2,10}$-kallidin. *Chem Ber* 97, 1818, 1964.
73. E Wünsch, K-H Deimer. On the synthesis of [15-leucine]human gastrin I. *Hoppe-Seyler's Z Physiol Chem* 353, 1255, 1972.
74. E Wünsch, E Jaeger, S Knof, R Scharf, P Thamm. On the synthesis of motilin, III Determination of the purity and characterization of [13-norleucine]-motilin and [13-leucine]motilin. *Hoppe-Seyler's Z Physiol Chem* 357, 467, 1976.
75. N Fujii, H Yajima. Total synthesis of bovine pancreatic ribonuclease A. Parts 1 to 6. *J Chem Soc Perkin Trans 1* 789, 1981.
76. S Sakakibara. Synthesis of large peptides in solution. *Biopolymers (Pept Sci)* 37, 17, 1995.
77. NL Benoiton, YC Lee, R Steinauer, FMF Chen. Studies on the sensitivity to racemization of activated residues in couplings of *N*-benzyloxycarbonyldipeptides. *Int J Pept Prot Res* 40, 559, 1992.

2.25 ON THE ROLE OF 1-HYDROXYBENZOTRIAZOLE AS AN EPIMERIZATION SUPPRESSANT IN CARBODIIMIDE-MEDIATED REACTIONS

1-Hydroxybenzotriazole increases efficiency in carbodiimide-mediated reactions by preventing *N*-acylurea formation by protonating the *O*-acylisourea (see Section 2.12), and experimental data have shown that HOBt facilitates the aminolysis of 2-alkyl-5(4*H*)-oxazolones and suppresses epimerization during the coupling of peptides. Epimerization is caused by the formation and enolization of the oxazolone. HOBt has an effect on both of these processes. It suppresses formation of the oxazolone by protonating the *O*-acylisourea (Figure 2.26) that enhances its electrophilicity, thus accelerating its consumption, while at the same time generating the oxybenzotriazole anion that is an additional acceptor of the electrophile. Similarly, HOBt suppresses enolization of the oxazolone by protonating it (Figure 2.26), while at the same time generating the oxybenzotriazole anion that is a good acceptor of the electrophiles. It must be borne in mind, however, that the reaction of HOBt with the 5(4*H*)-oxazolone of a peptide is not intantaneous; it takes several minutes for the reaction to go to completion. HOBt also increases efficiency in the coupling of *N*-alkoxycarbonylamino acids. A possible intermediate in the reactions of *N*-alkoxycarbonylamino acids is the 2-alkoxy-5(4*H*)-oxazolone (see Section 1.10). Using the same reasoning, it can be postulated that the reactivity of any 2-alkoxy-5(4*H*)-oxazolone that is formed is enhanced by its protonation. Thus, a beneficial effect of HOBt in couplings of *N*-alkoxycarbonylamino acids can be attributed to protonation

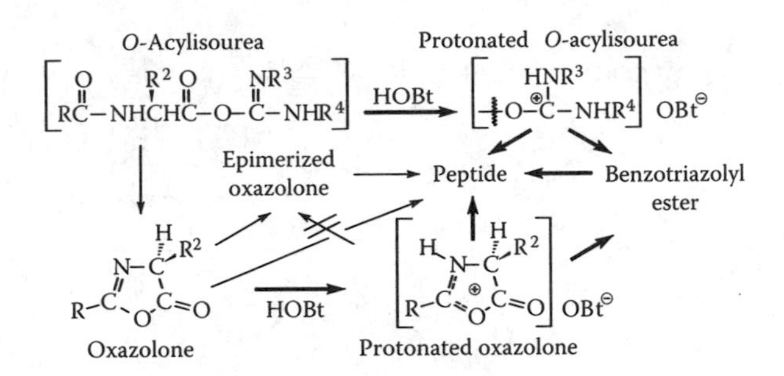

FIGURE 2.26 On the role of 1-hydroxybenzotriazole as an epimerization suppressant in carbodiimide-mediated reactions.

of the oxazolone that facilitates its consumption if it is formed. That the acidic nature of HOBt is involved in its beneficial effects is substantiated by the fact that camphorsulfonic acid, a similar noncarboxylic acid, catalyzes the aminolysis of oxazolones formed from *N*-peptidylaminoisobutyric acid.

The existence of a protonated oxazolone has been demonstrated indirectly by a simple experiment. When *p*-nitrophenol was added to an excess of 2-alkoxy-5(4*H*)-oxazolone in dichloromethane, a yellow color appeared. The color persisted until all the *p*-nitrophenol had been consumed by the oxazolone. The anion of *p*-nitrophenol is yellow. The explanation for the color of the mixture is the presence of the *p*-nitrophenoxide anion that was generated by abstraction of the proton by the oxazolone. In summary, protonation of the *O*-acylisourea suppresses the side reaction of oxazolone formation as well as the side reaction of *N*-acylurea formation and accelerates its consumption by enhancing its reactivity and generating an additional good nucleophile that consumes it. Protonation of the oxazolone suppresses epimerization by preventing its enolization and also increases the rate at which it is consumed.[4,68,78,79]

4. DH Rich, J Singh. The carbodiimide method, in E Gross, J Meienhofer, eds. *The Peptides: Analysis, Synthesis, Biology*, Academic, New York, 1979, Vol 1, pp 241-261.
68. P Wipf, H Heimgartner. Coupling of peptides with C-terminal α,α-disubstituted α-amino acids *via* oxazol-5(4*H*)-one. *Helv Chim Acta* 69, 1153, 1986.
78. P Wipf, H Heimgartner. 2. Synthesis of peptides containing α,α-disubstituted α-amino acids by the azirine/oxazolone method. *Helv Chim Acta* 73, 13, 1990.
79. NL Benoiton. 2-Alkoxy-5(4*H*)-oxazolones and the enantiomerization of *N*-alkoxycarbonylamino acids. *Biopolymers (Pept Sci)* 40, 245, 1996.

2.26 MORE ON ADDITIVES

Additives increase efficiency in carbodiimide-mediated reactions by preventing intermediates from undergoing side reactions and by transforming them into activated esters that become the precursors of the peptide products. *N*-hydroxysuccinimide and 4-hydroxy-3-oxo-3,4-dihydrobenzotriazine are the best nucleophiles or acceptors of activated species. They trap oxazolones before they have time to isomerize. The popularity of HONSu as an additive has diminished considerably during the

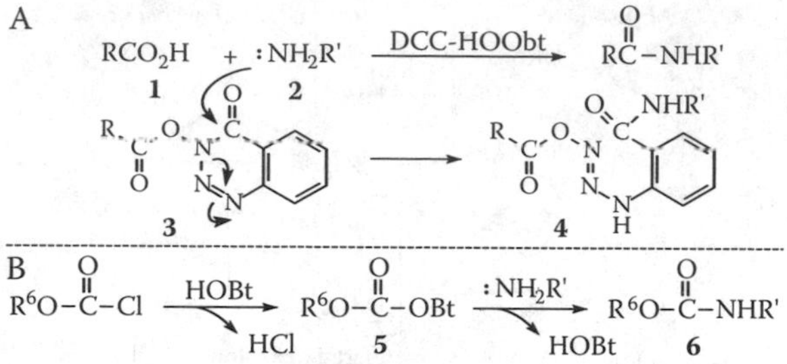

A

$$\underset{\mathbf{1}}{RCO_2H} + \underset{\mathbf{2}}{:NH_2R'} \xrightarrow{\text{DCC-HOObt}} RC(=O)-NHR'$$

$$\mathbf{3} \longrightarrow \mathbf{4}$$

B

$$R^6O-C(=O)-Cl \xrightarrow[\text{HCl}]{\text{HOBt}} \underset{\mathbf{5}}{R^6O-C(=O)-OBt} \xrightarrow[\text{HOBt}]{:NH_2R'} \underset{\mathbf{6}}{R^6O-C(=O)-NHR'}$$

FIGURE 2.27 More on additives. In a carbodiimide-mediated reaction between acid 1 and amine 2, addition of HOObt can lead to the side reaction of aminolysis at the carbonyl of the activating moiety of ester 3, generating addition product 4. Addition of HOBt to a mixed-anhydride reaction containing unconsumed chloroformate generates mixed carbonate 5, leading to production of urethane 6.

last decade. 1-Hydroxybenzotriazole and HOObt are the popular additives, the choice between them being arbitrary. The newer 7-azabenzotriazole may be superior in some cases, but it also may not. There have been reports that HOBt and HOAt acted as bases instead of acids, thus promoting epimerization during coupling instead of suppressing it. The α-proton of a benzotriazole ester is exchangeable, whereas that of a succinimido ester is not, so the former ester is more sensitive to isomerization by base-catalyzed enolization (see Sections 2.2 and 8.14).

When HOBt and HOAt are used with phosphonium and uronium salt-based reagents, they are present as anions, and they suppress epimerization by trapping the *O*-acyloxyphosphonium, *O*-acyluronium, and oxazolone intermediates as the activated esters (see Section 2.21).

There can be a minor side reaction associated with the use of HOObt in carbodiimide-mediated reactions; namely, aminolysis at the carbonyl of the activating moiety of ester 3 (Figure 2.27), giving addition product 4. The reaction is negligible in most cases.

There is a claim that HOBt suppresses undesired aminolysis at the carbonate carbonyl of a mixed anhydride (Figure 2.25, path F). It is rarely used for this purpose, but if it is, it must be added only after the chloroformate has been consumed; otherwise, mixed carbonate 5 is formed, and it depletes the amino-containing component by acylating it, giving stable urethane 6 (Figure 2.27).[29,31–34,77]

30. E Wünsch, F Drees. On the synthesis of glucagon. X. Preparation of sequence 22-29. (*N*-hydroxysuccinimide) *Chem Ber* 99, 110, 1966.

31. JE Zimmerman, GW Callahan. The effect of active ester components on racemization in the synthesis of peptides by the dicyclohexylcarbodiimide method. *J Am Chem Soc* 89, 7151, 1967.

32. W König, R Geiger. A new method for the synthesis of peptides: activation of the carboxyl group with dicyclohexylcarbodiimide and 1-hydroxybenzotriazoles. *Chem Ber* 103, 788, 1970.

33. W König, R Geiger. A new method for the synthesis of peptides: activation of the carboxyl group with dicyclohexylcarbodiimide and 3-hydroxy-4-oxo-3,4-dihydro-1.2.3-benzotriazine. *Chem Ber* 103, 2034, 1970.
34. LA Carpino. 1-Hydroxy-7-azabenzotriazole. An efficient peptide coupling additive. *J Am Chem Soc* 115, 4397, 1993.
77. NL Benoiton, YC Lee, R Steinauer, FMF Chen. Studies on the sensitivity to racemization of activated residues in couplings of *N*-benzyloxycarbonyldipeptides. *Int J Pept Prot Res* 40, 559, 1992.

2.27 AN AID TO DECIPHERING THE CONSTITUTION OF COUPLING REAGENTS FROM THEIR ABBREVIATIONS

Abbreviations are used by scientists for simplifying the written and spoken word. However, for a combination of reasons, the abbreviations for the onium salt-based coupling reagents can be a nightmare for the novice as well as the specialist who is trying to follow a discussion that involves a few or several of these reagents. Reasons for the difficulties are the use of the same letter representing different words, lack of consistency in the nomenclature, the sequence of letters based on the names in another language (see Section 2.18), and so on. In an attempt to help myself through this maze of abbreviations, I have developed an abbreviated chemical equivalent for each that conveys immediately to me the constitution of the compound. The principal features of the equivalents are the use of Bt instead of B for benzotriazolyl-1-yl, aBt instead of A for 7-azabenzotriazolyl-1-y, cBt instead of C for 5-chlorobenzotriazol-1-yl, and P^+ and C^+ with appropriate *N*-substituents instead of P and U (Figure 2.28). Ester forms of compounds are indicated as $BtOX^+(...)$; amide oxide forms (see Section 7.17) can be represented as $OBtX^+(...)$. Other abbreviations are handled in an analogous manner.

HOBt	=	1-hydroxybenzotriazole	HOAt	=	1-HO-7-azabenzotriazole
B	=	Bt = benzotriazolyl	A	=	aBt = azabenzotriazolyl
HOCt	=	5-chloro-1-hydroxybenzotriazole	C	=	cBt = chlorobenzotriazolyl
P	=	phosphonium = P^+	U	=	uronium = $OC^+(N_2)$
Py	=	Pyr = pyrrolidino = $c(-NC_4H_8-)$	T	=	tetramethyl
H	=	hexafluorophosphate	*T	=	Tetrafluoroborate
BOP	=	$BtOP^+(NMe_2)_3{\cdot}PF_6^-$	AOP	=	$aBtOP^+(NMe_2)_3{\cdot}PF_6^-$
PyBOP	=	$BtOP^+(Pyr)_3{\cdot}PF_6^-$	PyAOP	=	$aBtOP^+(Pyr)_3{\cdot}PF_6^-$
HBTU	=	$BtOC^+(NMe_2)_2{\cdot}PF_6^-$	HATU	=	$aBtOC^+(NMe_2)_2{\cdot}PF_6^-$
HCTU	=	$cBtOC^+(NMe_2)_2{\cdot}PF_6^-$	*TATU	=	$aBtOC^+(NMe_2)_2{\cdot}BF_4^-$
*TBTU	=	$BtOC^+(NMe_2)_2{\cdot}BF_4^-$	HAPyU	=	$aBtOC^+(Pyr)_2{\cdot}PF_6^-$
HBPyU	=	$BtOC^+(Pyr)_2{\cdot}PF_6^-$	*TSTU	=	$SuOC^+(NMe_2)_2{\cdot}BF_4^-$
S	=	Su = succinimido			
BroP	=	$(NMe_2)_3P^+Br{\cdot}PF_6^-$	PyBroP	=	$(Pyr)_3P^+Br{\cdot}PF_6^-$
BOP-Cl	=	$c(-CO_2C_2H_4N-)_2POCl$ = *bis*(2-oxo-oxazolidino)phosphinic chloride			
EEDQ	=	*N*-ethoxycarbonyl-2-ethoxy-1,2-dihydroquinoline			
DPPA	=	$(PhO)_2PON_3$ = diphenyl phosphorazidate			
HOObt	=	HODhbt = 3-hydroxy-4-oxo-3,4-dihydrobenzotriazine = HOBt with a carbonyl inserted between the aromatic ring and the hydroxylamine nitrogen	carbodiimide	=	CDI
			DCC	=	$(c\text{Hex})_2$CDI
			EDC	=	Et,Me_2NPrCDI
			DIC	=	DIPCDI = iPr_2CDI

FIGURE 2.28 An aid to deciphering the constitution of coupling reagents from their abbreviations. HBPyU and HAPyU correspond to incorrect names of the compounds because U = uronium = OC^+N_2 includes the nitrogen atoms of the pyrrolidine rings. The substitutents on each nitrogen are tetramethylene.

3 Protectors and Methods of Deprotection

3.1 THE NATURE AND PROPERTIES DESIRED OF PROTECTED AMINO ACIDS

The union of two amino acids to form a peptide requires suppression of the reactivities of the functional groups that are not incorporated into the peptide bond (see Section 1.5). This is achieved by combining each group with another compound in a manner that allows removal of thc added moieties at will. These moieties are referred to as protecting groups. The starting materials for peptide synthesis are thus protected amino acids. Several characteristics are essential or desirable for a good protector. First and foremost, it must be removable, preferably with ease and by a mechanism that does not lead to side reactions. The suppression of reactivity should be complete and last throughout the synthesis, and the products generated by the protecting moiety must be separable from the target molecule. Both preparation of the amino acid derivative and removal of the protector must take place with preservation of the chiral integrity of the residue. Finally, the derivativc should be crystalline and stable during storage and obtainable by a process that is not too laborious or expensive. In some cases, compromises are accepted by choice or necessity.

The functional groups located in peptides are presented schematically in Figure 3.1. The amino and carboxyl groups are the most prevalent, followed by hydroxyl groups. The natures of the moieties that are most commonly used for protecting these and sulfhydryl functions are indicated within the rectangles. Note that the protectors are composed of or contain the alkyl group of an alcohol and that the combination of reactants involves the elimination of a molecule of water. Carboxyl groups are protected as esters, hydroxyl and sulfhydryl groups are protected as ethers, and amino groups are protected as urethanes that incorporate an oxycarbonyl as well

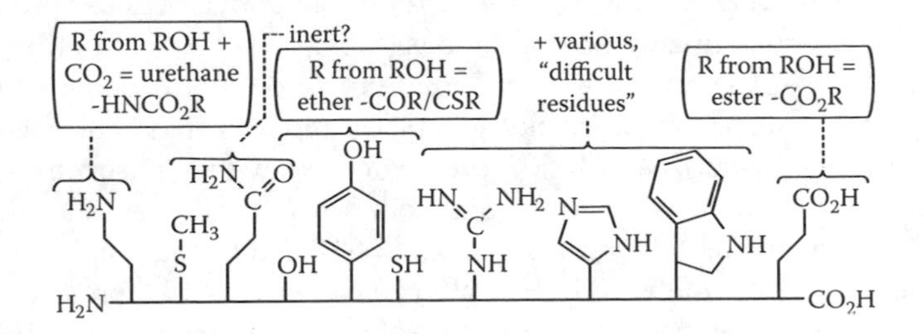

FIGURE 3.1 Functional groups and the nature of the moieties used for their protection. Protectors incorporate the alkyl of an alcohol.

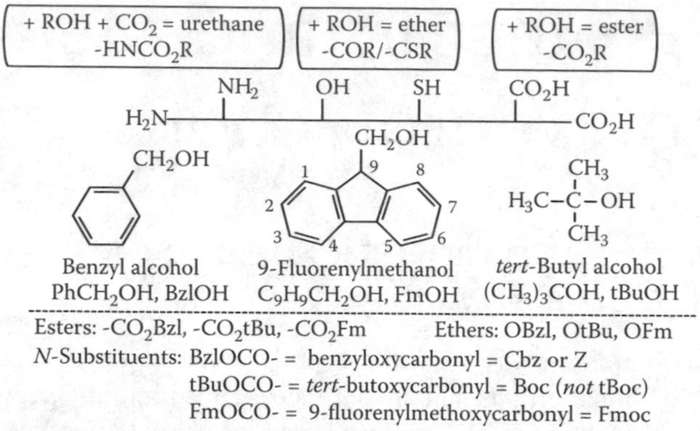

FIGURE 3.2 The alcohols from which protectors are derived, and their abbreviated designations.

as the alkyl group of the alcohol (see Section 1.6). Sometimes the corresponding alkyl halide and not the alcohol is employed to effect the protection, or some other type of protector is used. The methylthio and carboxamide groups are inert during peptide-bond formation, and the guanidino, imidazole, and indole groups are of such varied nature that they are not routinely protected by the same approaches.

3.2 ALCOHOLS FROM WHICH PROTECTORS DERIVE AND THEIR ABBREVIATED DESIGNATIONS

The common protecting groups are derived from a very limited number of alcohols, whose structures appear in Figure 3.2. Benzyl-based substituents emerged in the 1930s, *tert*-butyl-based substituents in the late 1950s, and 9-methylfluorenyl-based substituents in the 1970s. Designation of esters and ethers is straightforward. Bzl for benzyl is sometimes replaced by Bn by other organic chemists; Bz is reserved for benzoyl. The amino protectors were formerly referred to as carboalkoxy; hence, Cbz for carbobenzoxy. This nomenclature is outdated, but the abbreviation persists. An additional abbreviation, Z, is used for benzyloxycarbonyl, in honor of the codiscoverer of the protecting group, Leonidas Zervas. The abbreviation for the *N*-substituent containing *tert*-butyl is Boc and not t-Boc, the t- being superfluous because normal or secondary butyl and isobutyl substituents are not detached by the reagents that release *tert*-butyl-based protectors. Note that nitrogen protection involves a carboxylate linked to a nitrogen atom, whereas an ester is a carboxylate linked to a carbon atom. A fourth alcohol, $CH_2{=}CHCH_2OH$, has surfaced over the last decade, providing *O*-allyl (All) and allyloxycarbonyl (Aloc) protectors. Other alcohols common in peptide chemistry are substituted hydroxamates, HONR, in which the nitrogen atom is part of a ring. In cases in which the ring is linked to the oxygen atom through the nitrogen atom, the last letter of the name of the substituent is changed to -o; for example, esters formed with *N*-hydroxysuccinimide and *N*-hydroxypyrrolidine are named succinimido and pyrrolidino esters, respectively.

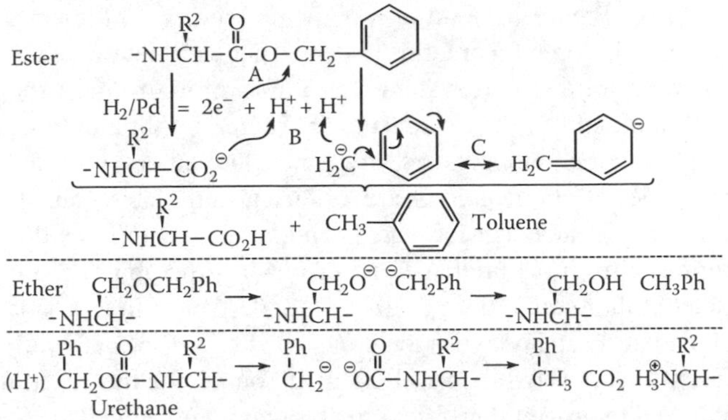

Ester $-NHCH(R^2)-C(=O)-O-CH_2-C_6H_5$ (A)

H_2/Pd = $2e^- + H^+ + H^+$ (B)

$-NHCH(R^2)-CO_2^{\ominus}$ + $H_2C^{\ominus}-C_6H_5$ $\overset{C}{\longleftrightarrow}$ $H_2C=C_6H_5^{\ominus}$

$-NHCH(R^2)-CO_2H$ + $CH_3-C_6H_5$ Toluene

Ether $-NHCH(CH_2OCH_2Ph)-$ $\rightarrow$ $-NHCH(CH_2O^{\ominus})-$ $^{\ominus}CH_2Ph$ $\rightarrow$ $-NHCH(CH_2OH)-$ CH_3Ph

(H^+) $PhCH_2OC(=O)-NHCH(R^2)-$ $\rightarrow$ $PhCH_2^{\ominus}$ $^{\ominus}OC(=O)-NHCH(R^2)-$ $\rightarrow$ $PhCH_3$ CO_2 $H_3\overset{\oplus}{N}CH(R^2)-$

Urethane

FIGURE 3.3 Deprotection of functional groups by reduction. Hydrogenolysis of benzyl-based protectors.[1] Attack by electrons liberates the protector as the benzyl anion because the latter is stabilized by resonance. This is a simplified presentation of the reaction.

Substituents that serve as protectors are of value because they are removable. The critical feature of each protector is, therefore, how it is removed. Therefore, instead of discussing protectors on the basis of their constitution, the subject of protection is presented on the basis of the methods of deprotection, using the more common protectors as examples.

3.3 DEPROTECTION BY REDUCTION: HYDROGENOLYSIS

Oxidation involves a loss of electrons (leo = loss of electrons, oxidation); reduction involves a gain of electrons (ger = gain of electrons, reduction). An oxidation is always accompanied by a reduction, and vice versa. Deprotection by reduction entails attack at an atom by a pair of electrons (Figure 3.3, path A), resulting in displacement of the protecting moiety as the anion. Fission occurs because the anion formed is stabilized by resonance (delocalization). The classical example is the rupture of benzyloxy. The benzyl moiety is displaced as the benzyl anion that is stabilized by charge delocalization (path C), liberating the functional group as the oxy anion in the process. The two anions are neutralized by two protons (paths B), thus regenerating the carboxyl, hydroxyl, or amino groups that had been protected, and producing toluene from the benzyl moiety. Carbon dioxide is released in addition from the benzyloxycarbonyl protector. The products generated by the protector are inert and readily eliminated. A phenacetyl [Pac, $-CH_2C(=O)Ph$] ester is also cleavable by reduction because the benzoylmethyl anion is stabilized by equilibration with the anion of the enol form [$CH_2=C(O^-)Ph$]. Functionalities such as methoxy, ethoxy, cyclohexyloxy, and so forth are not cleaved by reduction because there is no stabilization of the anions that might be generated. There are several procedures for carrying out reductions, but the simplest is the use of hydrogen in the presence of palladium catalyst. Palladium, usually 5–10% adsorbed on charcoal, induces oxidation of the chemisorbed gas, thus initiating a complex reaction that is presented in a simplified way (Figure 3.3). Atmospheric pressure is sufficient to effect the reaction. However, it is common practice to perform the reduction under the pressure of a column of water in a closed system that allows monitoring of the reaction by measuring the volume of gas that is consumed. There is only a marginal difference in sensitivity to reduction among esters, urethanes, and ethers, with the latter being slightly more stable. Hydrogenations are often carried out in an alcohol in the presence of a mild acid such as acetic acid. The latter decomposes the carboxylate anion that originates from the urethane and protonates the amino group, thus eliminating its nucleophilicity and any possible secondary reaction. A side reaction of *N*-methylation can accompany hydrogenations carried out in methanol if a trace of oxygen is present (section 6.20). An alternative source for the reacting species is palladium-catalyzed hydrogen transfer from an organic molecule (see Section 6.21). The first peptides synthesized by coupling protected amino acids were obtained by hydrogenolysis of *N*-benzyloxycarbonyldipeptides.[1,2]

1. M Bergmann, L Zervas. A general process for the synthesis of peptides. *Ber B* 65, 1192, 1932.
2. KW Rosenmund, F Heise. Oxidative catalytic dehydrogenation of alcohols. V. Catalytic reduction of esters and aldehydes. *Ber 54B* 2038, 1921.

3.4 DEPROTECTION BY REDUCTION: METAL-MEDIATED REACTIONS

In the removal of protectors by hydrogenolysis, reduction is effected by the electrons that are released by the palladium-catalyzed oxidation of hydrogen gas to the positively charged hydrogen atoms. Other reducing procedures involve metals that undergo oxidation themselves to the positively charged ions releasing electrons in the process. The required protons are supplied from another source. The common examples are sodium in liquid ammonia and zinc in acetic acid. Sodium reacts with liquid ammonia to generate solvated metal cations and solvated electrons (Figure 3.4). Reduction of the protecting moiety then occurs by attack of electrons (path A), which is followed at the end by protonation of anions (path B) by protons from water. The reaction is usually employed for deprotection after completion of the synthesis of a peptide. Benzyl esters, ethers, thioethers, and urethanes are cleaved, as well as toluenesulfonamides, at arginine and histidine side chains. Nitro-protection at guanidino of arginine (see Section 6.11) is also removed by this reaction. Completion of the reaction is indicated by persistence of the blue color emitted by sodium in liquid ammonia. The reaction is known as a Birch reduction. The Birch reduction proved invaluable in the 1950s and, thereafter, for the synthesis of peptides containing sulfhydryl groups that were protected as the benzyl ethers. Sulfur poisons palladium catalyst, thus precluding the use of hydrogenation for deprotection of sulfur-containing compounds. The value of this new method of reduction for peptide work was recognized in 1955 by the award of the Nobel Prize in Chemistry to Vincent duVigneaud for achieving the first synthesis of a peptide hormone — the nonapeptide oxytocin which contains a disulfide bond.

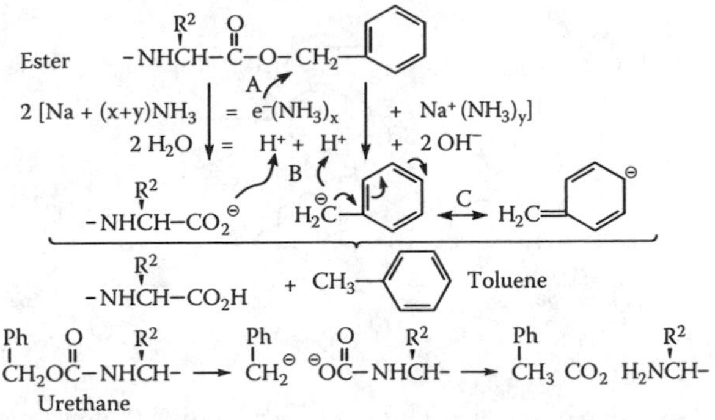

$$\text{Ester} \quad -NH\overset{R^2}{C}H-\overset{O}{\overset{\|}{C}}-O-CH_2-C_6H_5$$

$$2\,[Na + (x+y)NH_3 = e^-(NH_3)_x + Na^+(NH_3)_y]$$

$$2\,H_2O = H^+ + H^+ + 2\,OH^-$$

$$-NH\overset{R^2}{C}H-CO_2^{\ominus} \quad H_2C^{\ominus}-C_6H_5 \overset{C}{\longleftrightarrow} H_2C{=}C_6H_5^{\ominus}$$

$$-NH\overset{R^2}{C}H-CO_2H + CH_3-C_6H_5 \quad \text{Toluene}$$

$$\overset{Ph}{C}H_2O\overset{O}{\overset{\|}{C}}-NH\overset{R^2}{C}H- \longrightarrow \overset{Ph}{C}H_2^{\ominus}\ {}^{\ominus}O\overset{O}{\overset{\|}{C}}-NH\overset{R^2}{C}H- \longrightarrow \overset{Ph}{C}H_3\ CO_2\ H_2N\overset{R^2}{C}H-$$

Urethane

FIGURE 3.4 Deprotection of functional groups by reduction with sodium in liquid ammonia [du Vigneaud et al., 1930]. As in Figure 3.3, except reduction is effected by solvated electrons and protons are provided by water at the end of the reaction. Excess sodium is destroyed by NH_4Cl. This is a simplified presentation of the reaction. All benzyl-based protectors as well as –Arg(NO_2)–, –Arg(Tos)–, and –His(Tos)– are sensitive to sodium in liquid ammonia.

Another metal often employed for reductions is zinc. Zinc in the presence of acetic acid is oxidized to Zn^{++} with release of two electrons. Protons are provided by the acetic acid. The mixture is used in particular for cleavage of the benzoylmethoxy (PacO) [Ph(C=O)CH_2O–] and trichloroacetoxy [CCl_3C(=O)O–] groups of esters. The latter is cleavable by reduction because of the weak C–O bond resulting from the exceptionally strong electron-withdrawing effects of the chloro atoms. Zinc in acetic acid was originally employed for removing the nitro group from the side chain of arginine as well as cleaving nitroaromatic substituents such as nitrobenzyloxycarbonyl (see Figure 3.24). The mixture also removes allyl-based protectors (see Section 3.19) but not benzyloxycarbonyl (see Section 3.15 for the use of zinc to eliminate acid by reduction).[3,4]

3. V du Vigneaud, C Ressler, JM Swan, CW Roberts, PG Katsoyannis. The synthesis of oxytocin. *J Am Chem Soc* 76, 3115, 1954.
4. J Pless, S Guttmann. New results concerning the protection of the guanido group, in HC Beyermann, A van de Linde, W Massen van den Brink, eds. *Peptides. Proceedings of the 8th European Peptide Symposium*. North-Holland, Amsterdam, 1967, pp 50-54.

3.5 DEPROTECTION BY ACIDOLYSIS: BENZYL-BASED PROTECTORS

Acidolysis means lysis or scission of a bond by addition of the components H^+X^- of an acid to the atoms linked by the bond. Two decades after development of benzyloxycarbonyl as a protector for nitrogen (see Section 3.3), it was found that the protector could be removed by hydrogen bromide in acetic acid at room temperature and that the same reagent at 60°C cleaved benzyl esters. Acidolysis of these groups involves protonation of the oxygen atom of the carbonyl, followed by shift of electrons to give the positively charged carbon intermediate or carbenium ion that fragments because it is unstable (Figure 3.5). Fragmentation of the carbenium ion

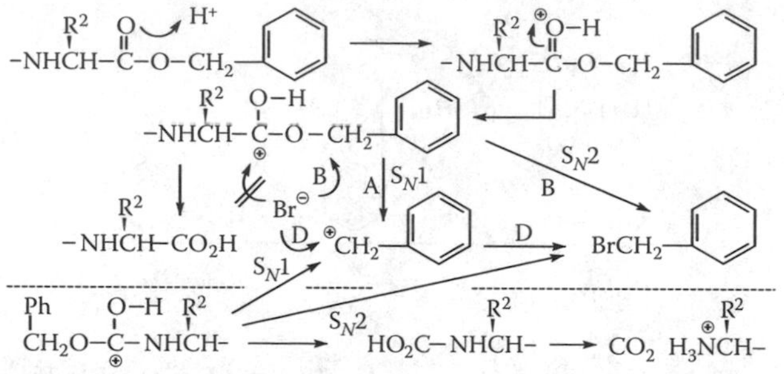

FIGURE 3.5 Deprotection of functional groups by acidolysis.[5] Protonation followed by carbocation formation during the removal of benzyl-based protectors by hydrogen bromide. Two mechanisms are involved in generating benzyl bromide from the protonated substrates.

occurs by two mechanisms, spontaneous rearrangement of electrons with release of the benzyl cation (path A), an S_N1 (substitution, nucleophilic, unimolecular) reaction with no other molecule being involved, and displacement of the benzyl cation by the bromide ion (path B), which is an S_N2 (substitution, nucleophilic, bimolecular) reaction in which cleavage is assisted by the incoming nucleophile. Both mechanisms give rise to the same product, benzyl bromide, which is lachrymatory.

In the case of the S_N1 reaction, the benzyl cation that is released is captured by the bromide anion (path D). Cleavage occurs because the leaving group can form a stable cation that is stabilized by delocalization of the positive charge. Methoxy, ethoxy, cyclohexyloxy, and so forth are not cleaved because they do not satisfy this condition. The reagent is obtained by bubbling hydrogen bromide through acetic acid that has been dried to prevent hydrolysis to an increase in weight of 38%. The sensitivity of protectors to HBr/AcOH is $PhCH_2OC(=O)NH$-(urethane) >> $PhCH_2OC(=O)$- (ester) >$PhCH_2OCH_2$- (ether). Only the urethane is cleaved at ambient temperature. The first peptide prepared by the solid phase method, namely, leucylalanylglycylvaline, was obtained on the basis of this selectivity. *N*-Benzyloxycarbonylamino acids were employed for the couplings, with the first residue attached to the resin as a benzyl ester (see Section 5.1). A variant of cleavage by the S_N2 mechanism involves participation of an added nucleophile such as thioanisole. The acid currently employed for debenzylations at the end of a synthesis is hydrogen fluoride (see Section 6.22). Other functional groups are sensitive to acid. The sensitivity to acidolysis of a moiety depends on the ease of protonation of the group and the ease of fragmentation of the carbenium ion, R_3C^+ (carbonium = R_4CH^+), which varies directly with the stability of the cation that is generated. Unfortunately, acidolysis has the unattractive feature — the moiety released is a good electrophile that has a tendency to react with any nucleophile in the vicinity (see Section 3.7).[5–7]

5. D Ben-Ishai, A Berger. Cleavage of *N*-carbobenzoxy groups by dry hydrogen bromide and hydrogen chloride. *J Org Chem* 17, 1564, 1952.
6. D Ben-Ishai. The use of hydrogen bromide in acetic acid for the removal of carbobenzoxy groups and benzyl esters of peptides. *J Org Chem* 19, 62, 1954.

7. Y Kiso, K Ukawa, T Akita. Efficient removal of *N*-benzyloxycarbonyl group by a "push-pull" mechanism using thioanisole-trifluoroacetic acid, exemplified by a synthesis of Met-enkephalin. *Chem Commun* 101, 1980.

3.6 DEPROTECTION BY ACIDOLYSIS: *tert*-BUTYL-BASED PROTECTORS

Benzyl-based protectors are cleavable by hydrogen bromide in acetic acid. *tert*-Butyl-based protectors, available since the late 1950s, are the second type of protectors that are sensitive to acid, succumbing to the weaker hydrogen chloride in an inert solvent (Figure 3.6). The reaction proceeds through the same carbenium ion formation after protonation, followed by fragmentation, in this case releasing the *tert*-butyl cation. However, only one mechanism, the S_N1 reaction, is involved in the breakdown of the carbenium ion (path A); the chloride counter-ion (path B) is not involved until after formation of the trisubstituted carbocation, which is trapped as *tert*-butyl chloride (path D). Some of the cation rearranges to isobutene (path E) before it is consumed by chloride. Acidolysis by hydrogen chloride was employed for the removal of *tert*-butoxycarbonyl groups when they first appeared, as well as *tert*-butyl from the esters that surfaced shortly after. When the solid-phase method of synthesis was developed, hydrogen chloride was employed for deprotection of the *tert*-butoxycarbonylamino-acid building blocks of the synthesis. The reagent now commonly used for acidolysis of *tert*-butyl-based protectors is trifluoroacetic acid-dichloromethane (1:1). The mechanism is the same except that the *tert*-butyl cation does not rearrange to isobutene; all of it is trapped as *tert*-butyl trifluoroacetate. As for benzyl-based protectors (see Section 3.5), the *tert*-butyl urethane is more sensitive to acidolysis than the ester and ether (see Section 6.22), and the *tert*-butyl cation is a good electrophile leading to undesired alkylations (see Section 3.7). *N*-*tert*-Butylamino-aromatics, especially *N*-*tert*-butyl,*N*-methylamino-aromatics, are also sensitive to acidolysis (see Section 8.14).[8–11]

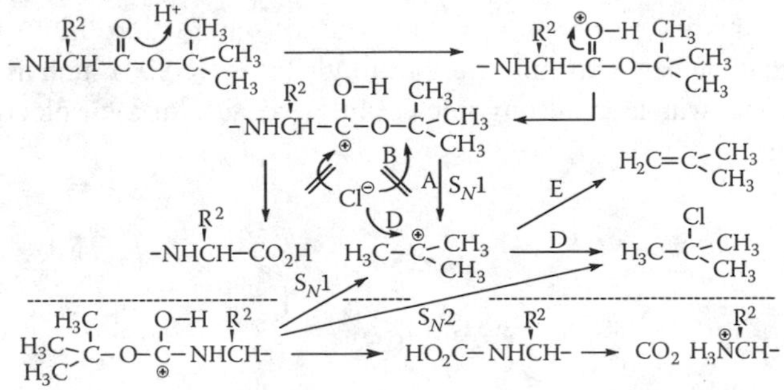

FIGURE 3.6 Deprotection of functional groups by acidolysis. Protonation followed by carbocation formation during the removal of *tert*-butyl-based protectors by hydrogen chloride.[8] One mechanism is involved in generating the *tert*-butyl cation, which is the precursor of two other molecules.

8. GW Anderson, AC McGregor. t-Butyloxycarbonylamino acids and their use in peptide synthesis. *J Am Chem Soc* 79, 6180, 1957.
9. R Schwyzer, W Rittel, H Kappeler, B Iselin. Synthesis of a nonadecapeptide with higher corticotropic activity. (*tert*-butyl removal) *Angew Chem* 72, 915, 1960.
10. RE Reid. Solid phase peptide synthesis. A study on the effect of trifluoroacetic acid concentration on the removal of the *tert*-butyloxycarbonyl protecting group. *J Org Chem* 41, 1027, 1976.
11. BF Lundt, NL Johansen, A Volund, J Markussen. Removal of *t*-butyl and *t*-butyloxycarbonyl protecting groups with trifluoroacetic acid. *Int J Pept Prot Res* 12, 258, 1978.

3.7 ALKYLATION DUE TO CARBENIUM ION FORMATION DURING ACIDOLYSIS

Acidolysis of protected amino acids releases the protectors as carbenium ions, which are good alkylating agents because of their electrophilicity. The carbenium ions will alkylate any nucleophile that is in the neighborhood. The possible acceptors found on the side chains of peptides are shown in Figure 3.7. Amino and imidazole groups are nucleophiles; however, in acidic media they are protonated so they are not acceptors of carbenium ions. The other functional groups are acceptors if the acid is not strong enough to protonate them (see Section 6.22). Alkylation occurs at the sulfur atoms of cysteine and methionine, the τ-nitrogen atom of histidine, *ortho* to the hydroxyl group of the ring of tyrosine, and at two positions of the indole ring of tryptophan, the imino nitrogen atom or adjacent tertiary carbon atom. An additional alkylating agent emerges when trifluoroacetic acid is employed for acidolysis (see Section 3.6), and this is *tert*-butyl trifluoroacetate. One way to try to avoid the alkylations is to swamp the acceptors with molecules of constitution similar to those intended to be protected. These molecules are known as scavengers. Examples of effective scavengers are ethylmethylsulfide to protect the thioether of methionine and 1,2-ethanedithiol to protect sulfhydryls of cysteine. 1,2-Ethanedithiol is also an effective scavenger of *tert*-butyl trifluoroacetate. Mixtures referred to as cocktails containing four or five scavengers are often used (see Section 6.22). Recent work has revealed that the best scavengers are trialkylsilanes (R_3SiH) such as triethyl- and triisopropylsilane, which are strong nucleophiles because of their electron-donating alkyl groups.[12–14]

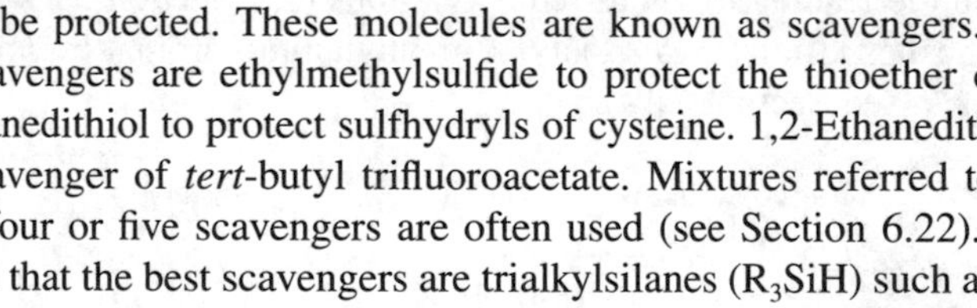

FIGURE 3.7 Potential sites of alkylation during acidolysis of protected functional groups.[11–13] Protonated groups are not alkylated.

12. S Guttmann, RA Boissonnas. Synthesis of benzyl *N*-acetyl-L-seryl-L-tyrosyl-L-seryl-L-methionyl-γ-glutamate and related peptides. (side-chain alkylation) *Helv Chim Acta* 41, 1852, 1958.
13. P Sieber, B Riniker, M Brugger, B Kamber, W Rittel. 255. Human Calcitonin. VI. The synthesis of calcitonin M. (side-chain alkylation) *Helv Chim Acta* 53, 2135, 1970.
14. DS King, CG Fields, GB Fields. A cleavage method which minimizes side reactions following Fmoc solid phase peptide synthesis. *Int J Pept Prot Res* 36, 255, 1990.

3.8 DEPROTECTION BY ACID-CATALYZED HYDROLYSIS

Although acidolysis means scission of a bond by the addition of H^+X^- (see Section 3.5), hydrolysis means scission of a bond by the addition of H^+OH^-, a process that nearly always requires the assistance of an acid or a base as a catalyst. Both amides and esters are cleavable by acid-catalyzed hydrolysis; as a consequence, it is rarely used for deprotection of functional groups. However, it is included in the discussion because it is related. Cleavage is initiated by the same mechanism as for acidolysis (see Section 3.5); that is, protonation of thc carbonyl followed by a shift of electrons to give the carbenium ion (Figure 3.8). The latter is relatively stable, however, and fragments only by the S_N2 mechanism on attack by water, and only if heat is applied. Acid-catalyzed hydrolysis can be a source of isomerization of amino acid residues during their release from a peptide (see Section 4.1). An example of acid-catalyzed deprotection was the removal of the ethyl and phthaloyl groups from N^α-benzyloxycarbonyl-N^β-phthaloyldiaminopropionic acid ethyl ester by hot 1 *N* hydrochloric acid after the side-chain ring had been opened to the *o*-carboxybenzamido substituent by aqueous sodium hydroxide. The urethane was not affected by the hot acid.[15]

15. L Benoiton. Conversion of β-chloro-L-alanine to N^α-carbobenzoxy-DL-diaminopropionic acid. *Can J Chem* 46, 1549, 1968.

3.9 DEPROTECTION BY BASE-CATALYZED HYDROLYSIS

Base-catalyzed hydrolysis is employed primarily for the liberation of carboxyl groups protected as esters. The reaction involves direct attack by the hydroxide anion

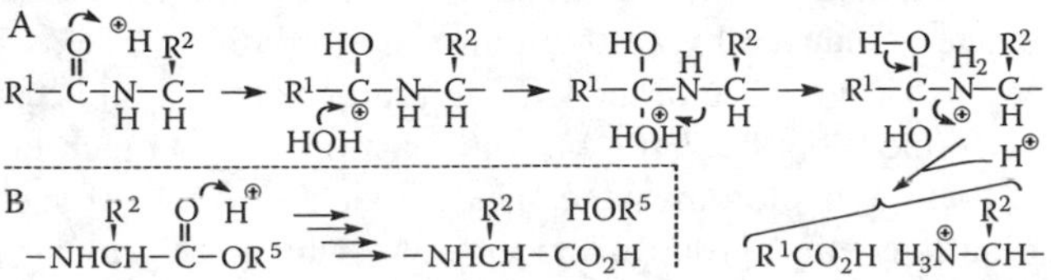

A: $R^1\text{–}C(=O)\text{–}NH\text{–}CHR^2\text{–} \rightarrow R^1\text{–}C^{\oplus}(OH)\text{–}NH\text{–}CHR^2\text{–} \;(+\,HOH) \rightarrow R^1\text{–}C(OH)(HOH^{\oplus})\text{–}NH\text{–}CHR^2\text{–} \rightarrow R^1\text{–}C(OH)_2\text{–}N^{\oplus}H_2\text{–}CHR^2\text{–} \rightarrow R^1CO_2H + H_3\overset{\oplus}{N}\text{–}CHR^2\text{–}$

B: $\text{–NHCHR}^2\text{–C(=O)–OR}^5 \;(+H^{\oplus}) \rightleftharpoons \text{–NHCHR}^2\text{–CO}_2\text{H} + \text{HOR}^5$

FIGURE 3.8 Deprotection of carboxyl groups by acid-catalyzed hydrolysis (A) of amides and (B) of esters. Protonation generates a relatively stable carbenium ion that usually requires heat to fragment it.

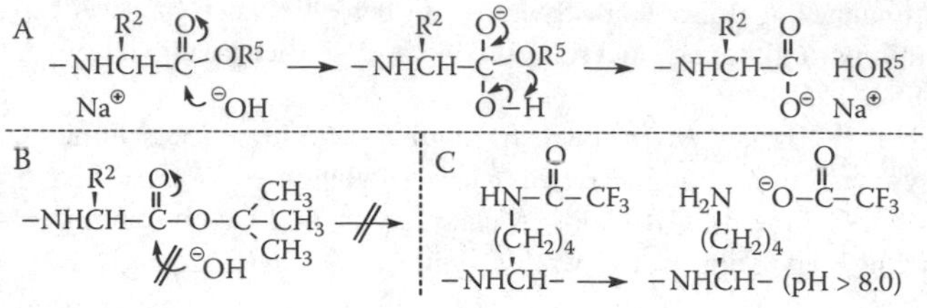

FIGURE 3.9 Deprotection of carboxyl groups by base-catalyzed hydrolysis of (A) esters and (C) trifluoroacetamides, involving direct attack by the hydroxide anion. (B) *tert*-Butyl esters are resistant to saponification.

at the carbon atom of the carbonyl to form a tetrahedral intermediate, which collapses to liberate the alkoxy substituent as the alcohol (Figure 3.9). The carboxyl group emerges as the sodium or other metal salt. Sodium salts of fatty acids are soaps, hence the term saponification to designate the base-catalyzed hydrolysis of esters. Methyl, ethyl, benzyl esters, and so on are subject to saponification; *tert*-butyl esters are not (Figure 3.9), because of hindrance or electronic effects. The more bulky the side chain R^2 of the residue, the more difficult it is to saponify the ester. Excess base or heat are not desirable, as they may cause isomerization by enolization (see Section 4.2). Saponification of esters of amino acid or peptide derivatives is usually effected in mixtures of organic and aqueous solvents to solubilize the reactants and products. The benzyloxy of a urethane is usually resistant to the conditions of saponification if there is no excess of base and the temperature is controlled. An amide undergoes the same reaction, liberating the carboxy-substituent –NHR as the amine NH_2R (Figure 3.9, C). Heat is usually required, though there are exceptions. The example shows the base-catalyzed hydrolysis of a trifluoroacetamide, which occurs as the pH is raised above neutral. The trifluoroacetyl group is attractive for protecting the side chain of lysine because it is so easy to remove it.[16]

16. A Neuberger. Stereochemistry of amino acids. *Adv Prot Chem* 4, 297, 1948.

3.10 DEPROTECTION BY *beta*-ELIMINATION

A protector can be removed by a process called *beta*-elimination if it contains a labile or "activated" hydrogen atom that is positioned *beta* to a good leaving group. The atom is labilized by an adjacent electron-withdrawing moiety such as substituted sulfonyl. Cleavage is induced by abstraction of the proton by base, which triggers a shift of electrons, with release of the leaving group as the anion because a double bond is readily formed (Figure 3.10, A). The reaction occurs much faster in dimethylformamide than in dichloromethane. The classical example of such a protector is the methanesulfonylethyl group of an ester, in which the leaving group is the carboxy anion (Figure 3.10). Note that the positioning of the activating and leaving groups implies their separation by two carbon atoms. A weak base creating a pH above neutral is sufficient to cause a β-elimination. Both the ester and methanesulfonylethoxycarbonyl (not shown) are cleaved by aqueous ammonia. The carboxamido

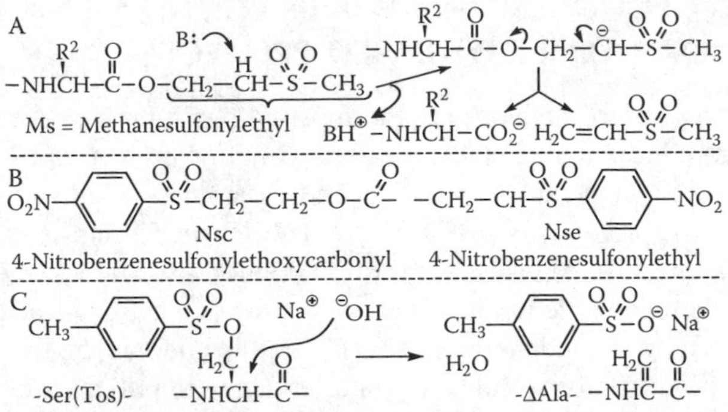

A
$-NHCH(R^2)-C(=O)-O-CH_2-CH(H)-S(O_2)-CH_3$ (B:) → $-NHCH(R^2)-C(=O)-O-CH_2-\overset{\ominus}{C}H-S(O_2)-CH_3$
Ms = Methanesulfonylethyl
$BH^{\oplus}$ $-NHCH(R^2)-CO_2^{\ominus}$ $H_2C{=}CH-S(O_2)-CH_3$

B
$O_2N-C_6H_4-S(O_2)-CH_2-CH_2-O-C(=O)-$ $-CH_2-CH_2-S(O_2)-C_6H_4-NO_2$
Nsc Nse
4-Nitrobenzenesulfonylethoxycarbonyl 4-Nitrobenzenesulfonylethyl

C
$CH_3-C_6H_4-S(O_2)-O-CH_2-CH(NH-)-C(=O)-$ $Na^{\oplus}$ $^{\ominus}OH$ → $CH_3-C_6H_4-S(O_2)-O^{\ominus}$ $Na^{\oplus}$ H_2O $H_2C{=}C(NH-)-C(=O)-$
-Ser(Tos)- -ΔAla-

FIGURE 3.10 Deprotection of functional groups by *beta*-elimination.[17] (A) Removal of a labile proton *beta* to a good leaving group leads to release of the protector as the didehydro compound. (B) Recently developed protectors (Samukov et al., 1988) also designated untraditionally as 4-nitrophenyl–. (C) Transformation of an *O*-protected serine residue into a dehydroalanine residue by *beta*-elimination.

anion of the latter collapses, releasing carbon dioxide and the amino group. These protectors have not found common use; however, the nitrobenzene equivalents (Figure 3.10, B) that were recently introduced show promise for the future. The 4-nitrobenzenesulfonylethoxycarbonyl group is less sensitive to base and less hydrophobic than the Fmoc group that is cleaved by the same mechanism (see Section 3.11); moreover, the addition product formed by its release by secondary amine is soluble in aqueous acid.

β-Elimination is encountered not only as a reaction for deprotection but also sometimes as a side reaction during the manipulation of *O*-substituted serines (Figure 3.10, C). In this case, the labile atom is the α-hydrogen of serine, the leaving group is the *O*-substituent along with the oxygen atom, and the double bond appears in the side chain of the amino acid, which is now dehydroalanine. Another example of this side reaction is formation of dehydroalanine residues during manipulation of *O*-dialkylphosphoserine residues. β-Elimination is avoided by use of the monoalkylphosphoserine derivatives. In contrast, the reaction can be employed for the preparation of peptides containing dehydroalanine by selecting an appropriate protector for the side chain of serine and effecting the elimination during or after chain assembly.[17–19]

17. CW Crane, HN Rydon. Alkaline fission of some 2-substituted dimethyl-ethylsulfonium iodides. *J Chem Soc* 766, 1947.
18. CGJ Verhardt, GI Tesser. New base-labile amino-protecting groups for peptide synthesis. *Rec Trav Chim Pays-Bas* 107, 621, 1988.
19. VV Samukov, AN Sabirov, PI Pozdnyakov. 2-(4-Nitrophenyl)sulfonylethoxycarbonyl (Nsc) group as a base-labile α-amino protection for solid phase peptide synthesis. *Tetrahedron Lett* 35, 7821, 1994.

3.11 DEPROTECTION BY *beta*-ELIMINATION: 9-FLUORENYLMETHYL-BASED PROTECTORS

9-Fluorenylmethyl-based moieties are the most common protectors removed by β-elimination. Introduced in the 1970s, 9-fluorenylmethyl protectors are stable to acid but sensitive to mild aqueous base and secondary amines. The C-9 hydrogen atom is activated by the aromaticity of the rings. The leaving group is the carboxylate anion, as in Figure 3.10, A, or the carboxamido anion (Figure 3.11). Abstraction of the proton by the hydroxide ion followed by a shift of electrons leads to rupture of the molecule, with generation of an exocyclic double bond on the three-ring moiety (path A). The product dibenzofulvene has a tendency to polymerize. In practice, a secondary amine is employed to remove the proton — one that is nucleophilic so that it also traps the deprotonated moiety that is released (path B). Piperidine is the amine of choice for this purpose: It deprotonates and further combines with the protecting moiety at the terminal carbon atom before the double bond can be formed. The addition product is not soluble in aqueous acid, so it cannot be disposed of by extraction. No gas is evolved during the deprotection, so the carbon dioxide must be bound to the piperidine as the salt. Diethylamine and DBU (see Section 8.12) cleave 9-fluorenylmethoxy-substituents without trapping the protector. The 9-fluorenylmethoxycarbonyl group is resistant to acidolysis, and, hence, it is orthogonal (see Section 1.5) to both benzyl and *tert*-butyl-based protectors. However, it is not orthogonal to benzyl-based protectors if the method of cleavage is reduction (see Section 6.21).[20]

20. LA Carpino, GY Han. The 9-fluorenylmethoxycarbonyl amino-protecting group. *J Org Chem* 37, 3404, 1972.

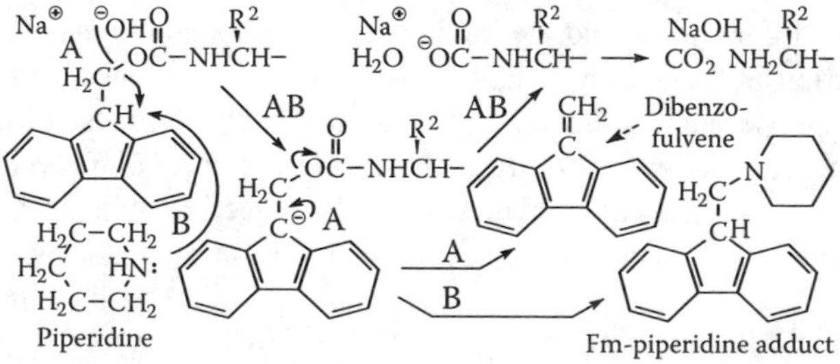

FIGURE 3.11 Removal of the 9-fluorenylmethoxycarbonyl group by *beta*-elimination (Carpino & Han, 1970). Deprotonation is achieved by hydroxide anion that generates dibenzofulvene or by piperidine that subsequently forms an addition product with the liberated moiety.

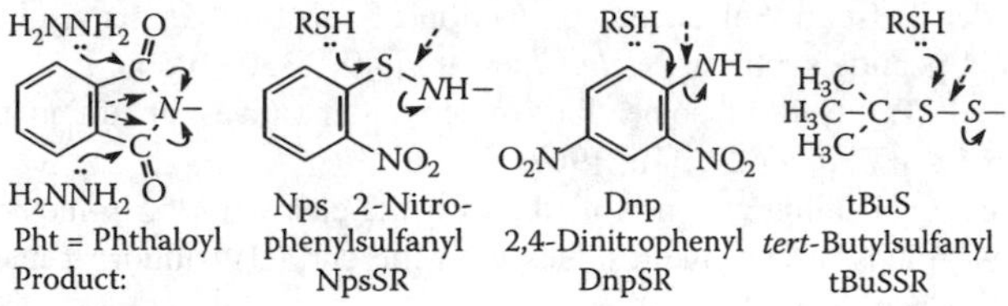

Pht = Phthaloyl | Nps 2-Nitrophenylsulfanyl | Dnp 2,4-Dinitrophenyl | tBuS *tert*-Butylsulfanyl

Product: NpsSR | DnpSR | tBuSSR

FIGURE 3.12 Protectors and their removal by displacement by a nucleophile. Protected atoms are indicated in italics, bonds that are cleaved are indicated by dashed arrows. Cleavage of phthalimido by hydrazine gives hydrazide $C_6H_4(CONH\text{-})_2$.

3.12 DEPROTECTION BY NUCLEOPHILIC SUBSTITUTION BY HYDRAZINE OR ALKYL THIOLS

There are some chemical bonds that are unstable to nucleophiles. Four protectors linked to functional groups by nucleophile-sensitive bonds appear in Figure 3.12. The nucleophiles, hydrazine or an alkyl thiol or other, displace the protecting moieties with which they combine, liberating the functional group. The three ring structures shown are protectors of nitrogen — the other is a protector of sulfhydryl. These substituents are employed primarily for side-chain protection, with the exception of 2-nitrophenylsulfanyl, which is used for N^{α}-protection. Thiolysis is employed for removing nitrophenylsulfanyl, 2,4-dinitrophenyl from the imidazole of histidine and *tert*-butylsulfanyl from the side chain of cysteine. N^{α}-Dinitrophenylamino acids are resistant to thiolysis. Thiolysis of alkylsulfanyl can be assisted by mercury or silver ions that coordinate with the protected sulfur atom, thus facilitating attack by the nucleophile. Hydrazinolysis is the usual reaction for cleaving phthalimido. 2-Nitrophenylsulfanyl is removable also by acidolysis that involves the usual protonation (see Section 3.5), but of the sulfur atom. The Nps–NH bond is much more sensitive to acid than the *tert*-butoxy group (see Section 6.22). Nps–amino acids undergo activation without forming an oxazolone because there is no carbonyl to attack the activated carboxyl group. The reaction of hydrazine with esters to give hydrazides (see Section 2.13) is the same as that described here, except that the nucleophile combines with the carboxyl group and not the protector. Several more-complex protectors recently developed for protection of the side chain of lysine are removed by assisted hydrazinolysis (see Section 6.4).[21–26]

21. L Zervas, D Borovas, E Gazis. New methods in peptide synthesis. I. Tritylsulfenyl and *o*-nitrophenylsulfenyl groups as N-protecting groups. *J Am Chem Soc* 85, 3660, 1963.
22. GH Nefkens, GI Tesser, RJ. Nivard. A simple preparation of phthaloyl amino acids via a mild phthaloylation. *Rec Trav Chim Pays-Bas* 79, 688, 1960.
23. A Fontana. F Marchiori, L Moroder, E Scoffone. New removal conditions of sulfenyl groups in peptide synthesis. *Tetrahedron Lett* 2985, 1966.

24. W Kessler, B Iselin. Selective deprotection of substituted phenylsulfenyl protecting groups in peptide synthesis. *Helv Chim Acta* 49, 1330, 1966.
25. S Shaltiel. Thiolysis of some dinitrophenyl derivatives of amino acids. *Biochem Biophys Res Commun* 29, 178, 1967.
26. M Fujino, O Nishimura. A new method for the cleavage of *S-p*-methoxy-benzyl and *S*-t-butyl groups of cysteine residues with mercury(II) trifluoroacetate. *J Chem Soc Chem Commun* 998, 1976.

3.13 DEPROTECTION BY PALLADIUM-CATALYZED ALLYL TRANSFER

Common protectors are derived from benzyl, *tert*-butyl, and 9-fluorenylmethyl alcohols (see Section 3.2). A fourth alcohol from which protectors of general utility are derived is allyl alcohol, CH_2=$CHCH_2OH$. Available since the 1950s, allyl-based protectors did not find favor until recently because the methods for their removal, such as the use of sodium in liquid ammonia, were not appealing. The situation changed, however, with development of allyl transfer reactions catalyzed by palladium(0) in 1980. Deprotection is achieved by palladium-catalyzed transfer of the π-allyl moiety. The metal, usually presented as palladium *tetrakis*-triphenylphosphine or *bis*triphenylphosphine dichloride, binds with the protector to form the π-allylpalladium complex (Figure 3.13). The two triphenylphosphine ligands increase the electrophilic nature of the complex. The -allyl moiety is transferred to an added nucleophile that attacks the complex on the face opposite to that of palladium. The palladium becomes the leaving group. Nucleophiles used to accept the allyl group have included morpholine, diethylamine, triphenylsilane, silylamines, and tributyltin hydride. The latter actually acts as a hydride donor, converting the allyl moiety to propene. Allyl protectors are stable to both acid and base and thus are orthogonal (see Section 1.5) to benzyl-, *tert*-butyl–, and 9-fluorenylmethyl-based protectors.[27–30]

27. BM Trost. New rules of selectivity: allyl alkylations catalyzed by palladium. *Acc Chem Res* 13, 385, 1980.
28. H Kuntz, C Unverzagt. The allyloxycarbonyl (Aloc) moiety — conversion of an unsuitable into a valuable amino protecting group for peptide synthesis. *Angew Chem Int Edn Engl* 23, 436, 1984.

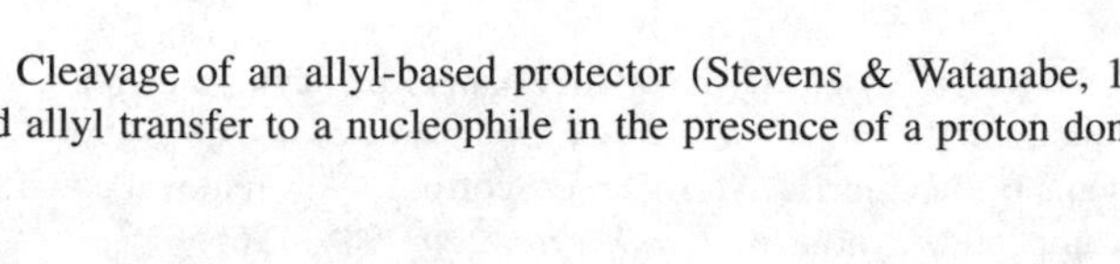

Aloc = Allyloxycarbonyl

FIGURE 3.13 Cleavage of an allyl-based protector (Stevens & Watanabe, 1950) by palladium-catalyzed allyl transfer to a nucleophile in the presence of a proton donor.[27,28]

29. O Dangles, F Guibe, G Balavoine, S Lavielle. A Marquet. Selective cleavage of the allyl and allyloxycarbonyl groups through palladium-catalyzed hydrostannolysis with tributyltin hydride. Application of the selective protection-deprotection of amino acid derivatives and in peptide synthesis. *J Org Chem* 52, 4984, 1987.
30. A Loffet, HX Zhang. Allyl-based groups for side-chain protection of amino-acids. *Int J Pept Prot Res* 42, 346, 1993.

3.14 PROTECTION OF AMINO GROUPS: ACYLATION AND DIMER FORMATION

Amino groups are usually protected as the alkoxycarbonyl derivatives. These are obtainable by the classical Schotten-Baumann reaction, involving acylation of the amino acid by the alkoxycarbonyl chloride, which is analogous to benzoylation using benzoyl chloride. The amino acid zwitter-ion is first converted to a nucleophile by base, the nucleophile attacks at the carbonyl of the reagent (Figure 3.14, path A), and the liberated acid is eliminated by additional base. The *N*-alkoxycarbonylamino acid that is produced and present as the sodium salt is then isolated by extraction into organic solvent after the addition of acid. Benzyloxycarbonyl- and 9-fluorenylmethoxycarbonyl-, but not *tert*-butoxycarbonylamino acids (see Section 3.16) were routinely prepared by this method until the 1980s. It then became apparent that the compounds, and in particular the Fmoc-derivatives, were contaminated by the corresponding *N*-alkoxycarbonyldipeptides. This was inferred and then proven after sequence analysis of peptides prepared by solid-phase synthesis (see Chapter 5) revealed two amino acid residues in which a single residue had been incorporated. The explanation was not difficult to unearth. The high activation of the acylating reagent because it is a chloride, combined with the fact that the product is completely ionized because of the strong base, induces formation of the mixed anhydride (path B), which undergoes aminolysis and consequently generates the protected dimer. The protected dimer has solubility properties similar to those of the protected amino acid, so it is not easy to purify the latter by crystallization. Interestingly, generation

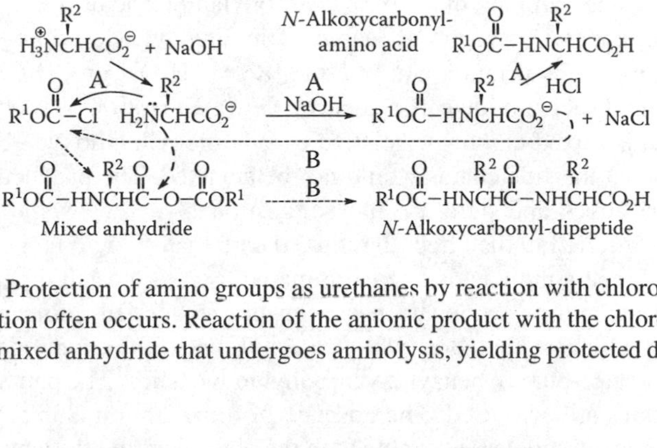

$$H_3\overset{\oplus}{N}CH(R^2)CO_2^{\ominus} + NaOH$$

$$R^1OC(=O){-}Cl \;+\; H_2\ddot{N}CH(R^2)CO_2^{\ominus} \xrightarrow[\text{NaOH}]{\text{A}} R^1OC(=O){-}HNCH(R^2)CO_2^{\ominus} + NaCl \xrightarrow[\text{A}]{\text{HCl}} R^1OC(=O){-}HNCH(R^2)CO_2H$$

N-Alkoxycarbonyl-amino acid

$$R^1OC(=O){-}HNCH(R^2)C(=O){-}O{-}C(=O)OR^1 \xrightarrow{\text{B}} R^1OC(=O){-}HNCH(R^2)C(=O){-}NHCH(R^2)CO_2H$$

Mixed anhydride — *N*-Alkoxycarbonyl-dipeptide

FIGURE 3.14 Protection of amino groups as urethanes by reaction with chloroformates (path A). A side reaction often occurs. Reaction of the anionic product with the chloroformate (path B) generates a mixed anhydride that undergoes aminolysis, yielding protected dimer. (Curtius, 1881).

of substituted dimers during acylation by the Schotten-Baumann procedure had been reported in the 1880s and then alluded to in a paper in the 1950s, but this obviously escaped the attention of peptide chemists for many years. A subsequent attempt to eliminate dimer formation by use of diisopropylethylamine instead of sodium hydroxide as the base showed that the tertiary amine prevented dimerization during acylations with benzoyl and ethoxycarbonyl chloride but only reduced it for reactions with benzyloxycarbonyl and 9-fluorenylmethoxycarbonyl chlorides. Methods of general application that avoid dimer formation are described in the next section.[31–35]

31. T Wieland, R Sehring. A new method of peptide synthesis. (dimer formation) *Ann Chem* 122, 1950.
32. C-D Chang, M Waki, M Ahmad, J Meienhofer, EO Lundell, JD Haug. Preparation and properties of N^{α}-9-fluorenylmethoxycarbonyl amino acids bearing *tert.*-butyl side chain protection. *Int J Pept Prot Res* 15, 59, 1980.
33. L Lapatsanis, G Milias, K Froussios, M Kolovos. Synthesis of *N*-2,2,2,-(trichloroethoxy carbonyl)-L-amino acids and *N*-(fluorenylmethoxycarbonyl)-L-amino acids involving succinimidoxy anion as a leaving group in amino acid protection. *Synthesis* 671, 1983.
34. NL Benoiton, FMF Chen, R Steinauer, M Chouinard. A general procedure for preparing a reference mixture and determining the amount of dimerized contaminant in *N*-alkoxycarbonylamino acids by high-performance liquid chromatography. *Int J Pept Prot Res* 27, 28, 1986.
35. FMF Chen and NL Benoiton. Diisopropylethylamine eliminates dipeptide formation during acylation of amino acids using benzoyl chloride and some alkyl chloroformates. *Can J Chem* 65, 1224, 1987.

3.15 PROTECTION OF AMINO GROUPS: ACYLATION WITHOUT DIMER FORMATION

Dimer formation occurs during acylation by the Schotten-Baumann procedure because the reagent is highly activated and the product is completely ionized, thus inducing reaction between the two (see Section 3.13). At the time when it became apparent that some samples of *N*-alkoxycarbonylamino acids were contaminated with protected dipeptide, a report appeared in the literature describing the acylation of hydroxyamino acids [Figure 3.15, path A, $R^2 = CH_2OH$ or $CH(CH_3)OH$] using 9-fluorenylmethyl succinimido carbonate ($R^1 = Fm$). The rationale was that the lesser activation of this mixed carbonate relative to 9-fluorenylmethoxycarbonyl chloride would ensure that the side chains would not be acylated. The products indeed were monoacyl derivatives and pure. At the suggestion of a reviewer of the paper, a statement was inserted to the effect that mixed carbonates could be generally useful for preparing Fmoc-amino acids uncontaminated by *N*-protected dipeptide. 9-Fluorenylmethyl succinimido carbonate has been the reagent of choice for preparing Fmoc-amino acids ever since. Benzyloxycarbonyl succinimido carbonate is similarly employed to prepare pure *N*-benzyloxycarbonylamino acids. The pentafluorophenyl-mixed carbonates are also used. The absence of dimerization is explainable on the basis that the reagents are less activated than the chlorides, and the acylations (Figure 3.15, path A) are carried out at lower pH, which helps to avoid reaction between

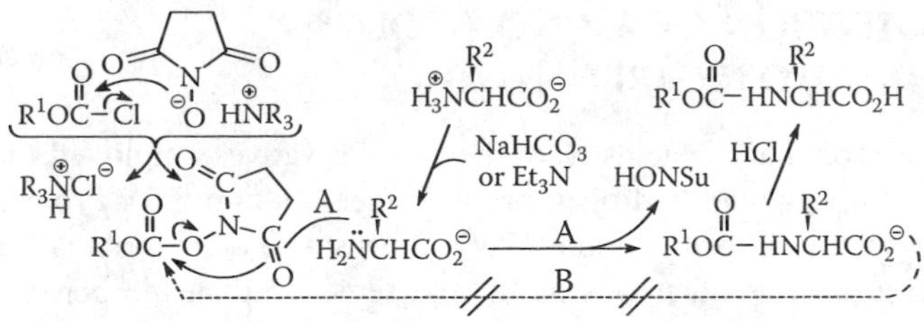

FIGURE 3.15 Protection of amino groups as urethanes by reaction with succinimido carbonates (path A).[33,36] The mixed carbonate is a weaker electrophile than the chloroformate. The *N*-alkoxycarbonylamino-acid anion does not react with the reagent (path B) in the presence of the weak base; hence no dimer is formed. R = triethyl or dicyclohexyl.

the acylating agents and the products (path B). The mixed carbonates are obtainable by reaction of the chlorocarbonate of one alcohol with the anion of the other alcohol.

A recently described approach involving zinc dust for eliminating acid allows acylation by 9-fluorenylmethoxycarbonyl chloride without dimer formation. The amino acid is dissolved in acetonitrile with the aid of hydrochloric acid, and zinc dust is added to destroy the acid and deprotonate the zwitter-ion, reducing the protons to gaseous hydrogen (Figure 3.16). Acylation is effected in the presence of zinc dust, which reduces the proton that is liberated by the reaction as soon it is formed. See Section 7.7 for another possible impurity in Fmoc amino acids.[34,36–39]

34. NL Benoiton, FMF Chen, R Steinauer and M Chouinard. A general procedure for preparing a reference mixture and determining the amount of dimerized contaminant in *N*-alkoxycarbonylamino acids by high-performance liquid chromatography. *Int J Pept Prot Res* 27, 28, 1986.
36. A Paquet. Introduction of 9-fluorenylmethoxycarbonyl, trichloroethoxycarbonyl, and benzyloxycarbonyl amino protecting groups into O-unprotected hydroxyamino acids using succinimidyl carbonates. *Can J Chem* 60, 976, 1982.
37. PBW Ten Koortenaar, BG Van Dijk, JM Peeters, BJ Raaben, PJH Adams, GI Tesser. Rapid and efficient method for preparation of Fmoc-amino acids starting from 9-fluorenylmethanol. *Int J Pept Prot Res* 27, 398, 1986.
38. DB Bolin, I Sytwu, F Humiec, J Meienhofer. Preparation of oligomer-free N^{α}-Fmoc and N^{α}-urethane amino acids. *Int J Pept Prot Res* 33, 353, 1989.
39. HN Gopi, VV Suresh Babu. Zinc-promoted simple synthesis of oligomer-free N^{α}-Fmoc-amino acids using Fmoc-Cl as an acylating agent under neutral conditions. *J Pept Res* 55, 295, 2000.

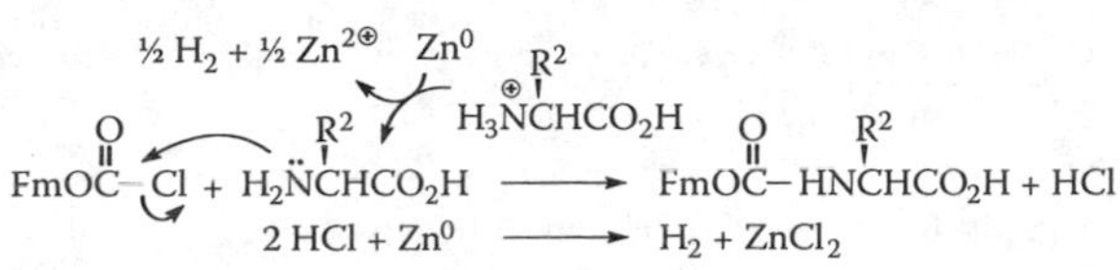

$$\text{FmOC(O)–Cl} + \text{H}_2\text{NCH(R}^2\text{)CO}_2\text{H} \longrightarrow \text{FmOC(O)–HNCH(R}^2\text{)CO}_2\text{H} + \text{HCl}$$

$$2\,\text{HCl} + \text{Zn}^0 \longrightarrow \text{H}_2 + \text{ZnCl}_2$$

FIGURE 3.16 *N*-Acylation at neutral pH employing zinc dust as a proton scavenger.[39] The zinc destroys the acid that is present or produced by reducing it to hydrogen.

3.16 PROTECTION OF AMINO GROUPS: *TERT*-BUTOXYCARBONYLATION

tert-Butoxycarbonylamino acids were the first derivatives employed that were sensitive to acid milder than hydrogen bromide (see Section 3.5). Protection by *tert*-butoxycarbonyl emerged in the late 1950s from studies on isophthalimides and on searches for a protector that was more acid sensitive than the benzyloxycarbonyl group. The substituent was employed to protect isophthalimide; it was shown to be removable by hydrogen chloride, trifluoroacetic acid, and hydrogen fluoride and suggested for use as a protector for amino acids. The protected amino acids were prepared in two other laboratories, employed successfully for the synthesis of peptides, and shown to be stable to hydrogenolysis and sodium in liquid ammonia. They had been obtained by acylation of the amino acids with a mixed carbonate, *tert*-butyl *p*-nitrophenyl carbonate (see Section 3.14). Boc-amino acids have never routinely been prepared by the Schotten-Baumann procedure (see Section 3.13) because *tert*-butoxycarbonyl chloride (Figure 3.17) does not possess the properties desired of a reagent; it decomposes at 10°C. The reagent of choice for a decade or more was *tert*-butoxycarbonyl azide (Figure 3.17), which gave high yields when used at mildly alkaline pH and not in large excess, but this was eventually withdrawn from the market because of its thermal and shock sensitivity. Several reagents, either mixed carbonates or similar, are now in current use for preparing Boc-amino acids. Two popular reagents appear in Figure 3.17. The substituted oxime is a mixed carbonate, and the other a pyrocarbonate, the anhydride of mono-*tert*-butyl carbonate. The acylations are carried out at mildly alkaline pH that is kept constant by use of a pH-stat and in dilute solution to avoid dimerization (see Section 3.13). It is notable that dimerization has rarely been associated with the preparation of Boc-amino acids, because *tert*-butoxycarbonyl chloride was hardly ever employed for their preparation.[13,40–46]

13. P Sieber, B Riniker, M Brugger, B Kamber, W Rittel. 255. Human Calcitonin. VI. The synthesis of calcitonin M. (side-chain alkylation) *Helv Chim Acta* 53, 2135, 1970.
40. LA Carpino, BA Carpino, PJ Crowley, PH Terry. *t*-Butyl azidoformate. *Org Syntheses* 15, 1964.

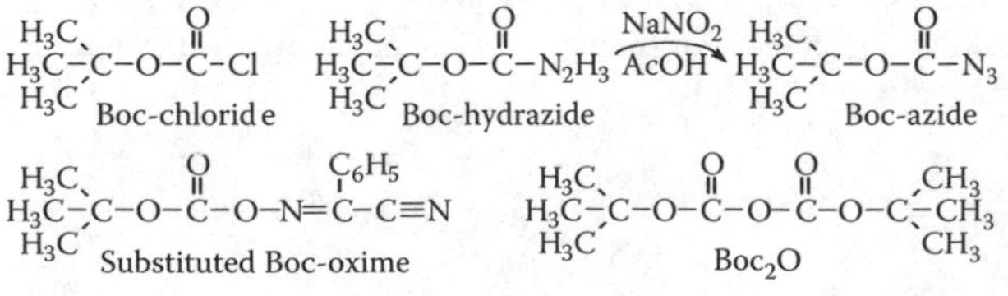

$(CH_3)_3C{-}O{-}C({=}O){-}Cl$ Boc-chloride

$(CH_3)_3C{-}O{-}C({=}O){-}N_2H_3$ Boc-hydrazide $\xrightarrow{NaNO_2,\ AcOH}$ $(CH_3)_3C{-}O{-}C({=}O){-}N_3$ Boc-azide

$(CH_3)_3C{-}O{-}C({=}O){-}O{-}N{=}C(C_6H_5){-}C{\equiv}N$ Substituted Boc-oxime

$(CH_3)_3C{-}O{-}C({=}O){-}O{-}C({=}O){-}O{-}C(CH_3)_3$ Boc_2O

FIGURE 3.17 Reagents for protection of amino groups as the *tert*-butoxycarbonyl derivatives. *tert*-Butyl chloroformate is rarely used because of its low boiling point. The oxime is 2-*tert*-butoxycarbonyloximino-2-phenylacetonitrile,[45] Boc_2O = di-*tert*-butyl dicarbonate, or di-*tert*-butyl pyrocarbonate.[46] (Tarbell et al., 1972; Pozdvev, 1974). Acylations are carried out at pH 9 to avoid dimerization.

41. LA Carpino. Oxidative reactions of hydrazines. II. Isophthalimides. New protective groups on nitrogen. *J Am Chem Soc* 79, 98, 1957.
42. FC McKay, NF Albertson. New amine-masking groups for peptide synthesis. *J Am Chem Soc* 79, 4686, 1957.
43. R Schwyzer, P Sieber, H Kappeler. On the synthesis of N-*t*-butyloxycarbonyl-amino acids. *Helv Chim Acta* 42, 2622, 1959.
44. E Schnabel. A better synthesis of tert-butyloxycarbonylamino acids through a pH-stat reaction. *Liebigs Ann Chem* 702, 189, 1967.
45. N Itoh, D Hagiwara, T Kamiya. A new *tert*-butoxycarbonylating reagent, 2-*tert*-butyl oxycarbonyloxyimino-2-phenylacetonitrile. *Tetrahedron Lett* 4393, 1975.
46. L Moroder, A Hallett, E Wünsch, O Keller, G Wersin. di-tert-Butyldicarbonat — an advantageous reagent for introduction of the tert-butyloxycarbonyl protecting group. *Hoppe-Seyler's Z Physiol Chem* 357, 1651, 1976.

3.17 PROTECTION OF CARBOXYL GROUPS: ESTERIFICATION

Carboxyl groups are usually protected as esters. This can involve esterification of the carboxyl group of an amino acid or that of an *N*-substituted residue. The approaches are different. Amino acids are esterified by acid-catalyzed reaction with alcohols (Figure 3.18, A), with the nature of the product dictating the nature of the catalyst. The acid protonates the carboxyl group, thus facilitating attack by alkoxy, which is followed by a release of water to produce the ester as the acid salt. The reaction is reversible; as a consequence, the water prevents the alkylation from going to completion, so some protocols recommend its removal with a separator (Figure 3.19) after it has been vaporized along with refluxing benzene. Hydrogen chloride is the classic catalyst for preparing methyl and ethyl esters. Use of the water separator when making the latter circumvents the need for dry hydrogen chloride and anhydrous alcohol. The rate of reaction is less for the hindered amino acids. The method is often attributed to Fischer, though it was used by Curtius in the 1880s. Because

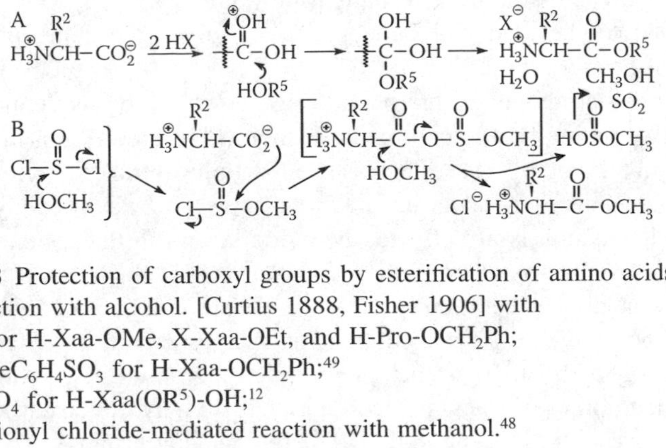

FIGURE 3.18 Protection of carboxyl groups by esterification of amino acids (A) by acid-catalyzed reaction with alcohol. [Curtius 1888, Fisher 1906] with
X = Cl for H-Xaa-OMe, X-Xaa-OEt, and H-Pro-OCH_2Ph;
X = 4-$MeC_6H_4SO_3$ for H-Xaa-OCH_2Ph;[49]
X = $^1/_2SO_4$ for H-Xaa(OR^5)-OH;[12]
and (B) by thionyl chloride-mediated reaction with methanol.[48]

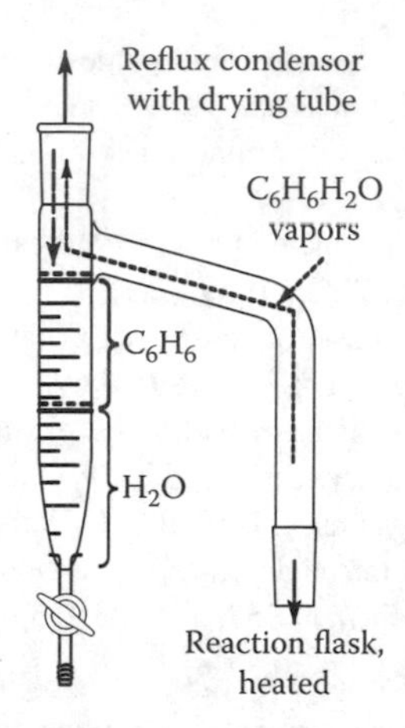

FIGURE 3.19 Dean-Stark water separator. Water is removed from the reaction medium by covaporization with benzene.

of the high boiling point of benzyl alcohol, benzyl esters are generated at the temperature of boiling benzene, with *p*-toluenesulfonic acid as the catalyst, except for proline benzyl ester. For convenience, the *p*-toluenesulfonates are often converted to the hydrochlorides by extraction of the acid into aqueous base, followed by addition of hydrogen chloride to a solution of the ester in organic solvent or evaporation of a solution of the ester in hydrochloric acid. Free esters can be obtained without the addition of a base by destroying the acid with zinc dust (see Section 3.15). Dicarboxylic amino acids are esterified exclusively on the side chain by use of concentrated sulfuric acid as a catalyst (see Section 6.24).

A variant that eliminates the production of water and that has proved effective for esterification of hydroxy and aromatic amino acids involves the use of thionyl chloride instead of acid. At a low temperature, the alcohol reacts with the chloride, generating methyl sulfinyl chloride, which produces the ester, probably through the mixed carboxylic acid-sulfinic acid anhydride (Figure 3.18, B). *p*-Toluenesulfonyl chloride added to the acid and benzyl alcohol serves the same purpose in the preparation of benzyl esters.

N-Substituted amino acids and peptides are esterified by nucleophilic substitution of an alkyl halide by the carboxylate anion (Figure 3.20, A), the latter being generated from the acid by a base such as triethylamine, tetralkylammonium hydroxide or monovalent alkali metal carbonates. In contrast to the acid-catalyzed reaction, the rate of this reaction is not affected by steric factors in the residue but depends on the extent of ionization of the carboxyl group because the anion is the displacing species. The crucial consideration is release of the anion from its counter-ion, which is best achieved by use of a large counter-ion such as cesium in an aprotic solvent such as dimethylformamide or hexamethylphosphoramide. Anhydrous salts are obtained by neutralizing the acid in alcohol with cesium hydrogen carbonate or other and removing the water by evaporation. The best leaving group X is iodide, though alkyl bromides are employed more frequently. All types of esters except *tert*-butyl are accessible by this approach, including 9-fluorenylmethyl and phenacyl esters

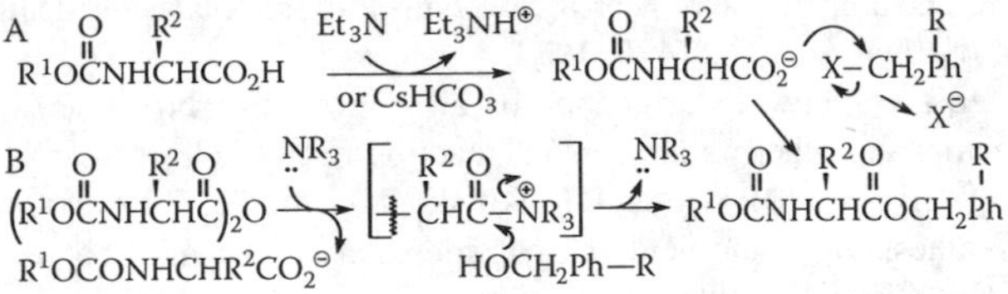

FIGURE 3.20 Protection of carboxyl groups by esterification of *N*-protected amino acids (A) by reaction of the anion with an alkyl halide or haloalkyl resin (R = resin) in dimethylformamide[51] and (B) by tertiary amine-catalyzed reaction of a symmetrical anhydride with hydroxymethylphenyl-resin (R = resin).[53] The intermediate is probably that depicted in Figure 3.19. Reaction (A) is applicable also to the carboxyl groups of peptides.

and benzyl esters with a variety of substituents such as nitro, methoxy, chlorodiphenyl (chlorotrityl), polystyrene, or other that constitute linkers for solid-phase synthesis (see Section 5.19). Another method for esterification to linkers is the 4-dimethylaminopyridine-catalyzed reaction with symmetrical anhydrides (Figure 3.20, B), which, however, is fraught with the danger of enantiomerization (see Section 4.16). Additional methods for attachment of a residue to linkers are presented in Section 5.22.

A more elaborate but general procedure for esterification involves reaction of the *N*-alkoxycarbonylamino acid with the alkyl chloroformate of the alcohol to be esterified in the presence of triethylamine and a catalytic amount of 4-dimethylaminopyridine (see Section 4.19) (Figure 3.21). The product probably arises by acylation of the alcohol by the acylpyridinum ion, both originating from decomposition of the mixed anhydride. The method can be used also to prepare activated esters (see Section 2.09), though the latter are usually obtained using the common coupling techniques (see Section 7.7).[47–57]

47. E Fischer. Synthesis of polypeptides XV. *Ber Deutsch Chem Ges* 39, 2893, 1906
48. M Brenner, W Huber. Determination of α-amino acids through alcoholysis of the methyl ester. *Helv Chim Acta* 36, 1109, 1953.
49. JD Cipera, RVV Nicholls. Preparation of benzyl esters of amino acids. *Chem Ind (London)* 16, 1955.
50. L Zervas, M Winitz, JP Greenstein. Studies on arginine peptides. I. Intermediates in the synthesis of N-terminal and C-terminal arginine peptides. (benzyl esters) *J Org Chem* 22, 1515, 1957.

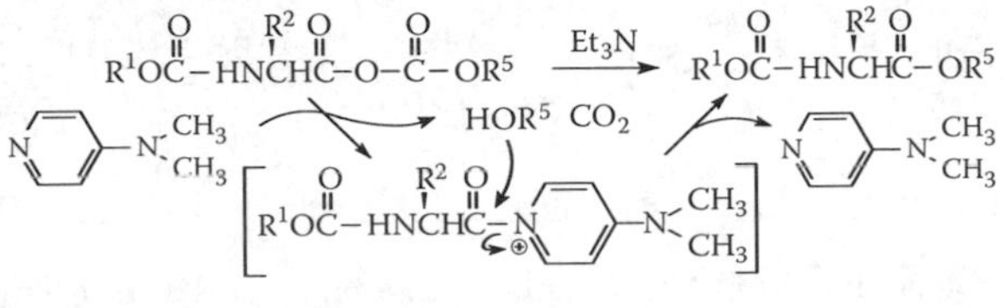

FIGURE 3.21 Esterification by decomposition of a mixed anhydride by triethylamine in the presence of 4-dimethylaminopyridine.[55] The active intermediate is probably the acylpyridium ion.

51. BF Gisin. The preparation of Merrifield-resins through total esterification with cesium salts. *Helv Chim Acta* 56, 1476, 1973.
52. PP Pfeffer, LS Silbert. Esterification by alkylation of carboxylate salts. Influence of steric factors and other parameters on reaction rates. *J Org Chem* 41, 1373, 1976.
53. S-S Wang, BF Gisin, DP Winter, R Makofske, ID Kulesha, C Tzougraki, J Meienhofer. Facile synthesis of amino acid and peptide esters under mild conditions via cesium salts. *J Org Chem* 42, 1286, 1977.
54. T Yamada, N Isono, A Inui, T Miyazawa, S Kuwata, H Watanabe. Esterification of *N*-(benzyloxycarbonyl)amino acids and amino acids using BF_3-etherate as catalyst. *Bull Chem Soc Jpn* 51, 1897, 1978.
55. S Kim, JL Lee, YC Kim. A simple and mild esterification method for carboxylic acids using the mixed carboxylic–carbonic anhydrides. *J Org Chem* 50, 560, 1985.
56. FMF Chen, NL Benoiton. Hydrochloride salts of benzyl esters from *p*-toluenesulfonate salts. *Int J Pept Prot Res* 27, 221, 1986.
57. K Ananda, VV Suresh Babu. Deprotonation of chloride salts of amino acid esters and peptide esters using commercial zinc dust. *J Pept Res* 57, 223, 2001.

3.18 PROTECTION OF CARBOXYL, HYDROXYL, AND SULFHYDRYL GROUPS BY *TERT*-BUTYLATION AND ALKYLATION

tert-Butyl esters cannot be prepared by the general methods of esterification (see Section 3.16) because of the unreactive nature of *tert*-butyl alcohol. Carboxyl groups of amino acids and *N*-alkoxycarbonylamino acids are *tert*-butylated by reaction with isobutene in organic solvent that is promoted by concentrated sulfuric acid (Figure 3.22, A). The strong acid is necessary to generate the *tert*-butyl cation by protonation of isobutene. The amino acid esters are conveniently crystallized as the hydrochlorides, after having been extracted into an organic phase from an alkaline solution. They are obtainable also by hydrogenolysis of the *N*-benzyloxycarbonylamino-acid *tert*-butyl esters. *N*-Protected amino acids and peptides can be alkylated also, using *tert*-butyl trichloroacetimidate in warm dichloromethane (Figure 3.22, B) This is a

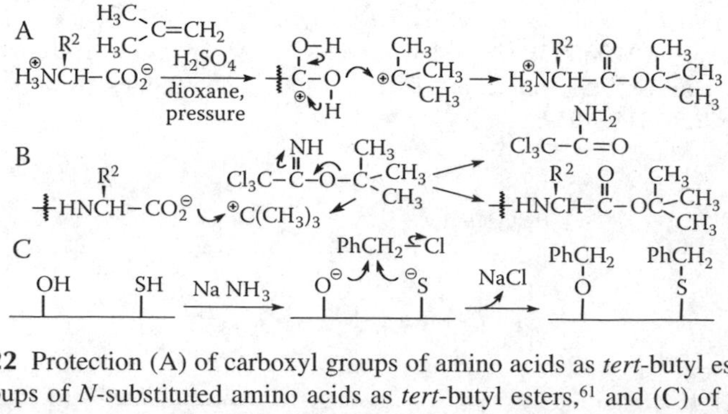

FIGURE 3.22 Protection (A) of carboxyl groups of amino acids as *tert*-butyl esters,[58] (B) of carboxyl groups of *N*-substituted amino acids as *tert*-butyl esters,[61] and (C) of phenolic and sulfhydryl groups as ethers. The amino acid esters are isolated as the hydrochlorides.

general reaction that also allows preparation of allyl and benzyl esters from the appropriate acetimidates. Hydroxyl and sulfhydryl groups are also alkylated by reactions A and B. Reaction with *tert*-butyl acetate and perchloric acid also esterifies *N*-protected amino acids. Benzyl ethers and thioethers are obtainable by displacement of chloride from benzyl chloride by oxy or thio anions generated by sodium in liquid ammonia (Figure 3.22, C). Modified benzyl (see Section 3.20) ethers and thioethers are obtained from the corresponding halides in alkaline solution.[58–62]

58. R Roeske. Amino acid tert-butyl esters. *Chem Ind (London)* 1121, 1959.
59. GW Anderson, FM Callahan. *t*-Butyl esters of amino acids and peptides and their use in peptide synthesis. *J Am Chem Soc* 82, 3359, 1960.
60. R Roeske. Preparation of t-butyl esters of free amino acids. *J Org Chem* 28, 1251, 1963.
61. A Armstrong, I Brackenridge, RFW Jackson, JM Kirk. A new method for the preparation of tertiary butyl ethers and esters. *Tetrahedron Lett* 29, 2483, 1988.
62. B Riniker, A Florsheimer, H Fretz, B Kamber. The synthesis of peptides by a combined solid phase and solution synthesis, in CH Schneider, AN Eberle, eds. *Peptides 1992. Proceedings of the 22nd European Peptide Symposium.* Escom, Leiden, 1993, pp 34-35.

3.19 PROTECTORS SENSITIZED OR STABILIZED TO ACIDOLYSIS

Shortly after introduction of the solid-phase method of peptide synthesis involving *tert*-butoxycarbonyl for temporary protection of α-amino groups (see Section 5.17), it became apparent that protection of side-chain functions as well as linkage to the resin by benzyl-based substituents was unsatisfactory. The benzyloxy bond was not stable enough to survive the successive treatments with acid that were required to remove the *tert*-butoxycarbonyl groups. There was a need for substituents that were more resistant to acidolysis. In the same vein, the necessity for strong acid to remove side-chain and carboxy-terminal protectors at the end of a synthesis led to development of substituents that were less resistant to acidolysis than benzyl substituents. Thus, there emerged protectors comprised of benzyl that is modified either to decrease or to increase the sensitivity to acidolysis, and as acidolysis involves protonation (see Section 3.5), this was achieved by adding functionalities that are either electron withdrawing, making protonation more difficult, or electron donating, making protonation easier. This is best illustrated by examples such as 4-nitro, which renders the benzyloxycarbonyl group stable to hydrogen fluoride for 12 hours, and 4-methoxy, which renders the benzyloxycarbonyl group sensitive to 10% trifluoroacetic acid, with a half-life of less than 1 minute (Figure 23). Addition of 2-chloro to benzyloxycarbonyl on the side chain of lysine reduces the rate of cleavage of the protector by 200 times. Modifications of the *tert*-butyl group also serve as examples of sensitized protectors. Replacement of one of the methyl groups by biphenyl (Figure 3.23) increases the sensitivity of the protector by thousands of times. In this case, the greater sensitivity resides in the increased stability of the cation that is generated by acidolysis (see Section 3.6). A cyclohexyl substituent is more stable

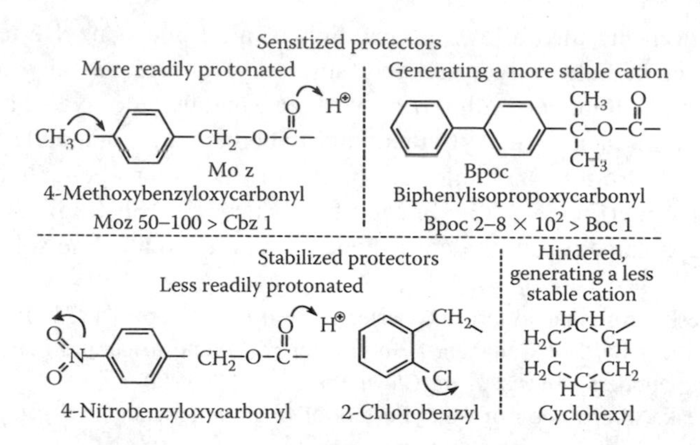

FIGURE 3.23 Protectors that are more sensitive or more stable to acidolysis than the parent protector. Electron-donating groups favor protonation and hence aid cleavage; electron-withdrawing groups disfavor protonation. The numbers indicate the relative ease of cleavage, which also depends on the stability of the carbenium ion that is released. Bpoc = 2-(biphenyl-4-yl)-prop-2-yloxycarbonyl.

than benzyl because it is more hindered at the protonating site and its removal generates a less-stable cation, which immediately isomerizes to the tertiary methylpentyl carbenium ion. In this regard, the ultimate stability to acidolysis and unreactivity of protectors seems to have been achieved in the 2,4-dimethylpent-3-yl (diisopropylmethyl) group (see Section 6.7). Protectors that are stabilized to acid are, however, more sensitive to nucleophiles. Acid-stabilized side-chain protectors that are popular today include 2-chlorobenzyl for tyrosine, and cyclohexyl-based protectors for tryptophan and the dicarboxylic amino acids. Derivatives that are sensitive to mild acid are often stored as the dialkylammonium salts to avoid decomposition. Both methoxybenzyl and nitrobenzyl protectors remain cleavable by hydrogenolysis (see Section 3.3), the latter by virtue of a 1,6-elimination resulting from a spontaneous shift of electrons from the nitrogen atom as the nitro group is reduced to the amino group (Figure 3.24), a reduction that can also be achieved by dithionite.

A protector of unique nature is the triphenylmethyl group, which is benzyl sensitized by two phenyl groups on the exocyclic carbon atom. When affixed to a hetero atom, the bond is cleavable by mild acid or hot acetic acid in part because of the very stable cation that is formed (Figure 3.25). N^{α}-Triphenylmethylamino

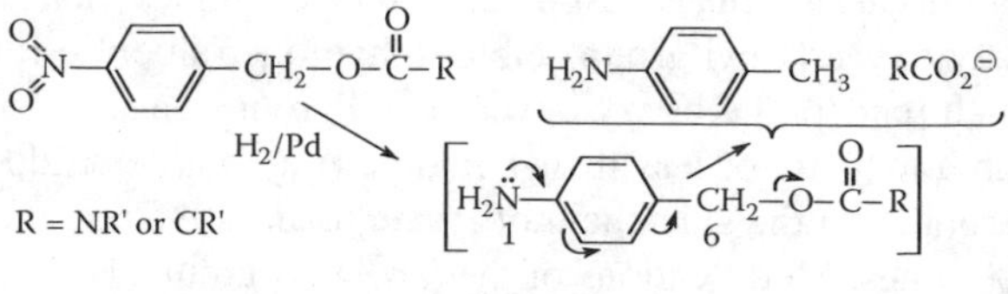

FIGURE 3.24 Cleavage of nitro-benzyl-based protectors by hydrogenolysis [Pless & Guttmann, 1960] or the action of dithionite ($S_2O_4^{2-}$).[71] A 1,6-elimination occurs as a result of a shift of electrons from the nitrogen atom of the arylamino group that is generated by reduction.

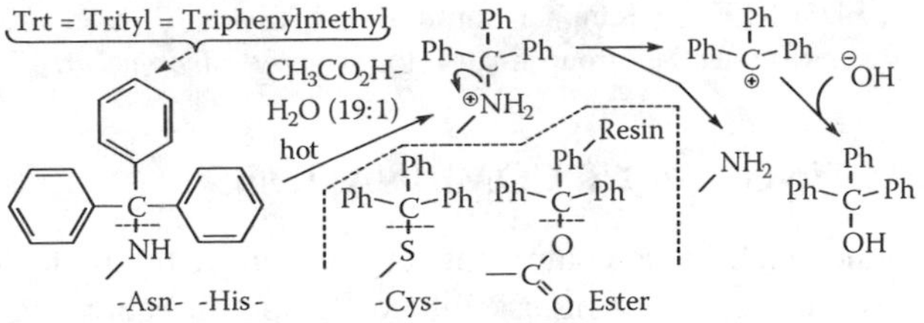

FIGURE 3.25 The triphenylmethyl group for protection[63,64] of side-chain and carboxy-terminal functional groups. The triphenylmethyl-heteroatom bond is sensitive to mild acid.

acids were evaluated for synthesis in the 1950s, but it was found that they couple with difficulty. The low reactivity can be attributed to hindrance by the bulky substituent as well as a suppressed activation at the carboxyl group. New interest in the triphenylmethyl group emerged decades later, when it was established to be effective for protection of the side-chain functional groups of asparagine, histidine, and cysteine (Figure 3.25).

The notion of stabilized and sensitized protectors is pertinent to solid-phase synthesis in particular. Practically all of the linkers through which carboxy-terminal residues are attached to the solid support are composed of benzyl that has been substituted with functional groups such as dialkoxy, dimethoxyphenyl, phenyl (benzhydryl), diphenyl (trityl), chlorodiphenyl (chlorotrityl, see Section 5.23), or other to modify the stability of the linking bond.[9,63–73]

9. R Schwyzer, W Rittel, H Kappeler, B Iselin. Synthesis of a nonadecapeptide with higher corticotropic activity. (*tert*-butyl removal) *Angew Chem* 72, 915, 1960.
63. G Amiard, R Heymes, L Velluz. On *N*-trityl-α-amino acids and their application in peptide synthesis. *Bull Soc Chim Fr* 191, 1955.
64. L Zervas, DM Theodoropoulus. *N*-Tritylamino acids and peptides. A new method of peptide synthesis. *J Am Chem Soc* 78, 1359, 1956.
65. GC Stelekatos, DM Theodoropoulus, L Zervas. On the trityl method of peptide synthesis. *J Am Chem Soc* 81, 2884, 1959.
66. H Schwarz, K Arakawa. The use of p-nitrobenzyl esters in peptide synthesis. *J Am Chem Soc* 81, 5691, 1959.
67. P Sieber, B Iselin. Peptide synthesis using the 2-(*p*-diphenyl)-isopropoxycarbonyl (Dpoc) amino protecting group. *Helv Chim Acta* 51, 622, 1968.
68. D Yamashiro, CH Li. Adrenocorticotropins. 44. Total synthesis of the human hormone by the solid-phase method. *J Am Chem Soc* 95, 1310, 1973.
69. BW Erickson, RB Merrifield. Use of chlorinated benzyloxycarbonyl protecting groups to eliminate Nε-branching at lysine during solid-phase peptide synthesis. *J Am Chem Soc* 95, 3757, 1973.
70. JP Tam, TW Wong, MW Reimen, FS Tjoeng, RB Merrifield. Cyclohexyl ester as a new protecting group for aspartyl peptides to minimize aspartimide formation in acidic and basic treatments. *Tetrahedron Lett* 4033, 1979.
71. E Guibe-Jampel, M Wakselman. Selective cleavage of *p*-nitrobenzyl ester with sodium dithionite. *Syn Commun* 12, 219, 1982.
72. S-S Wang, ST Chen, KT Wang, RB Merrifield. 4-Methoxybenzyloxycarbonyl amino acids in solid phase peptide synthesis. *Int J Pept Prot Res* 30, 662, 1987.

73. Y Nishiuchi, H Nishio, T Inui, T Kimura, S Sakakibara. N^{in}-Cyclohexyloxycarbonyl group as a new protecting group for tryptophan. *Tetrahedron Lett* 37, 7529, 1996.

3.20 PROTECTING GROUP COMBINATIONS

Success in peptide synthesis is contingent on removal of the protectors of the α-amino groups as the chain is being assembled without affecting the protectors of other functional groups (see Section 1.5). Various combinations of protecting groups are employed in the construction of a peptide chain. These combinations can be characterized as orthogonal systems, in which the α-protector is removed by a mechanism that is different from that used to deprotect the other functional groups, and systems that are not orthogonal, in which the other protectors are sensitive to acid that is much stronger than that used to remove the α-protectors (see Section 1.5). Examples of these systems appear in Figure 3.26. The carboxy terminus can be a simple ester or amide or one of these that is attached directly or indirectly to a resin. A third level of orthogonality can be introduced by incorporating on a side-chain functional group a protector that is not sensitive to the cleavage reagents indicated. An example is allyl/allyloxycarbonyl (see Section 3.13). Other protectors and methods of deprotection are occasionally employed for particular reasons. These include *o*-nitrobenzyl substituents that are removed by photolysis, benzyl-based protectors removed by electrolytic oxidation, esters removed by tertiary amine-catalyzed trans esterification, and trialkylsilylation instead of acid for the removal of *tert*-butoxycarbonyl groups.[67,74–79]

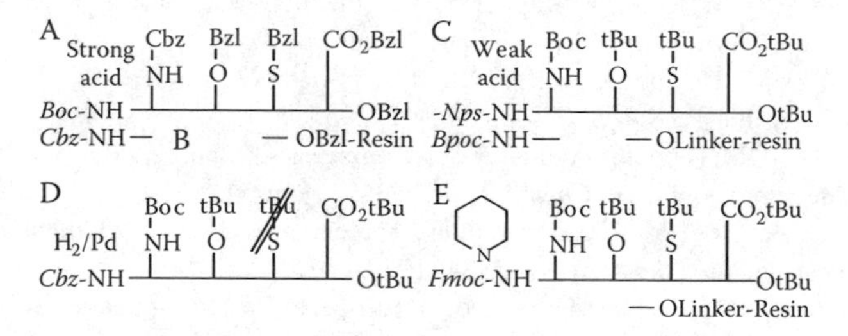

FIGURE 3.26 Combinations of protecting groups employed in synthesis. The protector written in italics is removed after each residue is incorporated into the chain. Protecting group combinations that are not orthogonal; all protectors are removed by acidolysis: (A) Boc/Bzl chemistry, the Boc group being removed by CF_3CO_2H-CH_2Cl_2 (1:1). Also employed in solid-phase synthesis. [Merrifield, 1963]. (B) Cbz/Bzl chemistry, the Cbz group being removed by 38% HBr in acetic acid.[6] (C) Nps/tBu [Moroder et al., 1978] and Bpoc/tBu [Sieber & Iselin, 1963] chemistries employing mild acid for deprotection of the α-amino groups. Protecting group combinations that are orthogonal: (D) Cbz/tBu chemistry [Schwyzer & Iselin, 1963] that is the most desirable protocol because removal of the Cbz group gives only inert volatile products. Catalytic hydrogenation fails in the presence of suflur-containing entities. (E) Fmoc/tBu chemistry used primarily in solid-phase synthesis.[77,78] Final deprotection is achieved for systems (C–E) by use of strong acid and stronger acid or reduction with sodium in liquid ammonia for system (A). Heated strong acid liberates functional groups of system (B).

67. P Sieber, B Iselin. Peptide synthesis using the 2-(*p*-diphenyl)-isopropoxycarbonyl (Dpoc) amino protecting group. *Helv Chim Acta* 51, 622, 1968.
74. R Schwyzer, B Rittel. Synthesis of intermediates for a corticotropic nonadecapeptide. I. N^{ε}-*tert*-Butoxycarbonyl-L-lysine, N^{α}(N^{ε}-*tert*-butoxycarbonyl-L-lysyl)-N^{ε}-*tert*-butoxycarbonyl-L-lysine, N^{ε}-*tert*-butoxycarbonyl-L-lysyl-L-prolyl-L-valylglycine and derivatives. *Helv Chim Acta* 44, 159, 1961.
75. R Schwyzer, P Sieber. The total synthesis of adrenocorticotrophic hormone. *Nature* 199, 172, 1963.
76. DH Rich, SK Gurwara. Preparation of a new *o*-nitrobenzyl resin for solid-phase synthesis of *tert*-butyloxycarbonyl-protected peptide acids. *J Am Chem Soc* 97, 1575, 1975.
77. C-D Chang, J Meienhofer. Solid-phase synthesis using mild base cleavage of N^{α}-fluorenylmethoxycarbonylamino acids, exemplified by a synthesis of dihydrosomatostatin. *Int J Pept Prot Res* 11, 246, 1978.
78. A Atherton, H Fox, D Harkiss, CJ Logan, RC Sheppard, BJ Williams. A mild procedure for solid phase peptide synthesis: Use of fluorenylmethoxycarbonylamino-acids. *J Chem Soc Chem Commun* 537, 1978.
79. KM Sivanandiaih, VV Suresh Babu, SC Shrankarama. Solid-phase synthesis of oxytocin using iodotrichlorosilane as Boc deprotecting reagent. *Int J Pept Prot Res* 45, 377, 1995.

4 Chirality in Peptide Synthesis

4.1 MECHANISMS OF STEREOMUTATION: ACID-CATALYZED ENOLIZATION

The objective of peptide synthesis is usually to prepare a single enantiomer. However, the amino acid enantiomers used in the synthesis are not always chirally stable during manipulation. Isomerization may occur at any step of a synthesis, be it the preparation of amino acid derivatives, coupling, or deprotection. The isomerization involves removal of the proton at the α-carbon atom of a residue, followed by a shift of the double bond of the adjacent carbonyl to the α-carbon, giving the enol form and eliminating the chirality. Reprotonation at the α-carbon atom generates the two possible configurations (Figures 4.1–4.5). The process is referred to as enolization and can be initiated by either an acid or base, with the latter more often being the culprit. Usually, though not always, it occurs at a residue that is substituted at both the amino and carboxyl groups. The affected residue may be a single one unattached to other residues or one that is incorporated into a chain (Figures 4.1–4.3), and in either case it may form part of a small cyclic structure that is usually the oxazolone (Figures 4.4, 4.5). The tendency for enolization to occur depends on the nature of the three substituents on the α-carbon atom, electron-withdrawing moieties on the β-carbon, and C-1 of the residue favoring the ionization. The least-encountered situation is that of acid-catalyzed enolization (Figure 4.1). Here enolization is initiated by protonation of the oxygen of the carbonyl, which induces migration of the electrons to neutralize the oxocation. Regeneration of the carbonyl from the enol produces the two isomeric forms of the residue. Examples of isomerization by this mechanism are the generation of D-isomers during the hydrolysis of peptides or proteins and during the exposure of *N*-substituted-*N*-methylamino acids to hydrobromic acid in acetic acid (see Section 8.14). The same protonation is involved in

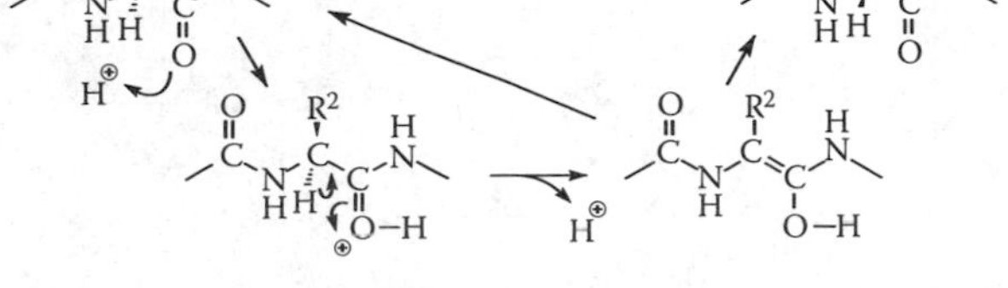

FIGURE 4.1 Enantiomerization of a residue (acid or amidated) by acid-catalyzed enolization.

the deprotection of functional groups by acidolysis (see Section 3.5,) but, fortunately, no enolization occurs during cleavage by the commonly used acidolytic reagents.[1–5]

1. A Neuberger. Stereochemistry of amino acids. *Adv Prot Chem* 4, 297, 1948.
2. JM Manning. Determination of D- and L-amino acid residues in peptides. Use of tritiated hydrochloric acid to correct for racemization during acid hydrolysis. *J Am Chem Soc* 92, 7449, 1970.
3. JR McDermott, NL Benoiton. *N*-Methylamino acids in peptide synthesis. III. Racemization during deprotection by saponification and acidolysis. *Can J Chem* 51, 2555, 1973.
4. H Frank, W Woiwode, G Nicholson, E Bayer. Determination of the rate of acidic catalyzed racemization of protein amino acids. *Liebigs Ann Chem* 354, 1981.
5. R Liardon, R Jost. Racemization of free and protein-bound amino acids in strong mineral acid. *Int J Pept Prot Res* 18, 500, 1981.

4.2 MECHANISMS OF STEREOMUTATION: BASE-CATALYZED ENOLIZATION

The most frequent cause of isomerization is direct removal of the α-proton by a base (Figure 4.2). The resulting unshared electron pair migrates to generate an equilibrium mixture of the carbanion and the oxoanion, with the double bond shifted to the α-carbon, thus eliminating the chirality. In the unique case of cysteine derivatives, however, the proton never gets completely detached but is carried from one side of the asymmetric center to the other by the base in a process known as isoracemization. Reprotonation generates the two configurations of the residue. Three different situations in which isomerization is caused by base-catalyzed enolization are encountered: first, the base is contained in the same molecule, with the classical case being the enantiomerization that occurs at activated N^{α}-alkoxycarbonylhistidine derivatives unprotected at the π-nitrogen of the imidazole ring of the side chain (see Section 4.3). Second, the base is introduced inadvertently, such as when the enantiomerization occurs at the activated residue of an acyl azide or activated ester that has been left in the presence of a tertiary amine. Particularly sensitive residues, because of the side chain electron-withdrawing groups, are

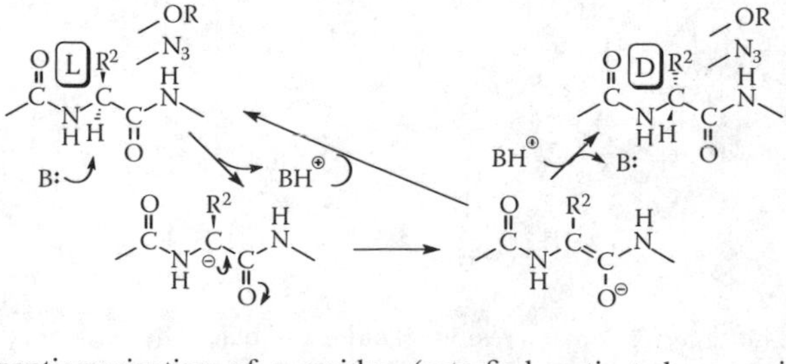

FIGURE 4.2 Enantiomerization of a residue (esterfied, activated, or amidated) by base-catalyzed enolization.

S-benzylcysteine, β-cyanoalanine, and β-carboxy-substituted aspartic acid. Third, the base is added intentionally. Here, isomerization can occur at the implicated residue such as that produced during the saponification of amino acid or peptide esters and during the aminolysis of Fmoc-proline chloride or aspartic acid activated at the β-carboxyl group (see Section 4.19). The isomerization also can occur at a distant residue such as that produced at the esterified, including resin-linked, carboxy-terminal cysteine or serine residues of a chain by the tertiary amine used to detach Fmoc-groups or to initiate onium salt-mediated coupling reactions (see Section 8.1) during chain assembly. Regardless, the residues most susceptible to base-catalyzed enolization are those in which the proton is abstracted from an α- or β-carbon that does not have an ionizable proton on the functional group linked to it, examples being $PgN(CH_3)$- or $-CO_2R$. Base-catalyzed enolization also occurs at a residue whose N-C^{α}-C=O atoms constitute part of the ring of an oxazolone (see Section 4.4) or a piperazine-2,5-dione (see Section 6.19), especially if prolyl is incorporated into the latter.[3,6–13]

3. JR McDermott, NL Benoiton. *N*-Methylamino acids in peptide synthesis. III. Racemization during deprotection by saponification and acidolysis. *Can J Chem* 51, 2555, 1973.
6. M Goodman, KC Stueben. Amino acid active esters. III. Base-catalyzed racemization of peptide active esters. *J Org Chem* 27, 3409, 1962.
7. B Liberek. Racemization during peptide synthetic work. II. Base catalyzed racemization of active derivatives of phthaloyl amino acids. *Tetrahedron Lett* 1103, 1963.
8. B Liberek. The nitrile group in peptide chemistry. V. Racemization during peptide synthesis. 4. Racemization of active esters of phthaloyl-β-cyano-L-alanine in the presence of triethylamine. *Acad Pol Sci Ser Sci Chim* 11, 677, 1963.
9. I Antanovic, GT Young. Amino-acids and peptides. Part XXV. The mechanism of base catalysed racemisation of the *p*-nitrophenyl esters of acylpeptides. *J Chem Soc C* 595, 1967.
10. J Kovac, GL Mayers, RH Johnson, RE Cover, UR Ghatak. Racemization of amino acid derivatives. Rate of racemization and peptide bond formation of cysteine active esters. *J Org Chem* 35, 1810, 1970.
11. GW Kenner, JH Seely. Phenyl esters for C-terminal protection in peptide synthesis. *J Am Chem Soc* 94, 3259, 1972.
12. Kisfaludy, O Nyeki. Racemization during peptide azide coupling. *Acta Chim (Budapest)* 72, 75, 1972.
13. F Dick, M Schwaller. SPPS of peptides containing C-terminal proline: racemization free anchoring of proline controlled by an easy and reliable method, in HLS Maia, ed. *Peptides 1994. Proceedings of the 23rd European Peptide Symposium*. Escom, Leiden, 1995, pp 240-241.

4.3 ENANTIOMERIZATION AND ITS AVOIDANCE DURING COUPLINGS OF *N*-ALKOXYCARBONYL-L-HISTIDINE

It had been established by midcentury that *N*-alkoxycarbonylamino acids do not isomerize during coupling. However, there emerged the puzzling observation that

N^{α}-substituted histidines, whether side-chain protected or not, underwent considerable enantiomerization during diimide-mediated reactions. Even Boc-L-histidine coupled by the azide method gave enantiomerically impure products. The results were attributed to intramolecular base-catalyzed proton abstraction and enolization (see Section 4.2). At the time, the position on the ring of substituents was not known, as evidenced by the designation of the popular substituent *im*-benzyl, which was ambiguous. Moreover, it was not helpful that the nitrogen atoms of the imidazole ring of histidine were designated in different ways by biochemists and chemists; namely, as 1,3 and 3,1. At least discussion was facilitated by the recommendation of the pertinent nomenclature committees that the nitrogen atom nearest to the chain (δ) should be designated *pros* ("near," abbreviated π) and the nitrogen atom farthest from the chain (ε) be denoted *tele* ("far," abbreviated τ). The α,β,γ,δ,ε designations of atoms (Figure 4.3) are those employed by x-ray crystallographers. There followed the suggestion by D. F. Veber that the side reaction might be caused by the *pros*-nitrogen atom of the imidazole ring, which remained unsubstituted in the derivatives. It is now known that substitution on the imidazole of histidine invariably occurs at the less hindered *tele*-nitrogen, and that isomerization indeed is caused by abstraction of the α-proton by the *pros*-nitrogen of the ring if it is left unprotected. The latter was demonstrated unequivocally by experiments with *tele*- and *pros*-substituted benzoylmethyl (phenacyl) derivatives of Cbz-histidine, with the latter being obtained from Cbz-His(τTrt)-OMe. *pros*-Substitution prevented isomerization during coupling; the *tele*-substituted derivative generated 30% of epimeric peptide (Figure 4.3). Further study established that for practical reasons, the *pros*-benzyloxymethyl derivative is the preferred derivative for couplings. Thus, substitution of the *pros*-nitrogen effectively suppresses enantiomerization during the coupling of N^{α}-alkoxycarbonyl-histidines. Furthermore, use of the *pros/tele* nomenclature eliminates the ambiguity of the previous 1,3/3,1 designations for the nitrogen atoms of the ring.[14–18]

14. GC Windridge, EC Jorgensen. Racemization in the solid phase synthesis of histidine containing peptides. *Intra-Sci Chem Rep* 5, 375, 1971.

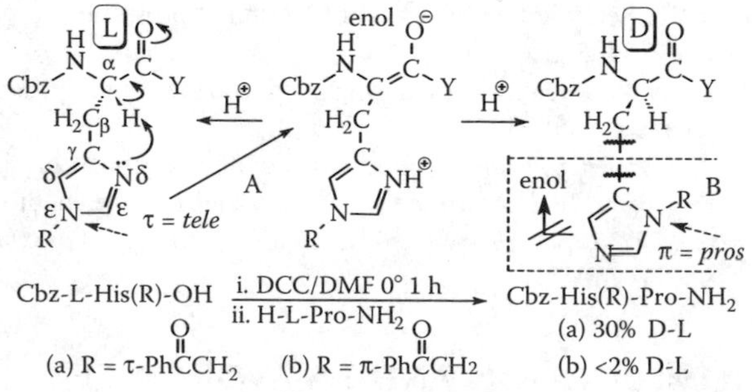

Cbz-L-His(R)-OH —(i. DCC/DMF 0° 1 h; ii. H-L-Pro-NH_2)→ Cbz-His(R)-Pro-NH_2

(a) R = τ-PhC(O)CH_2 (b) R = π-PhC(O)CH_2

(a) 30% D-L
(b) $<$2% D-L

FIGURE 4.3 Enantiomerization of activated N^{τ}-substituted N^{α}-benzyloxycarbonyl-L-histidine by enolization (A) promoted by the basic π-nitrogen atom of the imidazole ring that is in proximity to the α-hydrogen. Activation of the N^{π}-substituted derivative (B) proceeds without enolization. [Jorgenson, 1970; Weber, 1975; Jones et al.,1982]

15. GC Windridge, EC Jorgensen. 1-Hydroxybenzotriazole as a racemization-suppressing reagent for the incorporation of *im*-benzyl-L-histidine into peptides. *J Am Chem Soc* 93, 6318, 1971.
16. HC Beyerman, J Hirt, P Kranenburg, JLM Syrier, A van Zon. Excess mixed anhydride peptide synthesis with histidine derivatives. *Rec Trav Chim Pays-Bas* 93, 256, 1974.
17. AR Fletcher, JH Jones, WI Ramage, AV Stachulski. The use of the *N*(π)-phenacyl group for the protection of the histidine side chain in peptide synthesis. *J Chem Soc Perkin Trans* 1, 2261, 1979.
18. T Brown, JH Jones, JD Richards. Further studies on the protection of histidine side chains in peptide synthesis: use of the π-benzyloxymethyl group. *J Chem Soc Perkin Trans* 1, 1553, 1982.

4.4 MECHANISMS OF STEREOMUTATION: BASE-CATALYZED ENOLIZATION OF OXAZOLONES FORMED FROM ACTIVATED PEPTIDES

Oxazolones are readily formed from activated *N*-acylamino acids or peptides (see Sections 1.7, 1.9, and 2.23). Because of the strong tendency of the double bonds of the oxazolone to form a conjugated system, the α-proton is very labile and thus sensitive to base. The latter causes enolization, which eliminates the chirality (Figure 4.4). The lability of the α-proton is governed by electronic and conjugative effects at C-2, an electron donor such as methyl stabilizing the proton in contrast with phenyl, and steric effects at C-4, with the isopropyl of valine impeding release of the proton relative to isobutyl of leucine and then benzyl of phenylalanine. The effect of the latter is apparently anomalous and has been explained on the basis that the oxazolone from phenylalanine adopts a unique conformation resulting from stacking of the two rings, and this facilitates removal of the proton. As an example, the rates of racemization of the 2-phenyl-5(4*H*)-oxazolones from phenylalanine, leucine, and valine decrease in a ratio of 34:17:1 in dichloromethane. In tetrahydrofuran, in the presence of one equivalent of *N*-methylpiperidine, about 4 hours are required to

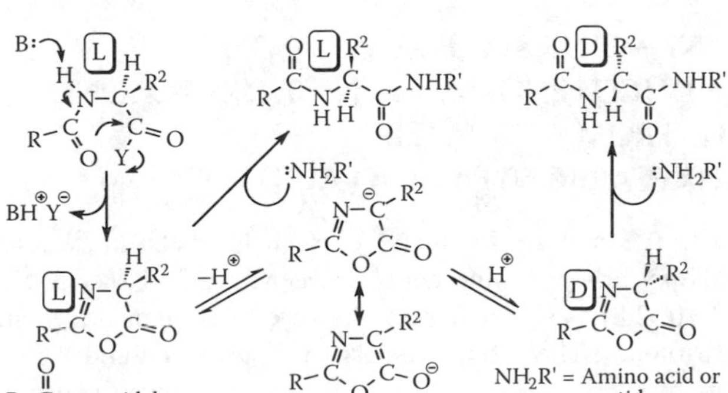

FIGURE 4.4 Enantiomerization of a residue by base-catalyzed enolization of the 2-alkyl-5(4*H*)-oxazolone formed during coupling of segments.

racemize Z-glycyl-L-valine. The rate of isomerization also depends dramatically on the nature of the base as well as the polarity of the solvent, being as much as 50 to 100 times greater in dimethylformamide than in dichloromethane. In the latter, *N*-methylmorpholine isomerized the oxazolone from a dipeptide more slowly than triethylamine, but in dimethylformamide it was the opposite, with the slowest isomerization occurring in the presence of diisopropylethylamine.

The double-bonded nitrogen atom of the oxazolone is slightly basic. As a result, oxazolones are not chirally stable on storage, undergoing an autocatalytic process called autoracemization — the result of one molecule abstracting the α-proton from a second molecule. Furthermore, oxazolones can isomerize by a different mechanism. The double bond can shift in the other direction from –CHR′-C(O-)=N-CHR^2– to –CR′=C(O-)-NH-CHR^2–, thus epimerizing the peptide at the second residue (see Section 7.23). This occurs in particular for oxazolones from aminoisobutyric acid that contain the sequence –CHR-C(O-)=N-$C(CH_3)_2$–.[6,9,19–23]

6. M Goodman, KC Stueben. Amino acid active esters. III. Base-catalyzed racemization of peptide active esters. *J Org Chem* 27, 3409, 1962.
9. I Antanovic, GT Young. Amino-acids and peptides. Part XXV. The mechanism of base catalysed racemisation of the *p*-nitrophenyl esters of acylpeptides. *J Chem Soc C* 595, 1967.
19. F Weygand, A Prox, W König. Racemisation of the second last carboxyl-containing amino acid in peptide synthesis. *Chem Ber* 99, 1446, 1966.
20. M Dzieduszycka, M Smulkowski, E Taschner. Racemization of amino acid residue penultimate to the C-terminal one during activation of N-protected peptides, in H Hanson, HD Jakubke, eds. *Peptides 1972*. North-Holland, Amsterdam, 1973, pp 103-107.
21. P Wipf, H Heimgartner. Coupling of peptides with C-terminal α,α-di-substituted α-amino acids via the oxazol-5(4*H*)-one. *Helv Chim Acta* 69, 1153, 1986.
22. M Slebioda, MA St-Amand, FMF Chen, NL Benoiton. Studies on the kinetics of racemization of 2,4-disubstituted-5(4*H*)-oxazolones. *Can J Chem* 66, 2540, 1988.
23. FMF Chen, NL Benoiton. Racemization of acylamino acids and the carboxy-terminal residue of peptides. *Int J Pept Prot Res* 31, 396.

4.5 MECHANISMS OF STEREOMUTATION: BASE-INDUCED ENOLIZATION OF OXAZOLONES FORMED FROM ACTIVATED *N*-ALKOXYCARBONYLAMINO ACIDS

Despite beliefs to the contrary for many years, it has been known since the early 1980s that *N*-alkoxycarbonylamino acids can generate 2-alkoxy-5(4*H*)-oxazolones (see Sections 1.10, 2.8). A major feature, however, distinguishes them from the 2-alkyl-5(4*H*)-oxazolones: They do not undergo isomerization under the normal operating conditions of synthesis. That being said, it must nevertheless be recognized that in the presence of base, the 2-alkoxy-5(4*H*)-oxazolones can and do enantiomerize. The 2-benzyloxy-5(4*H*)-oxazolone from Cbz-valine generated 7.5% of epimer when aminolyzed in the presence of 0.2 equivalents of triethylamine. The

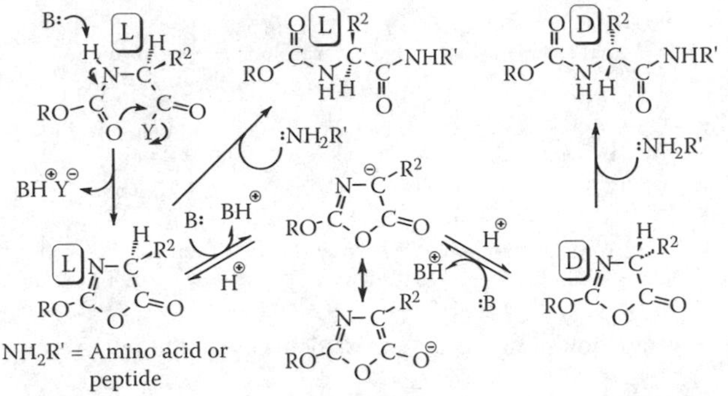

FIGURE 4.5 Enantiomerization of a residue by base-induced enolization of the 2-alkoxy-5(4*H*)-oxazolone formed during coupling of *N*-alkoxycarbonylamino acids.

isomerization can be attributed to the same base-catalyzed enolization (Figure 4.5), but the word "induced" has been inserted to emphasize the fact that the activated precursors do not isomerize when aminolyzed in the absence of added base. In contrast, the sensitivity of 2-alkoxy-5(4*H*)-oxazolones to tertiary amines is also indicated by the fact that they undergo autoracemization, as do the 2-alkyl-5(4*H*)-oxazolones (see Section 4.4), during storage, and surprisingly quickly, the oxazolone from *N*-ethoxycarbonyl-L-valine loses half of its optical activity after 6 days at 5°C.[24,25]

24. NL Benoiton, FMF Chen. 2-Alkoxy-5(4*H*)-oxazolones from *N*-alkoxycarbonylamino acids and their implication in carbodiimide-mediated reactions in peptide synthesis. *Can J Chem* 59, 384, 1981.
25. NL Benoiton, YC Lee, R Steinauer. Determination of enantiomers of histidine, arginine and other amino acids by HPLC of their diastereomeric *N*-ethoxycarbonyldipeptides. *Pept Res* 8, 108, 1995.

4.6 STEREOMUTATION AND ASYMMETRIC INDUCTION

Isomerization during couplings may occur by enolization of the activated residue (see Section 4.2) or by enolization of the oxazolone (see Sections 4.4 and 4.5) if it is formed or by both mechanisms. Several reaction steps are involved in this isomerization; namely, enolization of the activated residue at rate k_6, formation of the oxazolone at rate k_2, enolization of the oxazolone at rate k_4, and finally aminolysis of the isomerized intermediates at rates k_5 and k_7 (Figure 4.6). The desired product arises by aminolysis of the unisomerized intermediates at rates k_1 and k_3. Whether isomerization indeed occurs is dictated by the relative rates of these reactions. Examples of situations are presented in Figure 4.6 for couplings between two amino acid derivatives; the same would apply to couplings between segments. If the rates of aminolysis k_1 and k_3 are much greater than the rates of isomerization k_6 and k_4, no diastereoisomer will be generated. All rates are dependent on the natures of the two residues implicated in the coupling, as well as the natures of the amino and

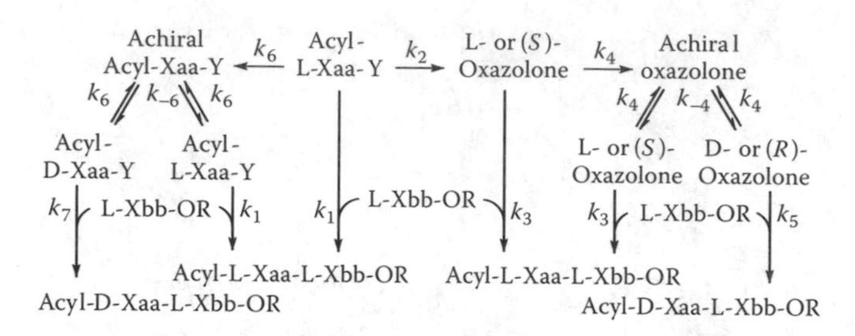

FIGURE 4.6 The generation of diastereoisomers depends on the relative rates of reactions. When

k_1	$>> k_2$,	there is no oxazolone formation;
k_1	$>> k_4 + k_{-4}$	there is no isomerization (e.g., if acyl = alkoxycarbonyl);
$k_4 + k_{-4}$	$>> k_3$	there is complete isomerization;
k_3	$> k_5$	equals a positive asymmetric induction (L-L > D-L);
k_3	$< k_5$	equals a negative asymmetric induction (D-L > L-L);
$k_6 + k_{-6}$	$> k_1$	equals isomerization by direct proton abstraction;
k_3	$= k_5$ & $k_1 = k_7$	means that the amino acid ester is not chiral.

carboxy substituents (near-neighbor residues for segments) and external factors such as temperature and polarity of the solvent (see Sections 4.12 to 4.16). In addition, a further complication arises because the two interacting molecules are chiral. A chiral molecule, in this case the amino acid ester, does not react at the same rates with two stereoisomers, be they enantiomers or epimers, so k_3 and k_5 are not equal, nor are k_1 and k_7. As a consequence, the diastereomeric products generated from the achiral intermediates are not formed in equal amounts. The greater the disparity in any two rates, the greater the difference between the amounts of the two isomeric products formed. This phenomenon is known as asymmetric induction. The direction of the induction is denoted as positive if more of the isomer containing the same configurations (-L-L- or -D-D-) is formed and negative if more of the isomer containing the two configurations (-D-L- or -L-D-) is formed. Again, the relative rates of aminolysis of the two isomers, and hence the magnitude of the asymmetric induction (percentage excess of one isomer), are dependent on the internal and external factors alluded to above. As an example, compare the +50% excess found for aminolysis in dichloromethane of the oxazolone from *Z*-L-alanyl-DL-valine by L-phenylalanine methyl ester with the 14% found for aminolysis in dimethylformamide of the oxazolone from *Z*-glycyl-DL-alanine by L-valine benzyl ester. There are various phenomena and reactions involved in producing unwanted isomers in couplings and a plethora of variables dictating the course of events, so except for the clear cases such as the reactions of *N*-alkoxycarbonylamino acids, in which $k_3 >> k_4$, the issue of chiral preservation during couplings is extremely complex and can be addressed only in generalizations.[26–31]

26. M Goodman, L Levine. Peptide synthesis *via* active esters. IV. Racemization and ring-opening reactions of optically active oxazolones. *J Am Chem Soc* 86, 2918, 1964.
27. F Weygand, W Steglich, X Boracio de la Lama. On the sterical course of the reaction of oxazol-5-ones with amino-acid esters. *Tetrahedron* Suppl 8, part 1, 9, 1966.

28. J Kovacs, GL Mayers, RH Johnson, RE Cover, UR Ghatak. Racemization of amino acid derivatives. Rates of racemization and peptide bond formation of cysteine esters. *J Org Chem* 35, 1810, 1970.
29. J Kovacs, EM Holleran, KY Hui. Kinetic studies in peptide chemistry. Coupling, racemization, and evaluation of methods useful for shortening coupling time. *J Org Chem* 45, 1060, 1980.
30. NL Benoiton, K Kuroda, FMF Chen. The dependence of asymmetric induction on solvent polarity and temperature in peptide synthesis. *Tetraheron Lett* 22, 3361, 1981.
31. NL Benoiton, YC Lee, FMF Chen. Studies on asymmetric induction associated with the coupling of *N*-acylamino acids and *N*-benzyloxycarbonyldipeptides. *Int J Pept Prot Res* 38, 574, 1991.

4.7 TERMINOLOGY FOR DESIGNATING STEREOMUTATION

Discussion of the isomerization process has proven to be a frustrating experience for peptide chemists. The reasons for this are twofold. First, the terms employed, namely, racemization and epimerization, describe the events from a different point of view — the former meaning the conversion of one isomer into an equimixture of two isomers, and the latter meaning the conversion of one isomer into another isomer. Second, although racemization applies to compounds with one stereogenic center and epimerization applies to compounds with two or more stereogenic centers, some amino acids belong to the first group, and others belong to the second group. The consequence of the latter is that often one cannot use the same term to describe the process that isomerizes a few amino acids, and the consequence of the former is that the numbers that quantify a certain amount of racemization are twice as large as those that quantify the same amount of epimerization. The unfortunate result has been that the term racemization is used by many to mean all changes in configuration, which is very different from its actual definition.

A further obstacle to exact phraseology arises when a unichiral substrate isomerizes during coupling and generates a mixture of two products that are not enantiomers but epimers. Achieving preciseness in expression has indeed become a difficult chore. My frustration with the situation in general prompted me to suggest the use of another term; namely, enantiomerizaton. The definition of enantiomerization has recently been revised by the nomenclature committees to signify the interconversion of enantiomers, implying mirror images containing one stereogenic center. Enantiomerization is consistent with epimerization — both mean the conversion of one isomer into another, thus eliminating the quantitation problem. Both are applicable (racemization is not) to the realistic case in which the conversion goes to greater than 50% because of a high negative asymmetric induction (see Section 4.6) that occurs during aminolysis. Finally, enantiomerization seems preferable to racemization for designating incomplete isomerization because it is more precise, partial racemization, meaning partial generation of an optically inactive product. A summary of definitions appears in Figure 4.7. To these definitions must be added diastereoisomers, which includes all stereoisomers that are not mirror images, and epimers, which are diastereoisomers that differ in configuration at one carbon atom only.

1 Achiral ⟶ 80% (*S*) + 20% (*R*) = 60% Enantiomeric excess (e.e.)
20% of minor enantionmer (max = 100%)

2 Racemization = conversion of enantiomer to a racemate (max = 50%)
L-Xaa ⟶ 80% L + 20% D = 40% racemized (L ⟶ DL)
20% enantiomerized (max = 100%)
Enantiomerization = interconversion of enantiomers (L ⟶ D)
(racemization is the unique case of enantiomerization progressing to 50%)

3 Epimerization = interconversion of epimers ((L-L ⟶ L-D; + L = L-D-L)
L-Xaa-L-Xbb + L-Xcc (the interconversion occurred before aminolysis)
⟶ 80% L-L-L + 20% L-D-L = 20% epimerized (can be >50%)

4 L-Xaa + L-Xbb (L ⟶ D; + L = D-L)
⟶ 80% L-L + 20% D-L = 20% enantiomerized? (can be >50%)
(the interconversion occurred before aminolysis)

FIGURE 4.7 Four different cases with pertinent terminology involving changes in isomeric composition generating a 4:1 mixture of products. max = theoretical maximum. The expression "enantiomerization of the residue" is applicable in all situations.

Enantiomeric excess applies to cases where the starting material of a reaction is not chiral, and hence is rarely used by peptidologists. Because it has proven to be very helpful, in this book enantiomerization is also employed to indicate a change in configuration sustained by a unichiral amino acid residue in a peptide, which is analogous to the use of epimerization to indicate a change in configuration sustained by a glycosidic residue in an oligosaccharide.[32,33]

32. NL Benoiton. Sometimes it is neither a racemisation nor an epimerisation but an enantiomerisation. A plea for preciseness in the use of terms describing stereomutations in peptide synthesis. *Int J Pept Prot Res* 44, 399, 1994.
33. JP Moss. Basic terminology of stereochemistry. *Pure Applied Chem* 68, 2193, 1996.

4.8 EVIDENCE OF STEREOCHEMICAL INHOMOGENEITY IN SYNTHESIZED PRODUCTS

Demonstrating that isomerization has or has not occurred during a synthesis is not straightforward. Other than by using x-ray crystallography, which is not practical, there is no way to prove the enantiomeric integrity of a peptide. One can only demonstrate the absence of isomers. This can involve demonstrating either the absence of D-residues in the peptide or the absence of diastereomeric peptides. Analysis for D-residues in a peptide after its hydrolysis is feasible, but interpretation of the data is fraught with uncertainties because all residues isomerize slightly to various extents during acid-catalyzed hydrolysis (see Section 4.23). D-Residues in a peptide can also be detected by an indirect method. The peptide is submitted to the action of a mixture of L-directed hydrolytic enzymes. Complete digestion indicates the absence of D-isomers; incomplete digestion indicates that D-residues may be present but unfortunately does not prove it. Diastereomeric contaminants in a peptide can often be detected by physical techniques such as nuclear magnetic resonance (NMR) spectroscopy or high-performance liquid chromatography (HPLC). NMR curves of intermediates or products may show double peaks that are caused by diastereoisomers. In contrast, NMR curves that do not show double peaks cannot be taken as proof that the compound is a single entity; they eliminate the possibility of contamination by a diastereoisomer only if the suspected diastereoisomer is available as reference compound. The same holds for HPLC. A minor peak closely following a major peak (see Section 4.11) on the HPLC profile of the compound may be the result of a diastereoisomer, especially if the two compounds responsible have identical amino acid compositions, but the only way to be certain is to chromatograph the reference compound. Fortunately, in particular for syntheses using segments, the possible sites of epimerization are known, and reference compounds with the appropriate D-residues can be made. Cochromatography of products with the suspected diastereoisomers is the recommended way of proceeding.[34]

34. R Riniker, R Schwyzer. Steric homogeneity of synthctic valyl[5] hypertension-II-aspartyl-ß-amide. *Helv Chim Acta* 44, 658, 1961.

4.9 TESTS EMPLOYED TO ACQUIRE INFORMATION ON STEREOMUTATION

Information on stereomutation has been acquired over the years by use of "racemization tests" composed of small peptides that bear the names of the senior persons of the laboratories where they have been developed. These are described in chronological order in Figure 4.8. Each test consists of an N^{α}-substituted amino acid Xaa that is coupled with an amino acid ester. Isomerization at residue Xaa generates a mixture of two diastereomeric peptides except for the first two tests. The Anderson test generates a racemate that crystallizes from solution and is quantified by weighing. The Young test produces a racemate that is quantified on the basis of the specific rotation of the peptide mixture. Diastereomeric peptides are determined by a variety of techniques. For the Weygand test, it is gas–liquid chromatography, for the Bodanszky test that generates D-alloisoleucine the latter is determined with an amino acid analyzer after hydrolysis of the peptide. The first truly representative test that involved an activated peptide with coupling between two chiral residues is the Izumiya test, which issued from a study of the chromatography of several dozen tripeptides. The epimers of glycylalanylleucine were found to be the easiest to

Author	Year	-Xaa-OH	+	H-Xbb-	Detection
a. Anderson	'52	Z-Gly-Phe-OH	+	H-Gly-OEt	Yd of D-L
b. Young	'63	Bz-Leu-OH	+	H-Gly-OEt	$[\alpha]_D$
c. Weygand	'63	Tfa-Val-OH	+	H-Val-OMe	GLC
d. Bodanszky	'67	Ac-Ile-OH	+	H-Gly-OEt	$H^+\Delta$; AAA
e. Izumiya	'69	Z-Gly-Ala-OH	+	H-Leu-OBzl	H_2/Pd; AAA
f. Weinstein	'72	Ac-Phe-OH	+	H-Ala-OMe	NMR, βCH_3
g. Davies	'75	Bz-Val-OH	+	H-Val-OMe	NMR, OCH_3
h. Various	'80+	Z-Gly-Xaa-OH	+	H-Xbb-OR	HPLC

FIGURE 4.8 "Racemization tests" employed for acquiring information on stereomutation. Couplings are carried out, and the isomeric content of the products is determined by a variety of techniques. AAA = amino acid analyzer; GLC = gas–liquid chromatography.

separate with the amino acid analyzer. Despite the value of this or similar tests for which quantitation was performed with an amino acid analyzer, these tests suffered from the drawback that reference compounds were required because two epimers do not give the same color yields when reacted with ninhydrin (see Section 5.4). Subsequently, there emerged the observation that the NMR spectra of some diastereomeric peptides show two separated peaks for methyl singlets. B. Weinstein developed the first test based on this fact, involving the peaks exhibited by the methyl protons of the side chain of the alanyl residue of Ac-L-Phe-L/D-Ala-OMe. This was followed by the test of Davies, which issued from the general phenomenon that the methoxy protons of Bz-L-Xaa-D/L-Xbb-OMe present as double peaks. The separation of peaks in these model peptides is attributed to shielding by the aromatic rings. Of all tests, those involving benzoylamino acids are the most sensitive because the aromaticity of the *N*-substituent is conducive to oxazolone formation, thus favoring isomerization. However, benzoylamino acids are less representative of couplings of peptides (see Section 4.13).

Use of these tests provided a host of information on the relative merits of different coupling methods and conditions of operation. However, perusal of Figure 4.8 reveals that such a variety of residues and *N*-substituents on the activated residues are implicated in the couplings that their use could not provide reliable information on either the relative tendency of residues (Xaa) to enantiomerize or the effect on the inversion at Xaa resulting from the nature of the aminolyzing residue (Xbb). The advent of HPLC as monitoring technique allowed this deficiency to be overcome. A series of tests involving Z-Gly-Xaa-OH coupled with H-Xbb-OR in which one residue could be varied while the other was constant was developed. It transpires that most small diastereomeric peptides can be separated by HPLC (see Section 4.11); the isomers also give the same response on the detector, which measures ultraviolet absorbance.[35–43]

35. W Anderson, RW Young. Use of diester chlorophosphites in peptide synthesis. *J Am Chem Soc* 74, 5307, 1952.
36. MW Williams, GT Young. Amino-acids and peptides. Part XVI. Further studies of racemisation during peptide synthesis. *J Chem Soc* 881, 1963.
37. F Weygand, A Prox, L Schmidhammer, W König. Gas chromatographic investigation of racemization in peptide synthesis. *Angew Chem* 75, 282, 1963.
38. M Bodanszky, LE Conklin. A simple method for the study of racemization in peptide synthesis. *Chem Commun* 773, 1967.
39. N Izumiya, M Muraoka. Racemization test in peptide synthesis. *J Am Chem Soc* 91, 2391, 1969.
40. B Weinstein, AE Pritchard. Amino-acids and Peptides. Part XXVIII. Determination of racemization in peptide synthesis by nuclear magnetic resonance spectroscopy. *J Chem Soc Perkin Trans* 1, 1015, 1972.
41. JS Davies, RJ Thomas, MK Williams. Nuclear magnetic resonance spectra of benzoyldipeptide esters. A convenient test for racemisation in peptide synthesis. *J Chem Soc Chem Commun* 76, 1975.
42. NL Benoiton, K Kuroda, ST Cheung, FMF Chen. Lysyl dipeptide and tripeptide model systems for racemization studies in amino acid and peptide chemistry. *Can J Biochem* 57, 776, 1979.

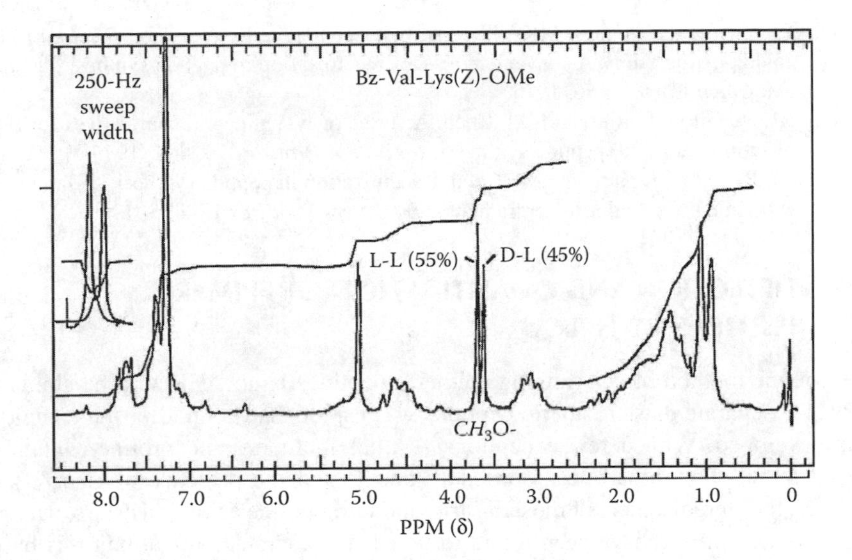

FIGURE 4.9 60-MHz H[1]–nuclear magnetic resonance spectrum of Bz-D/L-Val-L-Lys(Z)-OMe obtained by coupling Bz-L-Val-OH with H-L-Lys(Z)-OMe, using DCC in the presence of HOBt in dimethylformamide at 25°C.[45] See text for details of quantitation.

43. NL Benoiton, Y Lee, B Liberek, R Steinauer, FMF Chen. High-performance liquid chromatography of epimeric *N*-protected peptide acids and esters for assessing racemization. *Int J Pept Prot Res* 31, 581, 1988.

4.10 DETECTION AND QUANTITATION OF EPIMERIC PEPTIDES BY NMR SPECTROSCOPY

Epimeric peptides can be quantified by proton NMR spectroscopy if one or several protons of each isomer give rise to sharp peaks that are sufficiently separated. High-powered instruments are unnecessary. A typical 60-MHz spectrum of an isomerized benzoyldipeptide methyl ester appears in Figure 4.9. Quantitation was achieved by integrating from both directions the peaks of the methoxy protons obtained at an expanded sweep width and taking the average. A high percentage of –D-L-isomer was generated in this carbodiimide-mediated coupling despite the presence of 1-hydroxybenzotriazole because the activated component was a benzoylamino acid and the solvent was polar. The protocol employed to get the results was deemed so simple and potentially useful that it was described in the form of an experiment suitable for undergraduate students. Diastereoisomers of longer peptides have given rise to double peaks originating from protons of various natures.[40,41,44,45]

40. B Weinstein, AE Pritchard. Amino-acids and Peptides. Part XXVIII. Determination of racemization in peptide synthesis by nuclear magnetic resonance spectroscopy. *J Chem Soc Perkin Trans* 1, 1015, 1972.

41. JS Davies, RJ Thomas, MK Williams. Nuclear magnetic resonance spectra of benzoyldipeptide esters. A convenient test for racemisation in peptide synthesis. *J Chem Soc Chem Commun* 76, 1975.
44. NL Benoiton, K Kuroda, FMF Chen. A series of lysyldipeptide derivatives for racemization studies in peptide synthesis. *Int J Pept Prot Res* 13, 403, 1979.
45. NL Benoiton, K Kuroda, FMF Chen. Racemization in peptide synthesis. A laboratory experiment for senior undergraduates. *Int J Pept Prot Res* 15, 475, 1980.

4.11 DETECTION AND QUANTITATION OF EPIMERIC PEPTIDES BY HPLC

The routine method of determining epimeric peptides is now HPLC. The ability of HPLC to separate diastereomeric peptides was established soon after the technique was developed. With a few exceptions — that include some proline-containing peptides that show broad peaks probably because of the presence of *cis/trans* — isomers, the stereoisomers of most di-, tri-, and tetrapeptide acids and esters, whether *N*-substituted or not, have been found to be easily separated. Quantitation is based on measurement of ultraviolet absorbance at 208 nm (peptide bond) or 215 nm for aromatics and is straightforward because two epimers produce the same response in the detector. Isomers with residues of identical configuration are referred to as "positive" isomers, and isomers with a residue of the opposite configuration are referred to as "negative" isomers. Experience has revealed that in nearly all cases, the positive isomers emerge from a reversed-phase column before the negative isomers. The comportment is rationalized on the basis that the amino acid side chains that are responsible for the adsorption are closer together in negative isomers, and hence the latter are retained more strongly by the hydrophobic stationary phase. The Newman projections of the isomers show the proximities of the side chains of adjacent residues (Figure 4.10). A useful observation that issues from the order of emergence of the isomers is that more accuracy in analysis can be obtained by coupling residues of opposite configuration because the peak of the positive isomer that is generated by isomerization precedes the larger peak, and is thus less likely to be masked by it (Figure 4.10). When the coupling is between two L-residues, the

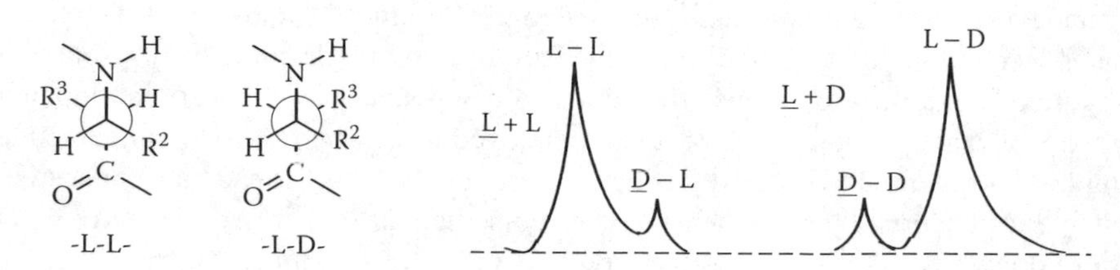

FIGURE 4.10 Newman projections and reversed-phase high-performance liquid chromatography profiles of epimeric peptides. The side chains are closer together in the negative isomers. This favors stronger interaction with the stationary phase, and thus later emergence from the column. Couplings of residues of opposite configuration yield an epimer that precedes the larger peak.

larger peak emerges first and can overlap the smaller second peak. In addition, the test is more sensitive because residues of opposite configuration lead to more isomerization because they react more slowly with each other than with residues of identical configuration. The first demonstration of the enantiomerization of *N*-alkoxycarbonylamino acids other than the histidines issued from the observation that the HPLC profile of a target peptide showed a second small peak, the two corresponding to products with identical amino acid compositions (see Section 4.19).[43,46–48]

43. NL Benoiton, Y Lee, B Liberek, R Steinauer, FMF Chen. High-performance liquid chromatography of epimeric *N*-protected peptide acids and esters for assessing racemization. *Int J Pept Prot Res* 31, 581, 1988.
46. J Rivier, R Wolbers, R Burgus. Application of high pressure liquid chromatography to peptides, in M Goodman, J Meienhofer, eds. *Peptides. Proceedings of the 6th American Peptide Symposium*. Halsted, New York, 1977, pp 52-56.
47. DJ Pietrzyk, RL Smith, WR Cahill. The influence of peptide structure on the retention of small peptides on reverse stationary phases. *J Liq Chromatog* 6, 1645, 1983.
48. R Steinauer, FMF Chen, NL Benoiton. Determination of epimeric peptides for assessing enantiomeric purity of starting materials and studying racemization in peptide synthesis using high-performance liquid chromatography. *J Chromatog* 325, 111, 1985.

4.12 EXTERNAL FACTORS THAT EXERT AN INFLUENCE ON THE EXTENT OF STEREOMUTATION DURING COUPLING

Experience has shown that there are several external factors that have an influence on the stereomutation that accompanies any coupling reaction. First among these is the polarity of the solvent, the more polar solvent producing the greater isomerization in the following order: tetrahydrofuran < dichloromethane << dimethylformamide < dimethyl sulfoxide. Temperature also has a dramatic effect, with a lower temperature suppressing isomerization. As an example, a sixfold decrease in epimer was generated by a 10˚C diminution in temperature from +5˚C to 5˚C for the dicyclohexylcarbodiimide-mediated coupling of *N*-acetylvaline, with the benzyl ester of side chain being protected lysine in dimethylformamide. The extent of stereomutation is dictated by the relative rates of 5(4*H*)-oxazolone formation and enolization of the activated components and the rates of aminolysis of the latter (see Section 4.6). Each rate can be affected differently by changes in temperature and solvent polarity, but very little is known about which rate constants are affected and how they are affected by changes in these parameters. Elevated temperature sometimes facilitates couplings; therefore, it must increase the rate constants for aminolysis, but this is usually accompanied by increased isomerization. The effect of the polarity of solvent can be rationalized on the basis that oxazolone formation is favored in the more polar solvent because it is a polar molecule. One thing that is known is that lower temperature and an increase in solvent polarity favor a positive asymmetric induction that contributes to the reduction in epimer formation. In some cases, in particular for reactions involving aminolysis of activated esters, a higher concentration of reactants will increase the rates of aminolysis.

Isomerization involves abstraction of a proton from one or more activated intermediates, so if tertiary amine is present, the nature and amount of the base has an influence on the results. The basicity of the amine and the hindrance around the nitrogen atom are significant (see Section 2.22). Abstraction of the proton is impeded by hindrance. In general, the more basic amine is more deleterious. However, if the base is hindered, then the effect of the basicity is diminished. Such is the case with diisopropylethylamine. Triethylamine is not hindered enough, so its use is rarely recommended. The worst case scenario is the presence of a strong base such as 4-dimethylaminopyridine that is unhindered (see Section 4.19). All this being said, it must be recognized that the basicity of an amine is not the same in different solvents. An additional factor that can have an enhancing or inhibitory effect on isomerization is salts of tertiary amines.[30,31,37,49,50]

30. NL Benoiton, K Kuroda, FMF Chen. The dependence of asymmetric induction on solvent polarity and temperature in peptide synthesis. *Tetraheron Lett* 22, 3361, 1981.
31. NL Benoiton, YC Lee, FMF Chen. Studies on asymmetric induction associated with the coupling of *N*-acylamino acids and *N*-benzyloxycarbonyldipeptides. *Int J Pept Prot Res* 38, 574, 1991.
37. F Weygand, A Prox, L Schmidhammer, W König. Gas chromatographic investigation of racemization in peptide synthesis. *Angew Chem* 75, 282, 1963.
49. W Williams, GT Young. Amino acids and peptides. XXXV. Effect of solvent on the rates of racemization and coupling of acylamino acid *p*-nitrophenyl esters. Base strengths of amines in organic solvents, and related investigations. *J Chem Soc Perkin Trans 1* 1194, 1972.
50. DS Kemp, S-W Wang, J Rebek, RC Mollan, C Banquer, G Subramanyam. Peptide synthesis with benzisoxazolium salts-II. Activation chemistry of 2-ethyl-7-hydroxybenzisoxazolium fluoroborate; coupling chemistry of 3-acyloxy-2-hydroxy-*N*-ethylbenzamides. *Tetrahedron* 30, 3955, 1974.

4.13 CONSTITUTIONAL FACTORS THAT DEFINE THE EXTENT OF STEREOMUTATION DURING COUPLING: CONFIGURATIONS OF THE REACTING RESIDUES

In any coupling, the rate of reaction between residues L-Xaa and L-Xbb will not be the same as the rate of reaction between residues D-Xaa and L-Xbb. As a consequence, if stereomutation occurs, the extents will not be the same. Limited data are available on the subject. Experiments with Z-Gly-Xaa-OH led to the conclusion that couplings between residues of identical configuration generated 25% less epimer than couplings between residues of opposite configuration. Thus, the rates of reactions must be greater for couplings between residues of identical configuration. The corollary is that if a -D-L- epimer is prepared for use as reference compound for confirming the nature of a suspected impurity in a synthesis (see Section 4.8), the HPLC profile of the product would show a larger minor peak than would be shown by the profile of the peptide that is under scrutiny. In contrast, some acylamino acids exhibit anomalous behavior. Less epimer is generated from benzoylamino acids in reactions between residues of opposite configuration.[51]

51. NL Benoiton, K Kuroda, FMF Chen. The relative susceptibility to racemization of L- and D- residues in peptide synthesis. *Tetrahedron Lett* 22, 3359, 1981.

4.14 CONSTITUTIONAL FACTORS THAT DEFINE THE EXTENT OF STEREOMUTATION DURING COUPLING: THE *N*-SUBSTITUENT OF THE ACTIVATED RESIDUE OR THE PENULTIMATE RESIDUE

The *N*-substituent of an activated residue is usually alkoxycarbonyl, acyl, or peptidyl. Activated *N*-alkoxycarbonylamino acids do not isomerize under normal circumstances (see Section 1.10), with the exception of Fmoc-proline chloride (see Sections 4.2 and 7.22). Isomerization of acyl- and peptidylamino acids is most often a result of oxazolone formation; hence the tendency to form oxazolones, along with their susceptibility to enolization, which is related to the electron-withdrawing property of the alkyl or aryl group of the substituent, has a significant effect on the extent of isomerization. In addition, the amount of oxazolone formed depends on the rates of coupling that are diminished substantially in the peptide. As example of the latter, aminolysis of activated esters of *Z*-glycylamino acids in tetrahydrofuran occurs much more slowly than aminolysis of the same esters of *Z*-amino acids. Benzoylamino acids are the most sensitive to enantiomerization, formylamino acids the least. Trifluoroacetylamino acids exhibit anomalous behavior because the oxazolone formed is of a different nature, 5(2*H*)- instead of 5(4*H*)-oxazolone (see Section 1.8). Acetyl and alkoxycarbonylaminoacylamino acids are of intermediate sensitivity. Illustrative data showing the influence of the nature of the *N*-substituent appear in Figure 4.11. Note also the variation depending on the method of coupling. When

Acid	Ester	DCC	DCC-HOBt	MxAn
For-Val-OH	H-Lys(Z)-OBzl	25	0.2	0
Z-Gly-Val-OH	H-Lys-(Z)-OBzl	11	5.0	1.5
Ac-Val-OH	H-Lys(Z)-OBzl	27	7.0	42
Tfa-Val-OH	H-Lys(Z)-OH	25	16	20
Bz-Val-OH	H-Lys(Z)-OMe	31	25	
Z-Gly-Val-OH	H-Phe-OEt	2.0	2.5	
Z-Ala-Val-OH	H-Phe-OEt	12	3.6	
Z-Leu-Val-OH	H-Phe-OEt	19	5.9	

FIGURE 4.11 Data showing the effect of the N^{α}-substituent of the activated residue on stereomutation.[31] Percentable -D-L- isomer formed in couplings in dimethylformamide at +5°C. Ester.HCl salts neutralized with *N*-methylmorpholine. MxAn = Mixed anhydride using $ClCO_2iPr$ with 5-minute activation time at –5°C. DCC = dicyclohexylcarbodiimide, HOBt = 1-hydroxybenzotriazole.

	Acid	Ester	MxAn	DCC	DCC-HOBt[a]
+	Z-*Leu*-Val-OH	H-Phe-OEt		18.9	5.9
++	Z-*Ala*-Val-OH	H-Phe-OEt		12.0	3.6
+++	Z-*Gly*-Val-OH	H-Phe-OEt		2.0	2.5
*	Z-Gly-Val-OH	H-Val-OEt	12.2		
**	Z-Gly-*Leu*-OH	H-Val-OEt	13.5		
***	Z-Gly-*Phe*-OH	H-Val-OEt	2.3		
*	Z-Gly-*Val*-OH	H-Val-OBzl	11.2	6.5	2.4
**	Z-Gly-*Leu*-OH	H-Val-OBzl	7.5	19.5	<0.1
***	Z-Gly-*Phe*-OH	H-Val-OBzl	6.8	16.5	0.7

FIGURE 4.12 Data showing that the penultimate residue, the activated residue, the alkyl group of the aminolyzing ester, and the method of coupling have an effect on the stereomutation observed in a coupling.[53] Percentage -D-L- isomer. Details as in Figure 4.11. At +5°C.

the *N*-substituent is a residue, the residue is referred to as the penultimate one. The nature of the side chain of the penultimate residue significantly affects the isomerization (Figure 4.12, pluses). In addition, the penultimate residue can isomerize, in particular when the activated residue is glycyl or aminoisobutyryl (see Sections 4.4 and 7.23). There exist derivatives of amino acids that are doubly substituted on the nitrogen atom and, hence, cannot form oxazolones. In these cases, the activated residue becomes more susceptible to enolization because there is no ionizable proton on adjacent atoms to prevent the equilibration (see Section 8.14). Examples are phthaloylamino acids and Schiff's bases of amino acids, which are chirally sensitive to tertiary amines. A useful observation that emerged from work on benzoylamino acids is that no isomerization occurs if they are coupled in dichloromethane using DCC assisted by 1-hydroxybenzotriazole.[30,44,52–53]

30. NL Benoiton, K Kuroda, FMF Chen. The dependence of asymmetric induction on solvent polarity and temperature in peptide synthesis. *Tetraheron Lett* 22, 3361, 1981.
44. NL Benoiton, K Kuroda, FMF Chen. A series of lysyldipeptide derivatives for racemization studies in peptide synthesis. *Int J Pept Prot Res* 13, 403, 1979.
52. J Kovacs, R Cover, G Jham, Y Hsieh, T Kalas. Application of the additivity principle for prediction of rate constants in peptide chemistry. Further studies on the problem of racemization of peptide active esters, in R Walter, J Meienhofer, eds. *Peptides: Chemistry, Structure and Biology*. Ann Arbor, MI, 1975, pp 317-324.
53. NL Benoiton, YC Lee, R Steinauer, FMF Chen. Studies on sensitivity to racemization of activated residues in couplings of *N*-benzyloxycarbonyldipeptides. *Int J Pept Prot Res* 40, 559, 1992.

4.15 CONSTITUTIONAL FACTORS THAT DEFINE THE EXTENT OF STEREOMUTATION DURING COUPLING: THE AMINOLYZING RESIDUE AND ITS CARBOXY SUBSTITUENT

Enantiomerization of the activated residue is affected by the nature of the aminolyzing residue as well as the nature of its carboxy substituent. The more hindered the incoming nucleophile, the slower the coupling rate, and hence the greater the danger for isomerization. In apparent disaccord with this is the observation that more

Z-Gly-Xaa-OH	+	H-Xbb-OR	H-Xbb-Gly-OR	H-Xbb-Gly$_2$-OR
DCC-HOBt	Phe	2.4	5.4	11
DCC-HOBt	Val	7.6	8.7	22
DCC-HONSu	Val	16.6	9.0	35
MxAn -5^0	Phe	1.2	8.7	5.0
DCC-HOBt	Leu	2.4	6.0	

FIGURE 4.13 Data showing the effect of the carboxy substituent of the aminolyzing residue on stereomutation.[43] Percentage -D-L- isomer formed in couplings in dimethylformamide at 23°C. Ester.HCl salts neutralized with *N*-methylmorpholine. MxAn using $ClCO_2iBu$ with 2-minute activation time. Xbb = Lys(Z); when Xaa = Leu, Xbb = Leu

isomerization occurred when proline was the aminolyzing residue (see Section 7.22). The latter may be related to the fact that proline produces a negative induction, whereas noncyclic residues produce a positive induction. Illustrative data showing the effect of the aminolyzing component appear in Figure 4.13, and the entries for Z-glycyl-L-valine appear in Figure 4.12. The data show that benzyl esters may generate more (Figure 4.12, ***) or fewer (Figure 4.12, **) epimers than ethyl esters, and replacement of the ester by one or two glycyl residues generally increased the numbers (Figure 4.13). Increasing the length of the aminolyzing peptide will diminish the rate of coupling. Regardless, it must be recognized that the results vary, depending on the coupling method employed. A unique situation arises when the incoming nucleophile is proline phenacyl ester. It transpires that the unactivated proline residue undergoes enantiomerization because of the formation of a Schiffs's base between the imino nitrogen and the carbonyl of the keto function of the ester. This unusual phenomenon was observed for HOBt-assisted EDC-mediated couplings of Boc-amino acids in dimethylformamide.[31,43,52–55]

31. NL Benoiton, YC Lee, FMF Chen. Studies on asymmetric induction associated with the coupling of *N*-acylamino acids and *N*-benzyloxycarbonyldipeptides. *Int J Pept Prot Res* 38, 574, 1991.
43. NL Benoiton, Y Lee, B Liberek, R Steinauer, FMF Chen. High-performance liquid chromatography of epimeric *N*-protected peptide acids and esters for assessing racemization. *Int J Pept Prot Res* 31, 581, 1988.
52. J Kovacs, R Cover, G Jham, Y Hsieh, T Kalas. Application of the additivity principle for prediction of rate constants in peptide chemistry. Further studies on the problem of racemization of peptide active esters, in R Walter, J Meienhofer, eds. *Peptides: Chemistry, Structure and Biology.* Ann Arbor, MI, 1975, pp 317-324.
53. NL Benoiton, YC Lee, R Steinauer, FMF Chen. Studies on sensitivity to racemization of activated residues in couplings of *N*-benzyloxycarbonyldipeptides. *Int J Pept Prot Res* 40, 559, 1992.
54. H Kuroda, S Kubo, N Chino, T Kimura, S Sakakibara. Unexpected racemization of proline and hydroxyproline phenacyl ester during coupling reactions with Boc-amino acids. *Int J Pept Prot Res* 40, 114, 1992.
55. JC Califano, C Devin, J Shao, JK Blodgett, RA Maki, KW Funk, JC Tolle. Copper(II)-containing racemization suppressors and their use in segment coupling reactions, in J Martinez, J-A Fehrentz, eds. *Peptides 2000. Proceedings of the 26th European Peptide Symposium*, EDK, Paris, 2001, pp 99-100.

4.16 CONSTITUTIONAL FACTORS THAT DEFINE THE EXTENT OF STEREOMUTATION DURING COUPLING: THE NATURE OF THE ACTIVATED RESIDUE

It has been known for years that the activated residues of acyl- and peptidylamino acids enantiomerize during coupling (1.9). However, the "racemization tests" available (see section 4.9) did not allow for a valid comparison of the tendency of residues to isomerize because they incorporated a variety of aminolyzing residues and *N*-substituents. Valid demonstration of the different sensitivities of residues was provided by classical work on the synthesis of insulin. It was found that a 16-residue segment with *O*-tert-butyltyrosine at the carboxy terminus produced 25% of epimer in HOBt-assisted DCC-mediated coupling in dimethylformamide, and the same segment with leucine at the carboxy terminus produced no epimer. Only when series such as Z-Gly-Xaa-OH coupled with valine benzyl ester became available was it possible to compare many residues with confidence. Unfortunately, it transpires that the issue is extremely complex.

First, the order of sensitivity depended on whether the solvent was polar or apolar. In dichloromethane, the order of chiral sensitivity was Ile/Val < Leu/Ala < Phe, whereas in dimethylformamide it was Leu/Ala < Phe < Val/Ile, with alanine occasionally showing apparently anomalous behavior. Second, the order of sensitivity depended on the method of coupling. The results for couplings in dimethylformamide by the mixed-anhydride method were significantly different from those for couplings mediated by reagents such as DCC, BOP (see Section 2.17), and TBTU (see Section 2.18) assisted by 1-hydroxybenzotriazole. There was also some variation in the order for couplings with the latter reagents. That being said, there are some generalizations that can be made on the basis of studies from a few laboratories. There is agreement that for reactions in polar solvents, after proline, the most stable residues in no particular order are asparagine, glutamine, leucine, protected lysine, and protected aspartic and glutamic acids. *O*-Benzylserine is intermediate in stability. Stability is less for valine and isoleucine and least for substituted histidines, arginines, and threonines. In line with these generalizations is the fact that activated asparagine and glutamine showed the lowest rates of 5(4*H*)-oxazolone formation, whereas substituted arginine showed the highest rate, and the oxazolones from the ω-esters of aspartic and glutamic acids were aminolyzed at the highest rates, and that from valine was aminolyzed at the lowest rate. It must be emphasized that because epimerization depends on so many internal and external factors, predicting the likely outcome of a coupling is an unenviable task.[53, 56–58]

53. NL Benoiton, YC Lee, R Steinauer, FMF Chen. Studies on sensitivity to racemization of activated residues in couplings of *N*-benzyloxycarbonyldipeptides. *Int J Pept Prot Res* 40, 559, 1992.
56. P Sieber, B Kamber, J Hartmann, A Jöhl, B Riniker, W Rittel. 4. Total synthesis of human insulin. IV Description of the final product. *Helv Chim Acta* 60, 27, 1977.

57. C Griehl, A Kolbe, S Merkel. Quantitative description of epimerization pathways using the carbodiimide method in the synthesis of peptides. *J Chem Soc Perkin Trans 2* 2525, 1996.
58. S Sakakibara. Chemical synthesis of proteins in solution. *Biopolymers (Pept Sci)* 51, 279, 1999.

4.17 REACTIONS OF ACTIVATED FORMS OF *N*-ALKOXYCARBONYLAMINO ACIDS IN THE PRESENCE OF TERTIARY AMINE

Up to 1980, formation of oxazolones was associated with the isomerization that acyl- and peptidylamino acids underwent during coupling (see Section 1.9). Activated *N*-alkoxycarbonylamino acids coupled without isomerizing; therefore, they were considered not to form oxazolones (see Section 1.19). 2-Alkoxy-5(4*H*)-oxazolones were unheard of. In 1973 Miyoshi had found that treatment of *Z*-amino acids with thionyl chloride or phosgene in the presence of two equivalents of tertiary amine produced compounds that showed infrared and NMR spectra consistent with a cyclic structure resulting from dehydration. The compounds also underwent aminolysis without generating a second isomer. In accordance with the prevailing wisdom of the time, Miyoshi concluded that the cyclic structures could not be oxazolones because the latter were considered to be chirally labile. He presented the compounds as 1-alkoxycarbonylaziridin-2-ones that contain a three-membered ring, formed by nucleophilic attack by the deprotonated nitrogen atom at the activated carbonyl of the *Z*-amino acid (Figure 4.14, path E). About 5 years later, prompted by skepticism expressed on the issue by a specialist in heterocyclic chemistry, Jones and Witty reinvestigated the reaction and demonstrated that the products obtained

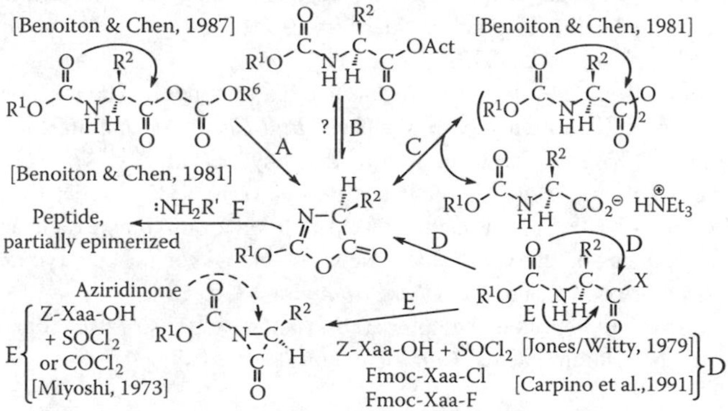

FIGURE 4.14 Reactions of activated *N*-alkoxycarbonylamino acids in the presence of tertiary amine. Acyl halides and mixed and symmetrical anhydrides generate 2-alkoxy-5(4*H*)-oxazolone in the presence of tertiary amine. Aminolysis of 2-alkoxy-5(4*H*)-oxazolone in the presence of Et_3N led to partially epimerized products. OAct = activating group.

by Miyoshi were in fact 2-alkoxy-(5*H*)-oxazolones (path D). The notion that oxazolones should isomerize during aminolysis was so deeply entrenched in the minds of peptidologists that it had induced Miyoshi to make a false conclusion. Shortly thereafter, symmetrical anhydrides were also shown to form oxazolones when left in the presence of tertiary amines, the reaction being reversible (path C). Later work revealed that mixed anhydrides are quickly and completely converted to oxazolones by tertiary amine (path A; see Section 2.8) and that the tertiary amine added to neutralize hydrohalide when Fmoc-amino-acid chlorides and fluorides are aminolyzed also gives rise to oxazolones (path D). Whether tertiary amine produces 2-alkoxy-5(4*H*)-oxazolones from activated esters (path B) has not been established; the equilibrium is likely very much in the direction of the activated ester because the oxazolones react with the phenols and hydroxylamines. The message to be gleaned is that activated forms of *N*-alkoxycarbonylamino acids, with the possible exception of the azides, can give rise to 2-alkoxy-5(4*H*)-oxazolones if tertiary amine is present during their aminolysis. This in itself is of no consequence; these oxazolones undergo aminolysis very quickly. However, it has been demonstrated that some 2-alkoxy-5(4*H*)-oxazolones were not chirally stable when aminolyzed in the presence of triethylamine (Figure 4.15). As much as 25% of other isomer was produced when the oxazolone from Z-L-valine was aminolyzed in the presence of half an equivalent of triethylamine. Few data on the issue are available, but the fact remains that there exists the danger that any 2-alkoxy-5(4*H*)-oxazolone produced from *N*-alkoxycarbonylamino acids during coupling might isomerize if tertiary amine is present.[24,59–64]

24. NL Benoiton, FMF Chen. 2-Alkoxy-5(4*H*)-oxazolones from *N*-alkoxycarbonylamino acids and their implication in carbodiimide-mediated reactions in peptide synthesis. *Can J Chem* 59, 384, 1981.
59. M Miyoshi. Peptide synthesis *via* *N*-acylated azidiridinone. I. The synthesis of 3-substituted-1-benzyloxycarbonylaziridin-2-ones and related compounds. *Bull Chem Soc Jpn* 46, 212, 1973.
60. M Miyoshi. Peptide synthesis *via* *N*-acylated aziridinone. II. The reaction of *N*-acylated aziridinone and its use in peptide synthesis. *Bull Chem Soc Jpn* 46, 1489, 1973.
61. JH Jones, MJ Witty. The formation of 2-benzyloxyoxazol-5(4*H*)-ones from benzyloxycarbonylamino-acids. *J Chem Soc Perkin Trans 1* 3203, 1979.
62. NL Benoiton, FMF Chen. Reaction of *N*-t-butoxycarbonylamino acid anhydrides with tertiary amines and carbodi-imides. New precursors for 2-t-butoxyoxazol-5(4*H*)-one and *N*-acylureas. *J Chem Soc Chem Commun* 1225, 1981.
63. FMF Chen, NL Benoiton. The preparation and reactions of mixed anhydrides of *N*-alkoxycarbonylamino acids. *Can J Chem* 65, 619, 1987.

Oxazolone from	Triethylamine (equiv) 0.0	0.2	0.5
Boc-L-Valine	<0.2	3	
Z-L-Valine	<0.2	7.5	25

FIGURE 4.15 Enantiomerization during aminolysis of 2-alkoxy-5(4*H*)-oxazolones in the presence of Et_3N.[24] Percentage -D-L- Epimer formed in reaction with H-Lys(Z)-OBzl.HCl/*N*-methylmorpholine in CH_2Cl_2.

64. LA Carpino, HG Chao, M Beyermann, M Bienert. ((9-Fluorenylmethyl)oxy)-carbonylamino acid chlorides in solid-phase peptide synthesis. *J Org Chem* 56, 2635, 1991.

4.18 IMPLICATIONS OF OXAZOLONE FORMATION IN THE COUPLINGS OF *N*-ALKOXYCARBONLYAMINO ACIDS IN THE PRESENCE OF TERTIARY AMINE

A deduction with critical implications issues from the discussion in Section 4.17. The facts are that in the presence of tertiary amine, activated *N*-alkoxycarbonylamino acids generate 2-alkoxy-5(4*H*)-oxazolones and that 2-alkoxy-5(4*H*)-oxazolones can enantiomerize when aminolyzed in the presence of tertiary amine. The inescapable conclusion is that activated *N*-alkoxycarbonylamino acids can enantiomerize when coupled in the presence of tertiary amine. This does not mean that isomerization does occur if a tertiary amine is present during a coupling but, rather, that the possibility cannot be disregarded (see Sections 7.21 and 8.1). That being said, it must be stressed that it is the tertiary amine that creates the danger — there is no danger introduced by the presence of salts of tertiary amines.

4.19 ENANTIOMERIZATION IN 4-DIMETHYLAMINOPYRIDINE-ASSISTED REACTIONS OF *N*-ALKOXYCARBONYLAMINO ACIDS

4-Dimethylaminopyridine (DMAP) is a basic tertiary amine endowed with exceptional catalytic properties because it exists as a resonating hybrid (Figure 4.16). It is particularly effective for assisting esterification reactions involving nucleophilic substitution at the carbonyl of an anhydride. When the shortcomings of solid-phase synthesis technology employing Boc/Bzl chemistry and polystyrene resin led to development of polyamide resins with Fmoc for temporary protection, the first residue was attached to the hydroxymethyl group of the linker by DMAP-catalyzed acylation with a symmetrical anhydride (see Section 2.5). Extra peaks in the HPLC profiles of target peptides, corresponding to peptides with identical amino acid

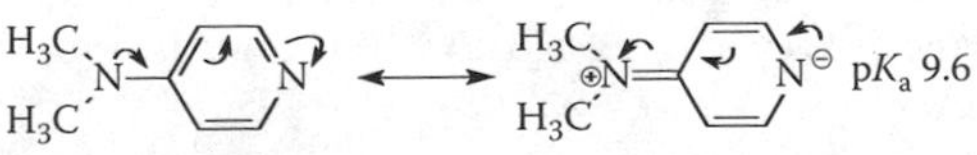

H_3C, H_3C N ⟷ H_3C, H_3C ⊕N N⊖ pK_a 9.6

FIGURE 4.16 4-Dimethylamipyridine has caused enantiomerization when used as catalyst for acylation of resin-bound functional groups.

Esterification[66]:

$(ROCO\text{-}Xaa)_2O + HOCH_2Ph$--- 1–5% -DXaa-

Aminolysis[67]:

Boc-Phe-OH + DCC + H-Glu(OBzl)-OCH_2Ph--- 1.7% -DPhe-

Esterification at β-CO_2H:[69]

Boc-Asp-OFm + $Me_2HCN{=}C{=}NCHMe_2$ + $HOCH_2Ph$--- 17% -DAsp-OFm compositions (see Section 4.8), evoked suspicions that isomerization had occurred during the esterification reactions. Investigation of the reaction with model compounds confirmed that up to 6% enantiomerization occurred during reaction of a symmetrical anhydride with the hydroxymethyl group of the linker-resin in dichloromethane or dimethylformamide in the presence of an equivalent of DMAP (Figure 4.16). At about the same time, it was shown that the addition of DMAP to a carbodiimide-mediated coupling of Boc-phenylalanine with γ-benzyl-protected glutamate attached to a resin led to minor but significant epimer formation. These were the first demonstrations of the isomerization of *N*-alkoxycarbonylamino acids other than histidine (see Section 4.3). The deleterious effects of DMAP on the acylation of hydroxymethyl groups have since been reported by many researchers. Thus, DMAP can lead to isomerization when used to assist both esterification and amide-bond forming reactions. The critical factor is the time of contact of the activated component with the base. Leaving Boc-phenylalanine anhydride with DMAP for 2 minutes before its aminolysis generated 17% of epimer. The results can be attributed to the formation and isomerization of the 2-alkoxy-5(4*H*)-oxazolones (see Section 4.17). It transpired that this new phenomenon of the isomerization of *N*-alkoxycarbonylamino acids was rationalizable on the basis of information acquired several months before its discovery. An additional and unusual case in which catalysis by DMAP proved disastrous is the carbodiimide-mediated esterification of the β-carboxyl group of Boc-Asp-OFm to a linker-resin, where 17% of D-isomer was generated. Here, the isomerization must be attributed to base-catalyzed enolization of the activated carboxyl group (see Section 4.1). In contrast, there are situations in which DMAP is effective and not deleterious in catalyzing the esterification of *N*-alkoxycarbonylamino acids — in particular when the activated component is an Fmoc-amino-acid fluoride and the solvent is not polar.[65–69]

65. W. Steglich, G Höfle. N,N-Dimethyl-4-pyridinamine, a very efficient acylation catalyst. *Angew Chem Int Edn Engl* 8, 981, 1969.
66. E Atherton, NL Benoiton, E Brown, RC Sheppard, B Williams. Racemisation of activated, urethane-protected amino-acids by *p*-dimethylaminopyridine. Significance in solid-phase synthesis. *J Chem Soc Chem Commun* 336, 1981.
67. SS Wang, JP Tam, BSH Wang, RB Merrifield. Enhancement of peptide coupling reactions by 4-dimethylaminopyridine. *Int J Pept Prot Res* 18, 459, 1981.
68. D Granitza, M Beyermann, H Wenschuh, H Haber, LA Carpino, GA Truran, M Bienert. Efficient acylation of hydroxy functions by means of Fmoc amino acid fluorides. *J Chem Soc Chem Commun* 2223, 1995.
69. M-L Valero, E.Giralt, D Andreu. Optimized Asp/Glu side chain anchoring in synthesis of head-to-tail cyclic peptides by Boc/OFm/benzyl chemistry on solid phase, in R Ramage, R Epton, eds. *Peptides 1996. Proceedings of the 24th European Peptide Symposium*, Mayflower, Kingswinford, 1998, pp 857-858.

4.20 ENANTIOMERIZATION DURING REACTIONS OF ACTIVATED *N*-ALKOXYCARBONYLAMINO ACIDS WITH AMINO ACID ANIONS

Activated forms of *N*-alkoxycarbonylamino acids that are stable enough to be isolated, such as mixed and symmetrical anhydrides and activated esters, possess a unique characteristic. They can be used for acylation of an amino acid without having to protect the latter's carboxyl group. The charged amino group of the zwitter-ion is made nucleophilic by removal of the proton by the addition of base. Attempts to use 2-ethoxy-5(4*H*)-oxazolones for acylating unprotected amino acids led to the surprising revelation that the oxazolones were not chirally stable under the conditions employed (see Section 4.5). This prompted an investigation of the chiral stabilities of the activated forms alluded to above during reaction with amino acid anions generated by various bases. Results showed that both types of anhydrides underwent substantial isomerization (0.7–15%) during aminolysis in dimethylformamide–water by an amino acid anion generated with sodium hydrogen carbonate (Figure 4.17). Less isomerization occurred when deprotonation was effected with sodium carbonate. Even activated esters, though to a lesser extent, behaved similarly. The isomerization could be minimized or suppressed by use of a 50% excess of sodium carbonate and a 50% excess of aminolyzing nucleophile. Succinimido esters are chirally the most stable of the activated forms examined. The apparently anomalous effect of the weaker base giving rise to more enantiomerization can be explained on the basis that the stronger base is more effective in converting the zwitter-ion into the aminolyzing nucleophile. However, if the aminolyzing component is a peptide instead of an amino acid, epimerization is not an issue because aminolysis can be effected without the addition of a base (see Section 7.21).[25,70–72]

25. NL Benoiton, YC Lee, R Steinauer. Determination of enantiomers of histidine, arginine and other amino acids by HPLC of their diastereomeric *N*-ethoxycarbonyldipeptides. *Pept Res* 8, 108, 1995.

Activated compound	1.0 $NaHCO_3$ 1.0 Valine	1.0 Na_2CO_3 1.0 Valine	0.75 Na_2CO_3 1.55 Valine
$(Z\text{-}Val)_2O$	8.2	<0.1	
$(Z\text{-}Phe)_2O$	6.3	<0.1	
Z-Phe-O-CO_2iBu	7.7	4.7	0.2
Z-Val-O-CO_2Et	35.5	9.8	4.4
Z-Phe-ONp	10.2	4.7	1.3
Boc-Leu-$OPhF_5$	13.3		0.2
Boc-Leu-ONp	5.7		0.7
Boc-Leu-ONSu	2.3		0.1

FIGURE 4.17 Enantiomerization data for reactions of activated *N*-alkoxycarbonyl-L-amino acids with an amino acid anion.[71,72] Percentage -D-D- peptide formed in reaction with H-D-Val-O^-·Na^+ in dimethylformamide-water (4:1) at 23°C.

70. NL Benoiton, FMF Chen. Activation and racemization in peptide bond formation, in GR Marshall, ed. *Peptides Chemistry and Biology. Proceedings of the 10th American Peptide Symposium*, Escom, Leiden, 1988, pp 152-155.
71. NL Benoiton, Y Lee, FMF Chen. Racemization during aminolysis of mixed and symmetrical anhydrides of *N*-alkoxycarbonylamino acids by amino acid anions in aqueous dimethylformamide. *Int J Pept Prot Res* 31, 443, 1988.
72. NL Benoiton, YC Lee, FMF Chen. Racemization during aminolysis of activated esters of *N*-alkoxycarbonylamino acids by amino acid anions in partially aqueous solvents and a tactic to minimize it. *Int J Pept Prot Res* 41, 512, 1993.

4.21 POSSIBLE ORIGINS OF DIASTEREOMERIC IMPURITIES IN SYNTHESIZED PEPTIDES

The objective of a synthesis is usually the preparation of an enantiopure peptide. However, the product sometimes contains one or more diastereomeric impurities. Possible actions and reactions that might have led to the generation of these impurities are as follows:

1. Use of nonenantiopure amino acid derivatives as starting materials. Notable examples have been serine derivatives that isomerized during *tert*-butylation and glutamate that isomerized during cyclodehydration to pyroglutamate.
2. Use of nonenantiopure protected amino acid-linker resins as starting materials. Examples have been products that were obtained by 4-dimethylaminopyridine-assisted esterification.
3. Use of histidine derivatives substituted at the τ-nitrogen. N^{π}-protection prevents isomerization during coupling (see Section 4.3), but N^{τ}-protection may not suppress it completely (see Section 6.10).
4. Esterification or aminolysis reactions catalyzed by 4-dimethylaminopyridine (see Section 4.19).
5. Aminolysis by amino acid anions (see Section 4.20).
6. Coupling of segments at other than Gly or Pro (see Section 4.4).
7. Prolonged contact of acyl azides or activated esters with tertiary amines.
8. Onium salt-mediated couplings of *N*-alkoxycarbonylamino acids involving a preactivation. Exposure of activated intermediates to tertiary amine can be deleterious (see Sections 4.18 and 7.20).
9. Onium salt-mediated couplings of *N*-alkoxycarbonylcysteines and serines. Activated forms of these compounds are particularly sensitive to the tertiary amines required to effect the couplings (see Section 8.1).
10. Onium salt-mediated couplings to the amino group of a peptide having an esterified carboxy-terminal cysteine residue. Esters of cysteine are sensitive to the base (see Section 8.1).
11. Treatment of compounds, especially esters (see Sections 3.11 and 4.20), *N*-substituted cysteine esters (see Section 8.1), and fully substituted *N*-methylamino-acid residues (see Section 8.14), by alkali.

12. Treatment of *N*-methylamino-acid derivatives by strong acid (see Section 8.14).
13. Storage of the peptide, even at temperatures close to 0°C.[73]

73. NF Sepetov, MA Krymsky, MV Ovchinnikov, ZD Bespalova, OL Isakova, M Soucek, M Lebl. Rearrangement, racemization and decomposition of peptides in aqueous solution. *Pept Res* 4, 308, 1991.

4.22 OPTIONS FOR MINIMIZING EPIMERIZATION DURING THE COUPLING OF SEGMENTS

The danger of epimerization during the coupling of segments exists for all cases, except when the activated residue is Pro and Gly, with a few exceptions (see Section 7.23). The obvious is to design a strategy that involves activation at these residues only. Options to try to minimize the side reaction for activation at other residues are as follows:

1. Devise a strategy that avoids activation at the more sensitive residues such as Val, Ile, Thr(Pg), His(Pg), and Arg(Pg) (Pg = protecting group; see Section 4.16).
2. Use an apolar solvent (see Section 4.12), though most of the time this is not possible because of the limited solubility of compounds in these solvents.
3. Use the acyl-azide method of coupling (see Sections 2.13 and 7.16) with particular care to avoid contact of the azide with tertiary amine.
4. Use a carbodiimide with additive such as HOBt, HOObt, or HOAt (see Sections 2.10 and 2.11), possibly in dimethylsulfoxide,[74] and furthermore with the addition of cupric ion. The latter is very effective in suppressing epimerization in carbodiimide-mediated amide-bond forming reactions (see Section 7.2).
5. Use an onium salt-based reagent such as PyBOP, T/HBTU (see Sections 2.18–2.21), PyAOP (see Section 7.19), or other with the corresponding additive and diisopropylethylamine or trimethylpyridine as tertiary amine without an excess. The additive may, however, promote epimerization.
6. Use the mixed-anhydride method in dimethylformamide with isopropyl chloroformate as the reagent and *N*-methylmorpholine, *N*-methylpiperidine, or trimethylpyridine as the base (see Sections 7.4 and 7.5).
7. Use the succinimido esters obtained from the acid by the mixed-anhydride method (see Section 7.8). This approach has been examined for segments of up to four residues.
8. Use the thio esters (-CO_2SR') obtained from the ROCO-Peptide-SR′-linker-resins assembled by solid-phase synthesis (see Section 7.10).[74]

74. K Barlos, D Gatos. 9-Fluorenylmethyloxycarbonyl/*t*butyl-based convergent protein synthesis. (dimethylsulfoxide as solvent) *Biopolymers (Pept Sci)* 51, 266, 1999.

4.23 METHODS FOR DETERMINING ENANTIOMERIC CONTENT

By definition, enantiomers differ by the direction in which a solution of the isomer deflects polarized light. The magnitude of the deflection, measured with a polarimeter, allows one to calculate the enantiomeric purity of the sample. The latter, however, requires that the specific rotation of the pure isomer is known; that the solution contains a known amount of the substance, which is chemically pure; and that the deflection is considerable so that the diminution in deflection caused by the other enantiomer is significant. Useful primarily for analyzing compounds containing one stereogenic center, measurement of optical rotation is seldom capable of detecting down to 1% of the other isomer and, consequently, is of limited value. A second characteristic that allows differentiation between enantiomers is their dissimilar behavior in the presence of another chiral molecule. This difference may be one of reaction or association (Figure 4.18). The classical example of the former is the interaction of enantiomers with enzymes. Enzymes are stereospecific biological catalysts that react with only one of two enantiomers, and this phenomenon was employed during the 1950s and 1960s to determine the enantiomeric purity of amino acids. L-Amino acids, obtained by enzymatic resolution of the racemates by selective hydrolysis of the *N*-acetyl or *N*-chloroacetyl derivatives by an L-directed acylase, were subjected to the action of hog kidney D-amino-acid oxidase (Figure 4.19) in a Warburg apparatus. Catalase was added to destroy the hydrogen peroxide generated, and quantitation was achieved by measuring the consumption of oxygen. L-Amino-acid oxidase from snake venom was employed to detect L-antipode in a D-amino acid. The method required 1 mmol of substrate, but it could detect down to 0.1% of the other isomer and was the method of choice for two decades. It was then superseded by the approach of converting the enantiomers into diastereomeric products (see Section 4.24) that are quantifiable by chromatographic techniques.

The dissimilar association of enantiomers with another chiral molecule (Figure 4.18) also allows their determination if the molecule is part of a chromatographic system. The unequal interactions result in different rates of migration of the enantiomers through the column. The chiral molecule may be a component of the mobile phase or the stationary phase of the system. Typical examples of the separation of

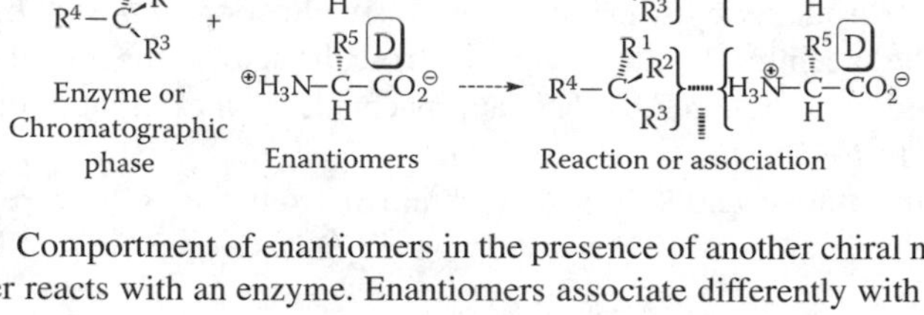

Enzyme or Chromatographic phase

Enantiomers

Reaction or association

FIGURE 4.18 Comportment of enantiomers in the presence of another chiral molecule. Only one enantiomer reacts with an enzyme. Enantiomers associate differently with another chiral molecule.

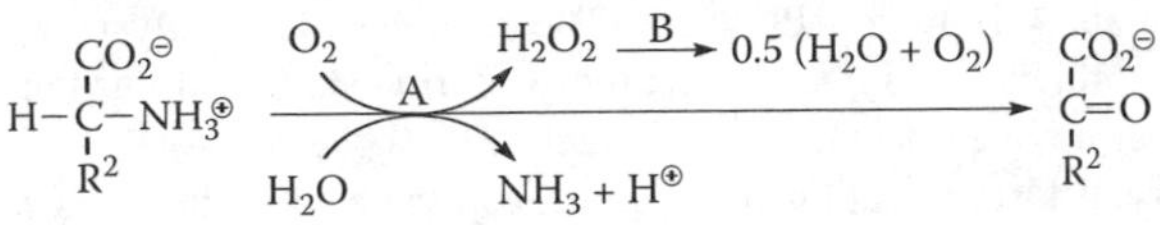

$$H{-}C(CO_2^{\ominus})(NH_3^{\oplus}){-}R^2 \xrightarrow[H_2O \;\; \rightarrow \;\; NH_3 + H^{\oplus}]{O_2 \;\; \xrightarrow{A} \;\; H_2O_2 \xrightarrow{B} 0.5\,(H_2O + O_2)} R^2{-}C(=O){-}CO_2^{\ominus}$$

FIGURE 4.19 Amino acid enantiomers are determined by reaction (A) with L- or D-amino-acid oxidase at pH 7–8.[75] Added catalase decomposes the hydrogen peroxide (B), which would otherwise oxidize the α-oxoacid. Quantitation is achieved by measuring oxygen consumption, which is 0.5 mol/mol of substrate.

enantiomers based on an association with a chiral phase are ion-exchange chromatography, using sodium acetate buffer containing cupric sulfate-proline (1:2) as eluent; reversed-phase HPLC, using the cupric salt of *N,N*-dipropyl-D-alanine as eluent; and gas-liquid chromatography on a *N*-lauroyl-L-valine *tert*-butyl amide or Chirasil-Val column. In the latter case, the enantiomers are chromatographed as the isopropyl esters of the *N*-pentafluoropropylamino acids. These methods are applicable to the analysis of several amino acids in a mixture, provided the pertinent peaks emerge separately. An analogous approach is analysis by capillary zone electrophoresis on a chiral phase.

An additional and serious obstacle arises when the amino acid to be analyzed is incorporated in a peptide. The residue must first be released by hydrolysis of the bonds connecting the amino acids. Unfortunately, acid-catalyzed hydrolysis of peptide bonds causes partial enantiomerization of the residue (see Section 4.1), with the extent depending on the nature of the residue as well as the nature of the adjacent residues — in effect, the sequence of the peptide. One must, therefore, make a correction for the amount of enantiomer that is generated during hydrolysis, but there is no simple method of doing this that is reliable. A valid correction can be made if the hydrolysis is carried out in the presence of deuterochloric acid, but interpretation of the data is too complicated for the nonexpert. In this regard, a technique for hydrolyzing a peptide or protein without isomerization is available. It involves heating it in 1*M* hydrochloric acid at 80–90°C for 15 minutes and treatment with pronase at 50°C for 6 hours, followed by treatment with a mixture of leucine amino-peptidase (enzyme commission 3.4.11.1) and peptidyl-D-amino-acid hydrolase (enzyme commission 3.4.13.17) for 24 hours. The aminopeptidase releases L-amino acids from the amino terminus of a chain, while the latter enzyme releases D-residues.[75–83]

75. A Meister, L Levintow, RB Kingsley, JP Greenstein. Optical purity of amino acid enantiomorphs. *J Biol Chem* 192, 535, 1951.
76. B Feibush. Interaction between asymmetric solutes and solvents. *N*-Lauroyl-valyl-*t*-butylamide as a stationary phase in gas liquid chromatography. *J Chem Soc Chem Commun* 544, 1971.
77. H Frank, GR Nicholson, E Bayer. Enantiomer labelling, a method for the quantitative analysis of amino acids. *J Chromatog* 167, 187, 1978.
78. PE Hare, E Gil-Av. Separation of D- and L- amino acids by liquid chromatography: use of chiral eluants. *Science* 204, 1226, 1979.

79. S Weinstein, MH Engel, PE Hare. The enantiomeric analysis of a mixture of all common protein amino acids by high-performance liquid chromatography using a new chiral mobile phase. *Anal Biochem* 121, 370, 1982.
80. DW Aswad. Determination of D- and L-aspartate in amino acid mixtures by high-performance liquid chromatography after derivatization with a chiral adduct and *o*-phthaldialdehyde. *Anal Biochem* 137, 405, 1984.
81. J Gerhardt, K Nokihara, R Yamamoto. Design and applications of a novel amino acid analyzer for D/L and quantitative analysis with gas chromatography, in JA Smith, JE Rivier, eds. *Peptides Chemistry and Biology. Proceedings of the 12th American Peptide Symposium*, Escom, Leiden, 1992, pp 457-458.
82. A D'Aniello, L Petrucelli, C Gardner, G Fisher. Improved method for hydrolyzing proteins and peptides without inducing racemization and for determining their true D-amino acid content. *Anal Biochem* 213, 290, 1993.
83. J Gerhardt, GJ Nichlson. Validation of a GC-MS method for determination of the optical purity of peptides, in J Martinez, J-A Ferentz, eds. *Peptides 2000. Proceedings of the 26th European Peptide Symposium*, EDK, Paris, 2001, pp 563-565.

4.24 DETERMINATION OF ENANTIOMERS BY ANALYSIS OF DIASTEREOISOMERS FORMED BY REACTION WITH A CHIRAL REAGENT

Interaction of enantiomers with another chiral molecule allows their separation on the basis of their dissimilar behaviors (Section 4.23). Reaction of enantiomers with another chiral molecule allows for separation of the products of the reaction by chromatography, and hence determination of the enantiomers because the products are diastereoisomers. A variety of chiral reagents and techniques is used for this purpose. The first popular method issued from knowledge of the chemistry of amino acid *N*-carboxyanhydrides (see Section 7.13) and familiarity with ion-exchange chromatography. Reaction of enantiomers with L-leucine *N*-carboxyanhydride (Figure 4.20) followed by separation, and analysis of diastereomeric dipeptides with an amino acid analyzer was developed into a method of determining enantiomers. Separation of diastereomeric dipeptides had previously been achieved on paper and Sephadex. L-Glutamic acid *N*-carboxyanhydride was used for derivatizing the hydrophobic and basic amino acids because the other reagent gave elution times that were too long. The method allows detection of one part in 1000 of the other isomer, but it suffers from the shortcoming that accurate quantitation requires the availability

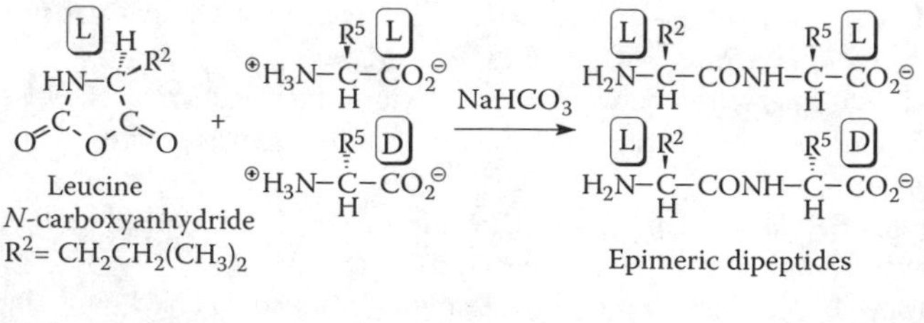

Leucine *N*-carboxyanhydride $R^2 = CH_2CH_2(CH_3)_2$ + $^{\oplus}H_3N\text{-}C(R^5)(H)\text{-}CO_2^{\ominus}$ (L) + $^{\oplus}H_3N\text{-}C(R^5)(H)\text{-}CO_2^{\ominus}$ (D) $\xrightarrow{NaHCO_3}$ $H_2N\text{-}C(R^2)(H)\text{-}CONH\text{-}C(R^5)(H)\text{-}CO_2^{\ominus}$ (L,L) + $H_2N\text{-}C(R^2)(H)\text{-}CONH\text{-}C(R^5)(H)\text{-}CO_2^{\ominus}$ (L,D)

Epimeric dipeptides

FIGURE 4.20 Conversion of enantiomers to diastereomeric products by reaction with an amino acid *N*-carboxyanhydride.[84]

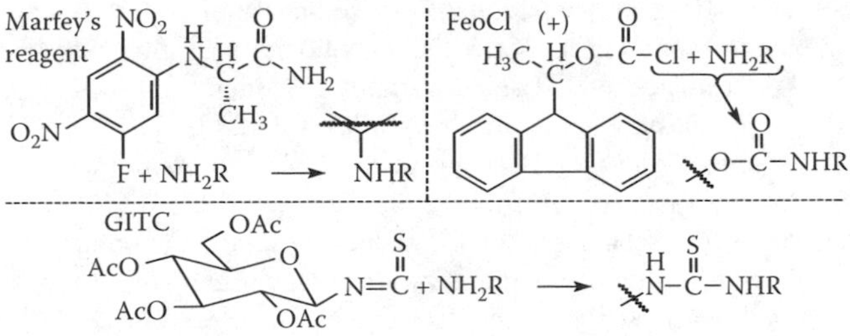

FIGURE 4.21 Chiral reagents and their reactions with enantiomeric mixtures NH_2R, generating diastereoisomers that are separable by high-performance liquid chromotography. Marfey's reagent = *N*-(2,4-dinitro-5-fluorophenyl)-L-alanine amide,[87] FeoCl = (+)-1-(9-fluorene)ethyl chloroformate,[88] GITC = 2,3,4,6-tetra-*O*-acetyl-β-D-glucopyranosyl isothiocyanate.[86]

of reference compounds because of the different color yields when ninhydrin reacts with diastereomeric peptides (see Section 4.9).

A second disadvantage is that the *N*-carboxyanhydride occasionally reacts further to generate tri- and tetrapeptides. This side reaction can be avoided by use of the succinimido ester of Boc-L-leucine as reagent, followed by removal of the Boc group. Other reagents commonly used for analysis of enantiomers as diastereomeric products appear in Figure 4.21. The diastereoisomers are separated by HPLC and determined by measurement of ultraviolet absorbance or fluorescence, the magnitudes of which are identical for the two isomers. GITC and FecCl are applicable also to the determination of secondary amino acids; namely, prolines and *N*-methylamino acids.

Amino acid derivatives can be examined for enantiomeric purity by the same procedures after removal of the protecting groups. Another approach is to couple them directly with another derivative to give protected dipeptides whose diastereomeric forms are usually easy to separate by HPLC (see Section 4.11). An *N*-protected amino acid is coupled with an amino acid ester, and vice versa. Use of soluble carbodiimide as reagent (see Section 1.16), followed by aqueous washes, gives clean HPLC profiles. It is understood that the derivative that serves as reagent must have been demonstrated to be enantiomerically pure.[43,84–89]

43. NL Benoiton, Y Lee, B Liberek, R Steinauer, FMF Chen. High-performance liquid chromatography of epimeric *N*-protected peptide acids and esters for assessing racemization. *Int J Pept Prot Res* 31, 581, 1988.
84. JM Manning, S Moore. Determination of D- and L-amino acids by ion-exchange chromatography as L-D and L-L- dipeptides. *J Biol Chem* 243, 5591, 1968.
85. AR Mitchell, SBH Kent, IC Chu, RB Merrifield. Quantitative determination of D- and L-amino acids by reaction with *tert*-butoxycarbonyl-L-leucine *N*-hydroxysuccinimide ester and chomatographic separation as D,L and L,L dipeptides. *Anal Chem* 50, 637, 1978.

86. T Kinoshita, Y Kasahara, N Nimura. Reversed-phase high-performance liquid chromatographic resolution of non-esterified enantiomeric amino acids by derivatization with 2,3,4,6-tetra-*O*-acetyl-β-D-glucopyranosyl isothiocyanate and 2,3,4 tri-*O*-acetyl-α-D-arabinopyranosyl isothiocyanate. *J Chromatog* 210, 77, 1981.
87. P Marfey. Determination of D-amino acids. II. Use of a bifunctional reagent 1,5-difluoro-2,4-dinitrobenzene. *Carlsberg Res Commun* 49, 591, 1984.
88. S Einarsson, B Josefsson, P Möller, D Sanchez. Separation of amino acid enantiomers and chiral amines using precolumn derivatization with (+)-1-(9-fluorenyl)ethyl chloroformate and reversed phase liquid chromatography. *Anal Chem* 59, 1191, 1987.
89. R Albert, F Cardinaux. RPHPLC resolution of enantiomeric *N*-methylamino acids by GITC (2,3,4,6-tetra-*O*-acetyl-β-D-glucopyranosyl isothiocyanate) derivatization, in JE Rivier, GR Marshall, eds. *Peptides. Chemistry Structure Biology. Proceedings of the 11th American Peptide Symposium*, Escom, Leiden, 1990, pp 437-438.

5 Solid-Phase Synthesis

5.1 THE IDEA OF SOLID-PHASE SYNTHESIS

Solid-phase synthesis means synthesis on a solid support. It is a technology that dates back to the early 1960s — an era when ion exchange on functionalized polystyrene beads was the prominent method for purification and analysis of small charged molecules. R. B. Merrifield, an immunologist working in the laboratory of D. W. Woolley at the Rockefeller Institute in New York, was required to synthesize analogues of biologically active peptides. In the process, Merrifield perceived that the approach of the day, made up of successive coupling and deprotection reactions in solution, followed by extractions at each step to eliminate unconsumed reactants and secondary products, was very labor intensive and repetitive, and he concluded that there was a need for a rapid and automatic method for the synthesis of peptides. He suggested that the synthesis be carried out with the first residue attached to an insoluble support (Figure 5.1), so that purification could be achieved by simple filtration instead of extraction. According to his proposal, a protected amino acid is anchored to an insoluble functionalized support by a bond that resists all chemistries employed during assembly of the peptide. The amino group is deprotected, and additional residues are introduced successively. Each reaction is followed by filtration, which removes unconsumed reactants and secondary products that are dissolved in the solvent. Final deprotection detaches the peptide from the support. Three advantages were envisioned by Merrifield: high yields achieved by forcing the reactions to completion, less manipulation and consequently less time, and minimized losses of material because reactions and purification would take place without removing the peptide support from the reaction vessel. Successful implementation of the method was announced in 1962, with the first publication appearing the following year. For the first 10 years or so, the method faced considerable opposition

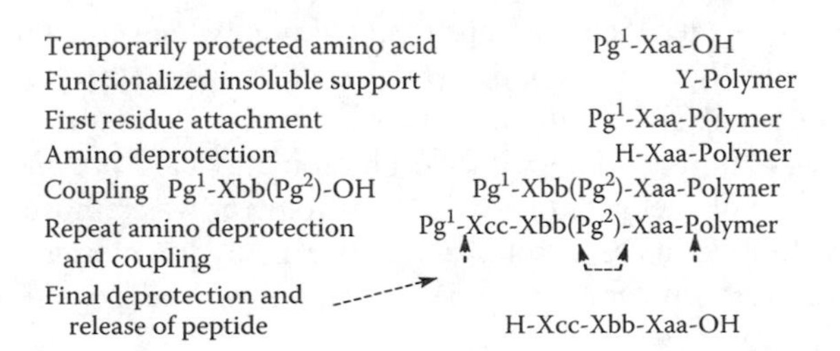

FIGURE 5.1 The idea of solid-phase synthesis as conceived by Merrifield in 1959. Protecting group Pg^1 is selectively removed after the addition of each residue.

by the traditionalists, but this resistance gradually dissipated. Twenty years of development and refinement by Merrifield and colleagues and others culminated in the award in 1984 of the Nobel Prize in Chemistry to Merrifield for development of a methodology for chemical synthesis on a solid matrix. The method has since found no end of applications. It is interesting to note that an analogous approach was employed by others in 1963 to synthesize a dipeptide, the amino-terminal residue being fixed to a polystyrene support through an amide bond.[1,2]

1. RB Merrifield. *Federation Proc* 21, 412, 1962.
2. RB Merrifield. Solid phase synthesis. I. The synthesis of a tetrapeptide. *J Am Chem Soc* 85, 2149, 1963.

5.2 SOLID-PHASE SYNTHESIS AS DEVELOPED BY MERRIFIELD

Initial attempts to develop the solid-phase method (see Section 5.01) involved attachment of the first residue to a functionalized copolymer of styrene and divinylbenzene, with the trade name Dowex 50, as the benzyl ester with benzyloxycarbonyl for protection of α-amino groups and selective deprotection by acidolysis with hydrogen bromide in acetic acid (see Section 3.5). The method was inefficient, however, because of the significant loss of the chain from the support at each step. The anchoring linkage was stabilized to acidolysis by nitration of the phenyl ring of the benzyl moiety (see Section 3.19), and this allowed the first successful synthesis of a peptide; namely, L-leucyl-L-alanylglycyl-L-valine. Unreacted amino groups were capped by acetylation with triethylammonium acetate after each coupling. The chain was detached from the resin by saponification. The structure was established by comparison with an authentic sample prepared in solution using 4-nitrophenyl esters (see Section 2.9) for peptide-bond formation. It was obvious, however, that the combination of two benzyl-based protectors was not going to be satisfactory. Fortunately, the *tert*-butoxycarbonyl protector (see Section 3.6) had just become available, and an improved methodology was developed. Figure 5.2 shows the scheme employed in 1964 by Merrifield to prepare bradykinin, which contains nine residues — the first biologically active peptide synthesized by his method. The first residue was attached to a polystyrene-divinylbenzene copolymer as the benzyl ester by reaction of Boc-N^{ω}-nitroarginine with the chloromethylated polymer (see Section 5.7) in the presence of triethylamine. The Boc-protector was removed by acidolysis with hydrogen chloride in acetic acid, and the amine hydrochloride that was generated was neutralized with triethylamine. Liberated *tert*-butoxycarbonyl produces isobutene and carbon dioxide.

The next residues were attached successively by dicyclohexylcarbodiimide-mediated coupling of Boc–amino acids with the free amino groups. The use of excess Boc–amino acid eliminated the need for capping after coupling. The last Boc-group and the benzyl-based side chain and carboxy-terminal protectors were removed at the end of the synthesis by acidolysis with hydrogen bromide in trifluoroacetic acid; the latter was used instead of acetic acid to avoid acetylation of hydroxymethyl side chains (see Section 6.6). Catalytic hydrogenolysis of the peptide removed the nitro

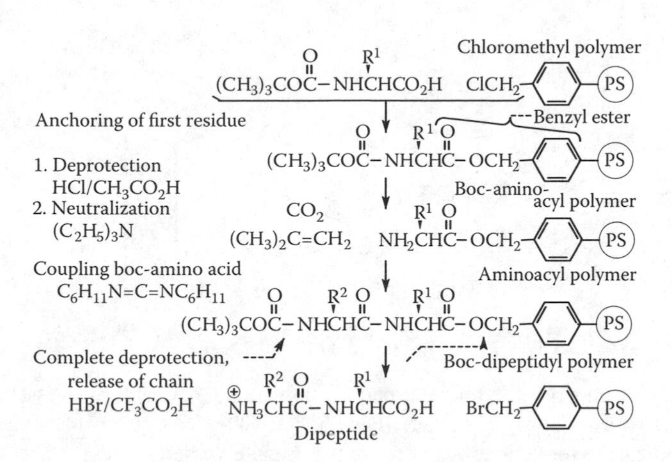

FIGURE 5.2 Synthesis of a peptide on a solid support according to Merrifield in 1964. PS = polystyrene. Initially, the N^{α}protector was benzyloxycarbonyl, removed by HBr in CH_3CO_2H, followed by final deprotection with the same reagent at 78°C. The current protocol employs CF_3CO_2H and HF, respectively.

groups. A 68% yield of bradykinin was obtained after purification of the crude product on a weakly acidic cation-exchange resin. Four days were required for the synthesis, and 4 more days were required to purify the product. All reactions except deprotection had been carried out in dimethylformamide. However, dichloromethane proved superior for couplings because less *N*-acyl-*N,N′*-dicyclohexylurea (see Section 2.2) is formed, and hence a smaller excess of Boc–amino acid is required. Peptide-bond formation employing activated esters had been examined, but the tactic was rejected on the basis that the reactions did not go to completion. It was later shown that activated esters are suitable for solid-phase synthesis if a solvent more polar than dichloromethane (see Section 7.6) is employed. Present-day protocol involves use of 50% trifluoroacetic acid in dichloromethane for deprotection of α-amino groups, and hydrogen fluoride for final deprotection and release of the chain from the support.[2–6]

2. RB Merrifield. Solid phase synthesis. I. The synthesis of a tetrapeptide. *J Am Chem Soc* 85, 2149, 1963.
3. RB Merrifield. Solid phase synthesis. II. The synthesis of bradykinin. *J Am Chem Soc* 86, 304, 1964.
4. RB Merrifield. Solid-phase synthesis. III. An improved synthesis of bradykinin. *Biochemistry* 3, 1385, 1964.
5. RB Merrifield. Solid phase synthesis. IV. The synthesis of methionyl-lysyl-bradykinin. *J Org Chem* 29, 3100, 1964.
6. RB Merrifield. Solid-phase peptide synthesis. *Endeavour* 3, 1965.

5.3 VESSELS AND EQUIPMENT FOR SOLID-PHASE SYNTHESIS

The unique feature of solid-phase synthesis is the elimination of unconsumed reactants and secondary products by filtration. The latter is possible because chain

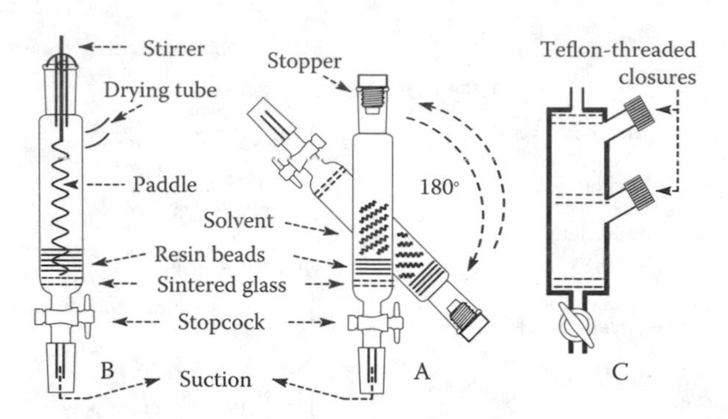

FIGURE 5.3 Reaction vessels for solid-phase synthesis. (A) 10–300-mL vessel (0.5–10 g of resin) affixed to a rotating-arc shaker; (B) 0.5–8-L (20–200 mm in diameter) cylindrical container with stirrer for up to 500 g of resin; (C) Ananth vessel with two 80-mL chambers.[10]

assembly is carried out on an insoluble support that is in the form of small beads, so the essential is a reaction vessel fitted at the bottom with a fritted glass disk and an outlet through which suction can be applied to remove the solvent containing the undesired components (Figure 5.3, A). The flow of liquid is controlled by a stopcock. Reactants and solvent are introduced through an upper entrance port that is closed by a stopper or screw cap. Agitation is achieved by fixing the vessel to a rotating-arc shaker. Simple movement of the support and not vigorous shaking is desired, so that the beads (see Section 5.7) are not damaged. Vessels of a variety of shapes and designs are available, including a two-chamber vessel (Figure 5.3, C) that allows synthesis of two different peptides at the same time. Some vessels have an inlet at the bottom for introducing a stream of nitrogen for agitation, with an appropriate outlet at the top. For synthesis on a larger scale, a vessel of increased diameter can be selected, into which is inserted a paddle connected to a mechanical stirrer for mixing the components (Figure 5.3, B). Monitoring of reactions can be achieved by withdrawing an aliquot of resin through the entrance port. Various arrangements of connected containers and three-way valves permit the delivery and removal of solvents and reagents without opening the vessels. An alternative involves synthesis under continuous-flow conditions (Figure 5.4), in which case the support is stationary, the solvent flows in a cycle through the support, and the reagents are injected into the moving solvent. Chain assembly employing the above systems is referred

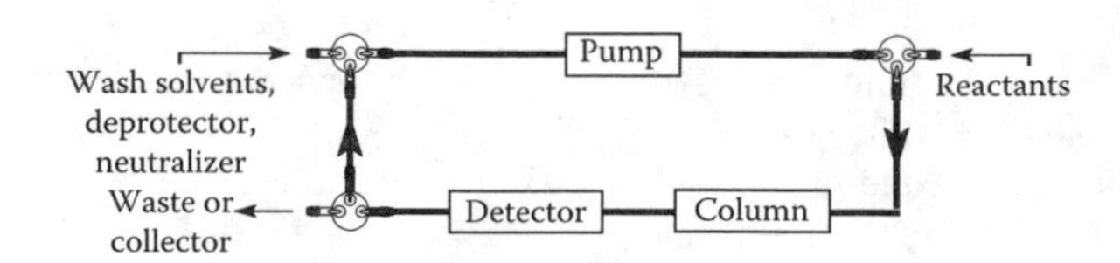

FIGURE 5.4 Schematic representation of a continous-flow system for the solid-phase synthesis of peptides. Solvent is forced through the system by a pump. The support is in the form of a column that is stationary. A reaction is monitored by measuring the change in absorbance of the solvent stream.

to as "manual" solid-phase synthesis. Instruments controlled by a programmer or computer produce peptides by "automated" synthesis.[2,7–10]

2. RB Merrifield. Solid phase synthesis. I. The synthesis of a tetrapeptide. *J Am Chem Soc* 85, 2149, 1963.
7. RB Merrifield, JM Stewart, N Jernberg. Instrument for automated synthesis of peptides. *Anal Chem* 38, 1905, 1966.
8. V Gut, J Rudinger. Rate measurement in solid phase peptide synthesis, in E. Bricas, ed. *Peptides 1968. Proceedings of the 9th European Peptide Symposium*, North-Holland, Amsterdam, 1968, pp 185-188.
9. TJ Lukas, MB Prystowsky, BW Erickson. Solid-phase peptide synthesis under continous-flow conditions. *Proc Natl Acad Sci USA* 78, 2791, 1981.
10. M Anantharamaiah, A Gawish, M Iqbal, SA Khan, CG Brouilette, JP Segrest. In CA Peeters, ed. *Peptides of Biological Fluids*, New York, Permagon, 1986, p 34.

5.4 A TYPICAL PROTOCOL FOR SOLID-PHASE SYNTHESIS

Solid-phase synthesis involves the combination of reagents with functional groups that are located on the surface and on the inside of beaded polymers. The beads are immersed in solvent containing the reagents, which approach the solvated sites by diffusion. Success is contingent on the sites being accessible to the reagents. Complete reaction is encouraged by use of a large excess of reagent; complete removal of soluble components is achieved by repeated washing and filtration. A typical protocol for the synthesis of a 14-mer on polystyrene resin using Boc/Bzl chemistry appears in Figure 5.5. The peptide resin is suspended in dichloromethane, then methanol, and then dichloromethane. The latter swells the resin beads, and the more polar methanol shrinks the beads (see Section 5.7). Alternating swelling and shrinking serves to expose reacting sites. Methanol is also used to remove reagents that are presented in dimethylformamide because it is miscible with the two other solvents.

	Min		Min
1. CH_2Cl_2 wash,	80 mL 3 X2	8. CH_2Cl_2 wash,	80 mL 3 X3
2. CH_3OH wash,	30 mL 3 X2	9. Boc-Xaa-OH (10 mmol) in 30 mL DMF + DCC (10 mmol) in DMF,	30 X1
3. CH_2Cl_2 wash,	80 mL 3 X2		
4. CF_3CO_2H-CH_2Cl_2 (1:1),	70 mL 10 X2	10. CH_3OH wash,	40 mL 3 X2
5. CH_2Cl_2 wash,	80 mL 3 X2	11. Et_2N-DMF (1:8),	70 mL 5 X2
6. Et_2N-DMF (1:8),	70 mL 5 X2	12. CH_3OH wash,	30 mL 3 X2
7. CH_3OH wash,	40 mL 3 X2	13. CH_2Cl_2 wash,	80 mL 3 X2

FIGURE 5.5 Schedule for the solid-phase synthesis of somatostatin, a 14-mer, on 10 g of resin reacted with 5 mequiv of the first amino acid, adapted from J. Rivier, *J. Am. Chem. Soc.* 96:2986, 1974. Min = time of mixing; X2 = two times; DCC = dicyclohexylcarbodiimide. Step 4 included 5% of $(CH_2SH)_2$ to prevent the oxidation of tryptophan. When the ninhydrin test on an aliquot after step 13 was negative, step 1 followed; when positive, steps 9–13 were repeated.

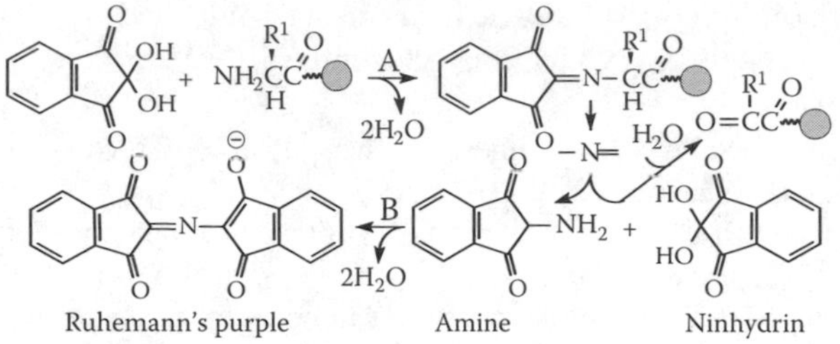

FIGURE 5.6 Reaction of ninhydrin (trioxohydrindene hydrate) with the amino group of a bound residue (A) generates the Schiff's base. Hydrolysis after shift of the double bond generates the aldehyde and another amine which reacts (B) with a second molecule of ninhydrin to give an equilibrium mixture of the anion depicted and its tetraoxo form with a maximum of absorbance at 570 nm.

Each wash is for a selected period of time and is carried out twice to completely effect the change in solvent. Step 4, followed by a dichloromethane wash, removes the Boc-group. Step 6 neutralizes the trifluoroacetate anion that is bound by ionic interaction. Step 9 is the coupling reaction. Step 11 serves to remove any Boc–amino acid that might be bound to the resin by adsorption. Step 13 is followed by a test for unreacted amino groups to verify that the coupling has gone to completion. An aliquot of beads is heated with a solution containing ninhydrin and other components. Two molecules of ninhydrin combine with an amino group with the liberation of three molecules of water (Figure 5.6) to produce a purple color, the intensity of which depends on the nature of the amino-terminal residue. The resin beads remain colorless if they do not contain cationic sites that combine with the anionic form of the chromophore (see Section 5.16). If the test is positive, a second coupling, steps 9–13, is effected, or the amino groups are permanently blocked or "capped" by acetylation or some other amide-bond forming reaction to prevent chain extension at these sites. The reagent for capping should not contain alcohol together with base so that fission by transesterification at the anchored residue is avoided. Final deprotection, which also releases the chain from the resin, is effected by strong acid. The synthesis described in Figure 5.5 was carried out manually on a scale up to 100 g of peptide-resin at the time when the solid-phase approach was still considered controversial. A multitude of protocols have emerged, a significant detail being the use of dichloromethane as the solvent for couplings.[11–13]

11. E Kaiser, RL Colescott, CD Bossinger, PI Cook. Color test for detection of free terminal amino groups in the solid-phase synthesis of peptides. *Anal Biochem* 34, 595, 1970.
12. JEF Rivier. Somatostatin. Total solid phase synthesis. *J Am Chem Soc* 96, 2986, 1974.
13. VK Sarin, SBH Kent, JP Tam, RB Merrifield. Quantitative monitoring of solid-phase peptide synthesis by the ninhydrin reaction. *Anal Biochem* 117, 147, 1981.

5.5 FEATURES AND REQUIREMENTS FOR SOLID-PHASE SYNTHESIS

There are several features that are characteristic of solid-phase synthesis. There is no loss of material accompanying the synthetic steps because the peptide is never taken out of the vessel. Secondary products issuing from the reagents and protectors do not affect the purity of the peptide because they are quantitatively removed after each reaction. Reactions can be encouraged to go to completion by the use of an excess of reagents. The question of the solubility of the peptide does not arise as it does for syntheses carried out in solution. No purification of intermediates is effected during the synthesis, and operations are repetitive and have been automated. However, monitoring the course of reactions is not straightforward, though some automated systems incorporate measurement devices, and synthetic impurities resemble the target molecule (see following), which makes them difficult to remove at the end of the synthesis.

The requirements for solid-phase synthesis are diverse. The support must be insoluble, in the form of beads of sufficient size to allow quick removal of solvent by filtration, and stable to agitation and inert to all the chemistry and solvents employed. For continuous-flow systems, the beads also must be noncompressible. Reactions with functional groups on beads imply reaction on the inside of the beads as well as on the surface. Thus, it is imperative that there be easy diffusion of reagents inside the swollen beads and that the reaction sites be accessible. Accessibility is facilitated by a polymer matrix that is not dense and not highly functionalized. A matrix of defined constitution allows for better control of the chemistry. Easier reaction is favored by a spacer that separates the matrix from the reaction sites. Coupling requires an environment of intermediate polarity such as that provided by dichloromethane or dimethylformamide; benzene is unsuitable as solvent.

Reactions in solid-phase synthesis should be complete to avoid the formation of mixtures of failure sequences, which creates a formidable purification challenge. As an example, a 99% yield at each of 30 steps of the synthesis of a 15-mer gives as product a 74% yield of target peptide and a 26% yield of peptides that lack one residue. This is assuming that all unreacted groups participate in the next step. When an amino group becomes unavailable for chain growth, the substance produced is referred to as a truncated peptide. Truncated peptides that resume growth give rise to peptides containing deletion sequences. Production of failure sequences resulting from incomplete coupling may be avoided by effecting a second acylation or by capping (see Section 5.4); however, there is no guarantee that this action will achieve the objective. Removal of reagents should be complete, achieved by repetitive washing with appropriate solvents. Successful solid-phase synthesis requires that there be no reactions on the side chains and main chain during assembly, including premature removal of protecting groups, which includes the carboxy-terminal protector. In view of the use of liberal amounts of solvents and amino acid derivatives in large excess, all materials should be pure, and the chiral integrity of the residues must be preserved throughout the synthesis.[6,14]

6. RB Merrifield. Solid-phase peptide synthesis. *Endeavour* 3, 1965.
14. S Hancock, DJ Prescott, PR Vagelos, GR Marshall. Solvation of the polymer matrix. Source of truncated and deletion sequences in solid phase synthesis. *J Org Chem* 38, 774, 1973.

5.6 OPTIONS AND CONSIDERATIONS FOR SOLID-PHASE SYNTHESIS

Solid-phase synthesis is a technology by which the synthesis of a peptide is simplified. The chemistry for synthesis on a solid support is the same as that for synthesis in solution, except that the protector of the carboxy terminus is linked to an insoluble support, either directly or indirectly. Peptides that are to be cyclized may be anchored through the functional group of a side chain (see Section 5.24). Chain assembly is by addition of single residues, which is the approach also employed for synthesis in solution. The nature of the appendage on the carboxy terminus provides the unique difference. The appendage consists of a polymer matrix attached to a protector. The nature of the linkage between the protector and the first residue, as well as the method employed to anchor the first residue, is dictated by the nature of the target peptide. The stability that is required of the anchoring bond depends on the conditions that are employed to remove the protectors on the amino terminus of the growing peptide chain, or vice versa. The consequence of the above is that the choice of one option limits the choices for the other options because most are interdependent. The variables include the type of support, the nature of the anchoring bond and how it is created, the method of coupling, the nature of the amino-terminus protector and the deprotecting reagent, and the method of final deprotection.

There are three major types of target molecules: peptides, which implies a free carboxy terminus; protected peptides, which usually implies a free carboxy terminus; and peptide amides. Resins are either microporous gels or composites. The more common gel is a polystyrene–divinylbenzene copolymer (see Section 5.7), an alternative is the polar polydimethylacrylamide (see Section 5.8), and recently made popular are the polyethyleneglycol–polystyrene adducts (see Section 5.9). The linkers (see Section 5.10) to which the peptides are bound are substituted benzyl alcohols for the synthesis of peptides and substituted benzyl amines for the synthesis of amides. Protectors are either *tert*-butoxycarbonyl for the amino termini with benzyl or variants for the side-chain functional groups or 9-fluorenylmethoxycarbonyl for the amino termini with *tert*-butyl for the side-chain functional groups (see Section 3.20). Bond formation is by use of carbodiimides or variations thereof, occasionally symmetrical anhydrides or activated esters, and onium salt-based reagents (see Section 5.15). Each coupling method has its attractive features and limiting implications. Some residues such as asparaginyl and glutaminyl may require special consideration. Selection among the options is often influenced by the experience of the operator and the traditions of the laboratory.

5.7 POLYSTYRENE RESINS AND SOLVATION IN SOLID-PHASE SYNTHESIS

Solid-phase synthesis is carried out with the first residue attached to a resin. The original and most popular resin is a copolymer of styrene and divinylbenzene (Figure 5.7). The matrix consists of a methylene chain with phenyl appendages at every second carbon atom, with cross links formed by fusion of phenyl rings with two chains. The resin, a porous gel, is in the form of beads, usually 38–75 μm in diameter (200–400 mesh: particles that pass through a 200-mesh screen and are retained by a 400-mesh screen), and occasionally 70–150 μm in diameter (100–200 mesh). Functionalization of the beads is achieved by derivatization after polymerization (Figure 5.7) or by including 4-methoxymethyl styrene in the polymerizing mixture. The latter approach produces a more defined product; functionalization generates positional isomers of which only 70% are *para*. The chloromethyl resin is referred to as Merrifield resin, and the functional groups are located primarily on the interior of the resin beads. The environment is heterogeneous, so some reaction sites are less accessible than others or are less likely to react with hindered incoming amino acid derivatives. Of primordial importance is the diffusion of reactants inside the beads. Rates of reactions are affected by rates of penetration of reagents, and reagents penetrate more freely when there is less obstruction and the sites are well solvated. In a swollen resin, there is no longer a phase boundary between the solvent and the resin. Experience has shown that a matrix with 1% cross links gives superior results to those of the 2% cross-linked material used in the earlier work. Maximum solvation occurs in a solvent of polarity similar to that of the matrix. The extent of swelling of a polystyrene resin is schematically depicted in Figure 5.8. The resin swells to three times its size in dimethylformamide and six times its size in the less polar solvent. The extent of solvation in dichlororomethane decreases during synthesis of the peptide as the material changes from hydrophobic polystyrene to the peptide-resin adduct that becomes increasingly more polar. Decreased solvation has a tendency to close reactive sites, and a gradual switch of solvent polarity during the

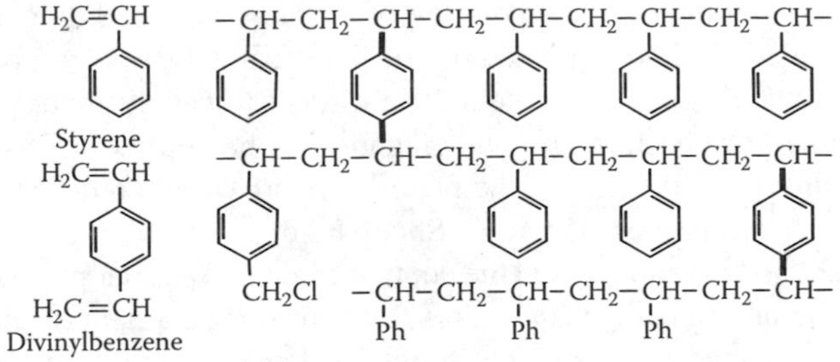

FIGURE 5.7 Merrifield resin obtained by polymerization of styrene in the presence of divinylbenzene followed by reaction with CH_3OCH_2Cl.[3] The amino-methyl resin is obtained by transformation or reaction of the copolymer with *N*-hydroxymethylphthalimide, followed by hydrazinolysis to liberate the amino groups.[31]

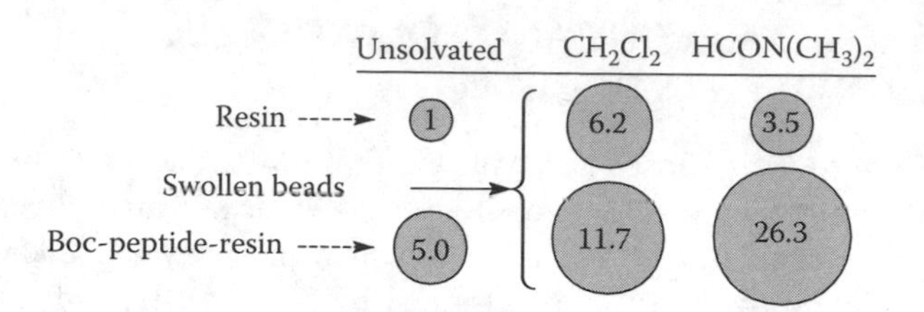

FIGURE 5.8 Relative volumes of a polystyrene resin and a peptide-resin adduct in the presence or absence of solvent molecules. The values are for a peptide of 40 residues (molecular weight = 5957 daltons), the first residue loaded at 0.95 mequiv/g of resin, giving a peptide to resin ratio of 4:1 by weight after assembly.[15]

synthesis serves to keep the reactive sites as accessible as possible. Alternating swelling–shrinking cycles helps to expose any buried sites. However, sites that are exposed by changes in solvation can give rise to deletion sequences (see Section 5.5). By the end of the synthesis (Figure 5.8), when 80% of the material is peptidic in nature, its volume is much greater in dimethylformamide than in dichloromethane. Dissolved substances are removed by filtration. A wash step with an alcohol that compresses the beads removes the last traces of reagents. The considerable increase in size of the insoluble component that occurs during a synthesis has to be kept in mind when selecting the size of a reaction vessel for a synthesis.[14,15]

14. S Hancock, DJ Prescott, PR Vagelos, GR Marshall. Solvation of the polymer matrix. Source of truncated and deletion sequences in solid phase synthesis. *J Org Chem* 38, 774, 1973.
15. VK Sarin, SBH Kent, RB Merrifield. Properties of swollen polymer networks. Solvation and swelling of peptide-containing resins in solid-phase peptide synthesis. *J Am Chem Soc* 102, 5463, 1980.

5.8 POLYDIMETHYLACRYLAMIDE RESIN

Despite the success achieved by synthesis on polystyrene supports, there were instances when the extent of incorporation of a residue dropped dramatically — the classical example being during assembly of the decapeptide corresponding to residues 65–74 of the acyl carrier protein. The difficulty was attributed to the loss of reactive amino groups resulting from aggregation of the peptide chains that occurred because the physical properties of the peptide chains were too different from those of the support. It was proposed by R. C. Sheppard that greater efficiency in synthesis could be expected if the support and the peptide chain had similar physical properties. There would be no dramatic changes in solvation, and hence no deleterious effects caused by such changes. An insoluble support suitable for synthesis was developed by extending the cross links of polyacrylamide by one carbon atom and by eliminating the hydrogen-bonding potential of the carboxamide groups. Polymerization of dimethylacrylamide in the presence of (1,2-*bis*-acrylamido)ethane gives polydimethylacrylamide, which consists of a methylene chain carrying dimethylformamide appendages at every second carbon atom, with cross links formed by the fusion of

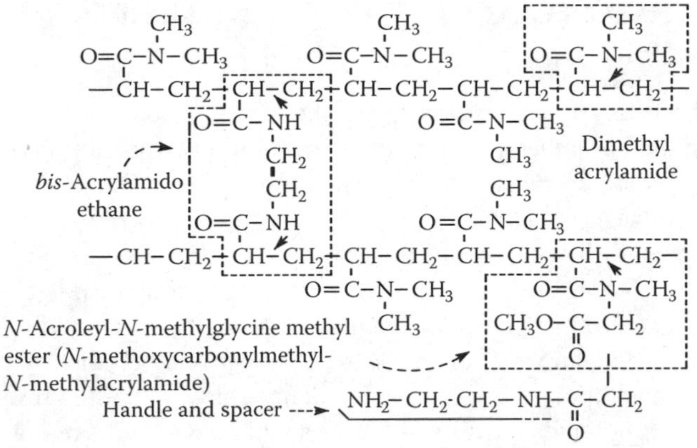

bis-Acrylamido ethane

Dimethyl acrylamide

N-Acroleyl-N-methylglycine methyl ester (N-methoxycarbonylmethyl-N-methylacrylamide)

Handle and spacer

FIGURE 5.9 Polydimethylacrylamide obtained by polymerizing dimethylacrylamide in the presence of cross-linking reagent *bis*-1,2-acrylamidoethane and the ester-bearing acrylamide, which provides the reactive functional group.[18] Reaction with diaminoethane produces a handle for attachment of a linker and subsequent chain assembly. The arrowheads indicate bonds that are double in the starting materials.

chains through two monomethylformamide appendages (Figure 5.9). A methoxycarbonyl group introduced by adding *N*-acryloylsarcosine methyl ester to the polymerizing mixture provides the functional group through which the target chain is anchored. The material is highly polar and so well solvated that it produces a transparent gel that is 10 times larger than the original material in dichloromethane, dimethylformamide, or methanol. With this resin as a support, the acyl carrier decapeptide 65–74 was obtained in a higher yield than with the polystyrene resin, using the same chemistry. Polydimethylacrylamide has provided an efficient alternative to polystyrene resin for the solid-phase synthesis of peptides. Variants exist in which dimethylamino has been replaced by 2-oxo-pyrrolidino and morpholino. The large expansion occurring in solvent has the undesirable implication that greater excesses of reagents are required to achieve good concentrations of reactants. Composite resins consisting of polydimethylacrylamide embedded within the pores of kieselguhr or polystyrene provide pressure-stable supports that are suitable for continous-flow systems.[16–20]

16. E Atherton, DL Clive, RC Sheppard. Polyamide supports for polypeptide synthesis. *J Am Chem Soc* 97, 6584, 1975.
17. R Arshady, E Atherton, MJ Gait, RC Sheppard. An easily prepared polymer support for solid phase peptide and oligonucleotide synthesis. Preparation of substance P and a nonadeoxyribonucleotide. *J Chem Soc Chem Commun* 423, 1979.
18. R Arshady, E Atherton, DL Clive, RC Sheppard. Peptide synthesis. Part 1. Preparation and use of polar supports based on poly(dimethylacrylamide). *J Chem Soc Perkin* 1, 529, 1981.
19. E Atherton, E Brown, RC Sheppard. A physically supported gel polymer for low pressure, continuous flow solid phase reactions. Applications to solid phase peptide synthesis. *J Chem Soc Chem Commun* 1151, 1981.
20. RC Sheppard. Continous flow methods in organic synthesis. *Chem Britain* 402, 1983.

5.9 POLYETHYLENEGLYCOL-POLYSTYRENE GRAFT POLYMERS

A different approach to synthesis surfaced shortly after solid-phase synthesis was developed. On the basis that synthesis on a solid support was not the ideal method because reactions are carried out in a heterogenous system that can be subject to steric effects, synthesis on a soluble support was explored by the group of E. Bayer. It was demonstrated that peptide chain assembly could be achieved efficiently in solution, with the first residue attached to the terminal group of polyethyleneglycol — a polymer that is soluble in dichloromethane or dimethylformamide. The reagents and products thus derived were separated from the peptide polymer by precipitating the adduct by the addition of ether. Kinetic studies showed that the reactivity of the amino group of a polymer-bound residue in solution was the same as that of the amino group of an amino acid ester; hence, the reaction kinetics are not controlled by diffusion, as is the case for heterogenous systems. The approach did not become competitive, however, because the manipulations necessary for separating the components were not amenable to automation. The study did lead to the development of a new type of carrier that possesses the favorable properties of each of the earlier supports.

Polyethyleneglycol–polystyrene graft polymers made up of polyoxyethylene chains attached to a polystyrene–divinylbenzene copolymer (Figure 5.10) exhibit the insolubility, mechanical stability, and inertness of hydrophobic polystyrene and maintain the flexibility and polarity of the polyethyleneglycol polymer. Peptide assembly is effected with the carboxy-terminal residues bound at the ends of the polyether chains. The rate of aminolysis of an activated derivative by a support-bound residue is the same as the rate of aminolysis by the corresponding amino acid ester. The resin exists in the form of beads that swell to two to three times their original size in the solvents that dissolve polyethyleneglycol. The swelling factor remains unchanged as a peptide chain is assembled. Maximum flexibility of the polar chains obtains at a molecular weight of 3000 Da, which defines the optimum spacer length. Beads of single size, which is preferable to resins with a wide distribution of sizes, are now available. Beads smaller (15 or 30 μm in diameter)

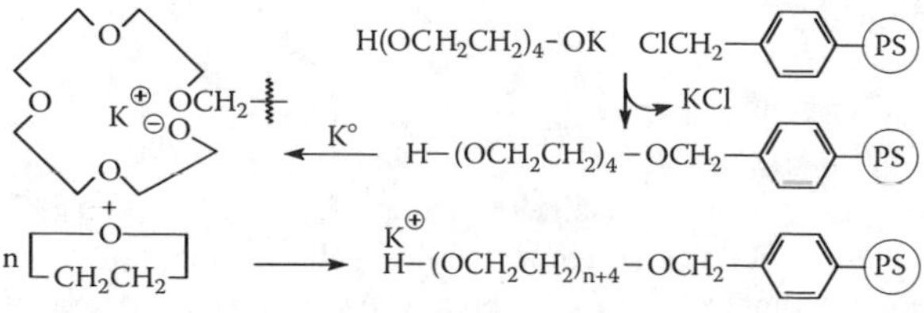

FIGURE 5.10 Synthesis of a polyethyleneglycol-polystyrene graft polymer by etherification of Merrifield resin using potassium tetra(oxyethylene) oxide, followed by extension of the chain by reaction of the potassium salt, which is present as the crown ether.[21] In several TentaGel resins, the connecting bond is an ethyl ether that is more acid-stable than the benzyl ether.

than usual (90 μm in diameter) allow for reduced coupling and deprotection times. The beads are insensitive to pressure and, therefore, are suitable for use in continuous-flow systems. The resin made up of 70% polyoxyethylene and 30% polystyrene matrix has been marketed as TentaGel (Tenta from tentacles) since 1989. Analogous gels where the two components are joined through other linkages have been developed. Some of these gels are sensitive to hydrogen fluoride at the interconnecting juncture. A polyethylene-polydimethylacrylamide copolymer that is stable to hydrogen fluoride has been developed.[21–25]

21. E Bayer, B Hemmasi, K Albert, W Rapp, M Dengler. Immobilized polyoxyethylene, a new support for peptide synthesis, in VJ Hruby, DH Rich, eds. *Peptides 1983. Proceedings of the 8th American Peptide Symposium*. Pierce, Rockford, IL, 1983, pp 87-90.
22. E Bayer, M Dengler. B Hemmasi. Peptide synthesis on the new polyoxyethylene-polystyrene graft copolymer, synthesis of insulin $B_{21\text{-}30}$. *Int J Pept Prot Res* 25, 178, 1985.
23. E Bayer, W Rapp. Polystyrene-immobilized PEG chains. Dynamics and applications in peptide synthesis, immunology, and chromatography, in JM Harris, ed. *Poly(Ethylene Glycol) Chemistry: Biotechnical and Biomedical Applications*. Plenum, New York, 1992, pp 326-345.
24. M Meldal. *Tetrahedron Lett* 33, 3077, 1992.
25. S Zalpsky, JL Chang, F Albericio, G Barany. Preparation and applications of polyethylene glycol-polystyrene graft resin supports for solid-phase peptide synthesis. *Reactive Polymers* 22, 243, 1994.

5.10 TERMINOLOGY AND OPTIONS FOR ANCHORING THE FIRST RESIDUE

Solid-phase synthesis involves assembly of a peptide chain whose carboxy-terminal residue is anchored to a support followed by release of the peptide from the support. Detachment of the peptide is contingent on the presence of an arrangement of interconnected atoms that renders the anchoring bond susceptible to the cleavage reagent employed. The structural moiety with which the carboxyl group of the first amino acid has reacted to form this combination of atoms is referred to as a "linker." The linker originates through elaboration of a functional group or "handle" on the support, or it may be a stand-alone molecule that is then secured to the support. The handle is affixed to the support either by derivatization of the resin or by including a molecule carrying the handle in the polymerizing mixture. The linker is bound to the support through the handle by a carbon-to-carbon, ether, or amide bond, all of which are stable to cleavage reagents. Stand-alone linkers are bifunctional molecules containing a carboxyl group. They are secured to supports by coupling with an amino group on the support. The resulting carboxamido group contributes to the lability/stability of the anchoring bond. Stand-alone linkers that incorporate an amino group must have the amino group protected before effecting the coupling. Spacers are sometimes inserted for different reasons between a linker and the support. The carboxy-terminal residue of the peptide is anchored to the linker by an ester or amide bond, depending on the nature of the target peptide (see Section 5.12). Examples of handles are methoxymethylphenyl ($CH_3OCH_2C_6H_4$–; see Section 5.7) and the methyl ester of sarcosine [$CH_3O_2CCH_2N(CH_3)$–; Figure 5.9]. Examples of linkers elaborated from a handle are 4-chloromethylphenyl ($ClCH_2C_6H_4$–), 4-hydroxybenzyloxybenzyl ($HOCH_2C_6H_4OCH_2C_6H_4$–; Figure 5.17), and chlorotrityl [$ClC(Ph_2)C_6H_4$–; Figure 5.17]. Examples of linkers that are stand-alone molecules are 4-hydroxymethylphenoxyacetic acid ($HOCH_2C_6H_4OCH_2CO_2H$; Figure 5.17) and 4-aminomethyl-3,5-dimethoxyphenoxypentanoic acid [$H_2NCH_2C_6H_2(OCH_3)_2O(CH_2)_4CO_2H$; Figure 5.18]. There are dozens of linkers available for synthesis, and nearly all of them, once acylated by a protected amino acid, provide a benzyl ester or a benzyl amide that has been sensitized to cleavage by acid by the presence of electron-donating moieties such as alkoxy, phenyl, or substituted phenyl. There are cases in which a peptide chain is bound to a support through two different linkers in series. This allows for versatility in synthesis. The distinction between designation of a moiety affixed to a support as a handle or linker is sometimes arbitrary.

The resins with their linkers are identified, unfortunately, by names that are inconsistent and confusing — a situation that has arisen primarily because the designations are based on diverse aspects of the resin such as the name of the developer (Wang resin, Rink resin; Figure 5.17), the trivial or chemical name of the linker (trityl resin; Figure 5.17), acronyms based on the chemical nature (PAM resin [Figure 5.15]; XAL [Figure 5.18]), or the property (SASRIN; Figure 5.20) or applicability of the linker (XAL, PAL; Figure 5.18).

There are three different approaches for anchoring the first residue to a support (Figure 5.11): first, the *N*-protected residue is anchored directly to a linker that has been elaborated on the support. Second, a stand-alone linker is coupled to the support, after which the protected residue is combined with the linker-resin. Third, the protected residue is combined with a stand-alone linker, and the resulting compound is then coupled to a handle on the support. The first option carries the disadvantage that if it is an ester that is formed, it is often difficult to force the

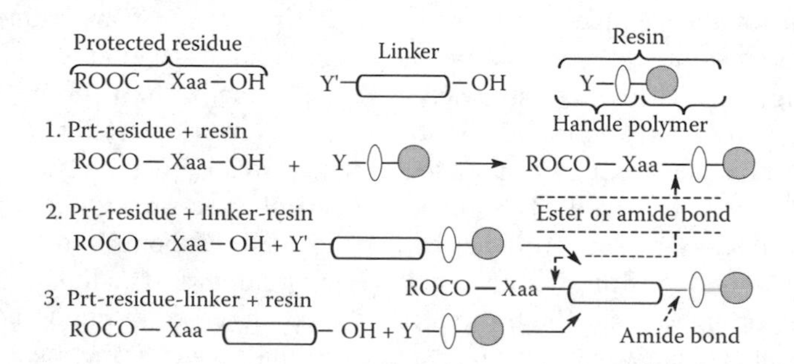

FIGURE 5.11 Three options for anchoring the first residue. Cleavage at the bond between Xaa and the support generates the product (see Figure 5.12). The word "resin" usually implies a polymer with a handle. For option 1, the handle serves as a linker. Options 2 and 3 allow insertion of a spacer or reference amino acid between the linker and the support. Option 3 allows purification of the *N*-acylaminoacyl linker and quantitative anchoring via an amide bond.

reaction to completion (see Section 5.23). The third option possesses the feature that the protected residue-linker is a small molecule that can be purified and then secured to the support by an amide bond that is easy to achieve quantitatively. The second and third options allow for the insertion of a spacer or a reference residue such as norleucine between the linker and the support. Analysis for the reference amino acid gives the amount of reactive sites (mequiv) per gram of resin, referred to as the "loading," that are available for use in synthesis. Spacers serve to distance the reactive sites from the matrix. The reagent of choice for attaching a linker to a support is a carbodiimide.

5.11 TYPES OF TARGET PEPTIDES AND ANCHORING LINKAGES

There are four types of target peptides: unsubstituted peptides that are sometimes referred to as peptide acids, protected peptides with a free carboxy terminus, peptide amides, and peptide alkyl thioesters (see Section 7.10), which have recently emerged. Amides are of interest because many naturally occurring peptides are amides. Protected peptides are of interest for use in synthesis in solution or for coupling to a resin-bound segment. The nature of the target peptide determines the nature of the bond through which the carboxy-terminal residue is anchored to the linker. With few exceptions, linkage is through a benzyl ester or a benzyl amide (Figure 5.12). The peptide acid or amide is detached by acidolysis, which releases the bound methylene atom as a carbenium ion (see Section 3.5). The susceptible bonds are cleavable because of the nature of the neighboring substituents (Figure 5.12). The acyl-oxime bond is cleaved by displacement by a nucleophile (see Section 5.21), and cyclic peptides are obtained by the same approach, except that the peptides are usually anchored to the support through the side chain of a multifunctional amino acid such as aspartic acid or serine (see Section 5.24).

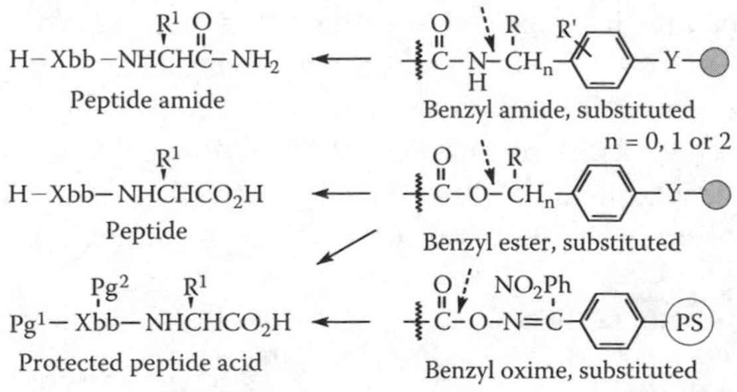

Peptide amide ← Benzyl amide, substituted

n = 0, 1 or 2

Peptide ← Benzyl ester, substituted

Protected peptide acid ← Benzyl oxime, substituted

FIGURE 5.12 The type of bond through which the first residue is anchored to the support is dictated by the nature of the target peptide. A peptide with a free carboxy terminus is produced by cleavage of a substituted benzyl ester or oxime. A peptide amide is produced by cleavage of a substituted benzylamide at the carboxamido–carbon bond. The natures of R, R′, and Y determine the stabilities of the pertinent bonds (dashed arrows). Amongst substituents sensitizing the bonds to acid are R = Ph, Ph_2, $(MeO)_2Ph$, $ClPh_2$; R′ = $(MeO)_2$; Y = OCH_2.

5.12 PROTECTING GROUP COMBINATIONS FOR SOLID-PHASE SYNTHESIS

The protecting group combinations employed in peptide synthesis are presented in Section 3.20. Of these, only two are routinely employed for solid-phase synthesis. The competing methods are Boc for N^{α}-protection with removal by acid and benzyl-based protectors for the side-chain functions, which implies a linker as stable to acid as a benzyl ester, and Fmoc for N^{α}-protection with removal by a secondary amine and *tert*-butyl-based protectors for the side-chain functions, which implies a linker of moderate sensitivity to acid (Figure 5.13). Removal of benzyl-based protectors and detachment of the chain is achieved either by very strong acid such a hydrogen fluoride or, occasionally, by reduction with sodium in liquid ammonia. If the peptide is simple enough, debenzylation and detachment can be achieved by catalytic hydrogenation. Removal of *tert*-butyl-based protectors and detachment of the chain is by strong acid such as trifluoroacetic acid.[4,26–28]

4. RB Merrifield. Solid phase synthesis. III. An improved synthesis of bradykinin. *Biochemistry* 3, 1385, 1964.
26. JM Schlatter, RH Mazur, O Goodmonson. Hydrogenation in solid phase peptide synthesis. I. Removal of product from the resin. *Tetrahedron Lett* 2851, 1977.
27. CD Chang, J Meienhofer. Solid-phase synthesis using mild base cleavage of N^{α}-fluorenylmethoxycarbonylamino acids, exemplified by a synthesis of dihydrosomatostatin. *Int J Pept Prot Res* 11, 246, 1978.
28. J Meienhofer, M Waki, EP Heimer, TJ Lambros, RC Makofske, C-D Chang. Solid phase synthesis without repetitive acidolysis. *Int J Pept Prot Res* 13, 35, 1979.

5.13 FEATURES OF SYNTHESIS USING BOC/BZL CHEMISTRY

The Boc/Bzl combination is not an orthogonal system (see Section 1.5). It suffers from the shortcoming that repeated acidolysis to remove the Boc groups causes slight loss of benzyl-based protectors, including the anchoring linker if it is an unmodified benzyl ester. To overcome these difficulties, stabilized protectors (see Section 3.19) based on 2-chlorobenzyl or cyclohexyl, and stabilized linkers such as hydroxymethylphenylacetamido (PAM-resin; see Section 5.17), have been developed. Removal of benzyl-based protectors generates carbenium ions that alkylate

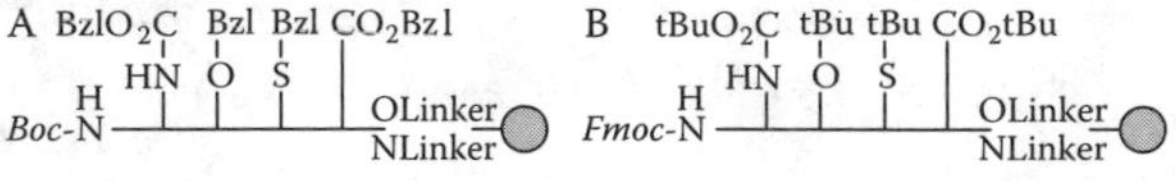

A $BzlO_2C$ Bzl Bzl CO_2Bzl HN O S *Boc*-NH OLinker NLinker

B $tBuO_2C$ tBu tBu CO_2tBu HN O S *Fmoc*-NH OLinker NLinker

FIGURE 5.13 Protecting group combinations employed in solid-phase synthesis. The protector written in italics is removed after each residue is incorporated into the chain. (A) Boc/Bzl[3] is not an orthogonal system — all substituents are removed by acidolysis. (B) Fmoc/tBu[27] [Atherton et al., 1978] is an orthogonal system – Fmoc is removed by β-elimination, other substituents by acidolysis. More suitable variants of the side-chain protectors are also used.

nucleophilic functions on side chains. Removal of stabilized protectors is even more troublesome because the stronger acid required induces cleavage by the S_N1 mechanism, which favors intramolecular over intermolecular side reactions (see Section 3.6) and compels the use of special equipment for effecting the reactions. The acid employed for deprotection leaves the amino group protonated; a neutralization step is required to convert the amino group to a nucleophile. Activated Boc–amino acids have a slight tendency to decompose in nonalkaline medium if they are not consumed quickly (see Section 7.15). This can lead to incorporation of two residues instead of one. No decomposition occurs in the basic milieu of *tert*-amine driven coupling reactions.

5.14 FEATURES OF SYNTHESIS USING FMOC/TBU CHEMISTRY

The Fmoc/tBu combination is an orthogonal system that was developed for solid-phase synthesis a decade and a half after Boc/Bzl chemistry. N^{α}-Deprotection is mediated by a base, so the amino group is available for the next coupling as soon as the protector is removed, and there is no loss of side-chain protectors during chain assembly. Final deprotection is by acid of moderate strength, so no special equipment is necessary, and the linker need not be especially stable to acid. On the other hand, *tert*-butyl-based protectors also generate carbenium ions that alkylate side chains, and the use of piperidine for Fmoc removal leads to side reactions because it is a good nucleophile and a strong base. It promotes aspartimide formation in sequences containing esterified aspartyl residues (see Section 6.13), giving rise to piperidides. It promotes piperazine-2,5-dione formation when the protector is removed from the second residue of a chain that is esterified to a linker-resin (see Section 6.19) and causes epimerization at *S*-protected cysteine residues that are esterified to a linker-resin. In addition, it is incompatible with the synthesis of *O*-acylserine-containing sequences because the acyl group shifts to the adjacent amino group as soon as the latter is liberated from the protector (see Section 6.6). Fmoc–amino acids are not ideal as starting materials. They are extremely hydrophobic, which tends to reduce the solubility of derivatives as well as hamper the deprotection reaction, and they have limited stability in the commonly used solvent dimethylformamide. The obstacles of piperazine-2,5-dione formation and epimerization at esterified cysteine are eliminated if the linker is 2-chlorotrityl chloride (see Section 5.23). For synthesis in solution, there is the additional unattractive feature that the piperidine adduct (see Section 3.11) is not easy to separate from the peptide. Fmoc–amino acids cost more than Boc–amino acids, but their use involves one step less, thus reducing consumption of solvent, which is a significant cost savings.[27–30]

27. CD Chang, J Meienhofer. Solid-phase synthesis using mild base cleavage of N^{α}-fluorenylmethoxycarbonylamino acids, exemplified by a synthesis of dihydrosomatostatin. *Int J Pept Prot Res* 11, 246, 1978.
28. J Meienhofer, M Waki, EP Heimer, TJ Lambros, RC Makofske, C-D Chang. Solid phase synthesis without repetitive acidolysis. *Int J Pept Prot Res* 13, 35, 1979.

29. E Atherton, CJ Logan, RC Sheppard. Peptide synthesis. 2. Procedures for solid-phase synthesis using N^{α}-fluorenylmethoxycarbonylamino-acids on polyamide supports. Synthesis of substance P and acyl carrier protein 65-74 decapeptide. *J Chem Soc Perkin Trans 1* 538, 1981.
30. E Atherton, M Caviezel, H Fox, D Harkiss, H Over, RC Sheppard. Peptide synthesis. Part 3. Comparative solid-phase synthesis of human β-endorphin on polyamide supports using t-butoxycarbonyl and fluorenylmethoxycarbonyl.

5.15 COUPLING REAGENTS AND METHODS FOR SOLID-PHASE SYNTHESIS

The reagents and methods employed for coupling in solid-phase synthesis are the same as for synthesis in solution, but a few are excluded because they are unsuitable. The mixed-anhydride method (see Section 2.6) and 1-ethoxycarbonyl-2-ethoxy-1,2-dihydroquinoline (see Section 2.15) are not used because there is no way to eliminate aminolysis at the wrong carbonyl of the anhydride. Acyl azides (see Section 2.13) are too laborious to make and too slow to react. The preparation of acyl chlorides (see Section 2.14) is too complicated for their routine use; this may be rectified, however, by the availability of triphosgene (see Section 7.13). That leaves the following choices, bearing in mind that a two to three times molar excess of protected amino acid is always employed.

A carbodiimide is added to the two reacting species. The urea generated from dicyclohexylcarbodiimide is insoluble and voluminous, so it is often replaced by diisopropylcarbodiimide, which generates a soluble urea. The soluble carbodiimide ethyl-(3-dimethylaminopropyl)-carbodiimide hydrochloride (see Section 1.16) is suitable but expensive. Efficiency of coupling is greater in dichloromethane than in dimethylformamide. There is also the option of adding 1-hydroxybenzotriazole to minimize the side reactions of *N*-acylurea (see Section 2.12), cyano (see Section 6.15), and aspartimide (see Section 6.13) formation.

The symmetrical anhydride is prepared using dicyclohexylcarbodiimide in dichloromethane, the urea and solvent are removed, and the anhydride is dissolved in dimethylformamide and added to the peptide-resin (see Section 2.5). The anhydride is a more selective acylating agent than the *O*-acylisourea and, thus, gives cleaner reactions than do carbodiimides, but twice as much amino-acid derivative is required, so the method is wasteful. It avoids the acid-catalyzed cyclization of terminal glutaminyl to the pyroglutamate (see Section 6.16) and is particularly effective for acylating secondary amines (see Section 8.15).

Activated esters (see Section 2.9) with 1-hydroxybenzotriazole as a catalyst are employed — pentafluorophenyl or 4-oxo-3,4-dihydrobenzotriazin-3-yl esters in particular for continuous-flow systems and special cases such as dicarboxylic amino acids. Other activated esters are not reactive enough. An alternative is preparation of benzotriazolyl esters using a carbodiimide followed by addition of the solution to the peptide-resin.

Phosphonium or uronium salts added to the two reacting species followed by the tertiary amine. The protected amino acid must be added before the reagent; otherwise, the latter will react with the amino groups. Bond formation is rapid, secondary products are soluble, and there is no major side reaction relating to the reagent. The alternatives are BOP (see Section 2.17), PyBOP (see Section 2.19), TBTU or HBTU (see Section 2.18), HATU (see Section 2.27), HAPyY (see Section 7.19), and similar. Azabenzotriazole-based reagents are expensive. There is the option of adding the corresponding auxiliary nucleophile (see Section 2.21), 1-hydroxybenzotriazole, or 1-hydroxy-7-azabenzotriazole; however, this can be counterproductive, as the activated ester formed may be less reactive than the intermediate generated from the reagent (see Section 2.20). Tertiary-amine driven reactions are especially well suited for Boc/Bzl chemistry because the basic milieu prevents decomposition of the activated form that occurs in nonalkaline medium (see Section 7.15) and eliminates the need for the neutralization step to dislodge Boc–amino acid that might be bound to the resin by adsorption (see Section 5.4).

5.16 MERRIFIELD RESIN FOR SYNTHESIS OF PEPTIDES USING BOC/BZL CHEMISTRY

The first peptides produced by solid-phase synthesis were obtained on Merrifield resin (see Section 5.7), using Boc/Bzl chemistry. Despite the success achievable, there are inherent imperfections associated with the chloromethylphenyl linker of Merrifeld resin. It is not straightforward to establish the number of reactive sites on the support. Quantitation requires determination of chloride by the Volhard or ninhydrin methods, neither of which is simple. Anchoring of the first residue by reaction of the cesium or trialkylammonium salt of the derivative with the chloromethyl group is an esterification reaction (Figure 5.14, A), which does not go to completion readily. Unreacted chloromethyl groups undergo quaternization by tertiary amines such as triethylamine and pyridine. The triethylamine used to neutralize the acid after deprotection produces an ion exchanger of trifluoroacetate (Figure 5.14, B). Pyridine from

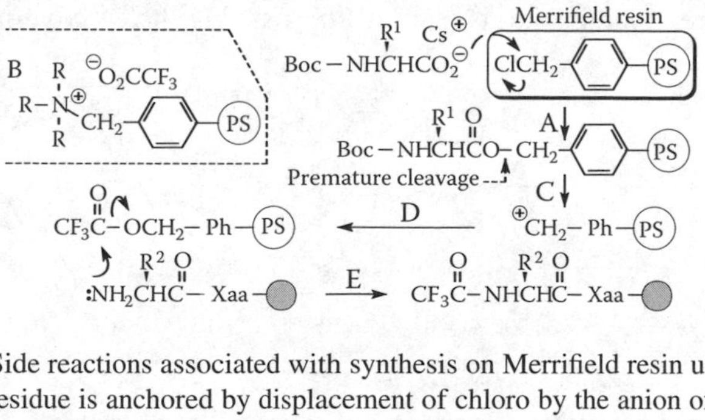

FIGURE 5.14 Side reactions associated with synthesis on Merrifield resin using Boc–amino acids. The first residue is anchored by displacement of chloro by the anion of the cesium salt of the Boc–amino acid (A). Unreacted chloromethyl groups can undergo quaternization by the tertiary amine employed for neutralization generating an exchanger of trifluoroacetate anion (B). Premature release of the peptide chain during treatment with CF_3CO_2H (C) generates trifluoroacetoxymethyl-resin (D) which leads to chain termination due to *O*-to-*N* transfer of the acyl group (E).

the hot ninhydrin solution used for assay of amino groups produces a blue color because of the binding of the anionic chromophore by the pyridinium cation (see Section 5.6). The color can be removed, however, by displacement of the anion by triethylamine hydrochloride in dichloromethane. Linkage through the benzyl ester is slightly sensitive to the deprotecting acid. Premature cleavage of the ester generates the benzyl-resin cation, which reacts with trifluoroacetate, with the acyl group then transferring to any amino group in the vicinity (Figure 5.14, D) as soon as it is deprotonated, thus terminating chain assembly after each deprotection step. These shortcomings prompted the Merrifield group to develop a modified linker, the use of which would eliminate the problems (see Section 5.17).[13]

13. VK Sarin, SBH Kent, JP Tam, RB Merrifield. Quantitative monitoring of solid-phase peptide synthesis by the ninhydrin reaction. *Anal Biochem* 117, 147, 1981.

5.17 PHENYLACETAMIDOMETHYL RESIN FOR SYNTHESIS OF PEPTIDES USING BOC/BZL CHEMISTRY

To overcome the deficiencies associated with chloromethyl resin (see Section 5.16), the Merrifield group developed the methyl-bearing phenylacetamidomethylpolystyrene support, better known as PAM resin, in which the first amino acid is bound through a benzyl ester that has been stabilized by inserting an acetamidomethyl group between the phenyl ring and the polymer (Figure 5.15). The *para*-insertion renders the new linkage 100 times more stable to acid than a benzyl ester. In addition, the work introduced the concept of first binding the protected residue with the linker before securing the two to the support (see Figure 5.11, 3). Connection to the support is accomplished by coupling the carboxyl group of the linker with the aminomethyl group of the polymer — a reaction that can be achieved quantitatively. The protected amino acid-linker combination is obtained by reacting the carboxy-protected linker

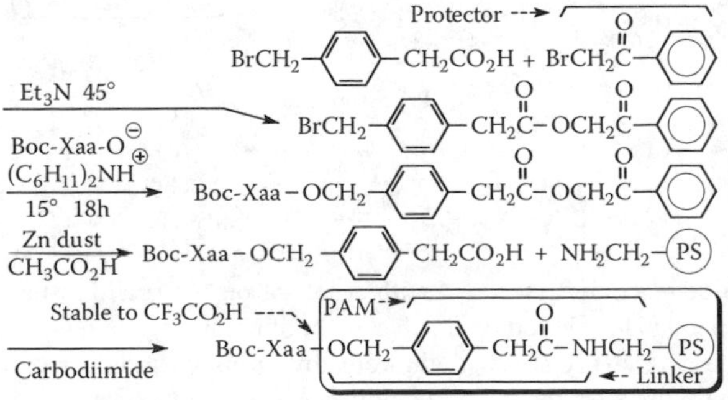

Protector
$BrCH_2-C_6H_4-CH_2CO_2H + BrCH_2C(=O)-C_6H_5$
Et_3N 45°
$BrCH_2-C_6H_4-CH_2C(=O)-OCH_2C(=O)-C_6H_5$
Boc-Xaa-$O^{\ominus}$ $(C_6H_{11})_2NH^{\oplus}$ 15° 18h
Boc-Xaa$-OCH_2-C_6H_4-CH_2C(=O)-OCH_2C(=O)-C_6H_5$
Zn dust / CH_3CO_2H
Boc-Xaa$-OCH_2-C_6H_4-CH_2CO_2H + NH_2CH_2-$(PS)
Stable to CF_3CO_2H
Carbodiimide
PAM
Boc-Xaa$-OCH_2-C_6H_4-CH_2C(=O)-NHCH_2-$(PS)
Linker

FIGURE 5.15 Synthesis of PAM (phenylacetamidomethyl) resin. (Merrifield et al., 1976). Use of PAM resin implies a Boc–amino acid anchored to oxymethylphenylacetamidomethyl-polystyrene through an ester linkage. The acetamido group renders the ester more stable to acid.

with the protected amino acid, followed by deprotection of the terminal carboxyl group (Figure 5.15). Use of a PAM resin usually implies purchase of the support to which a Boc–amino acid is already attached. In addition to the enhanced stability of the anchoring bond, a significant feature of the use of PAM resin is that the Boc–amino acid–linker combination is isolated and purified before it is fixed to the support. More recently the self-standing linkers 4-hydroxy- and 4-bromomethylphenylacetic acid became available. Coupling of these to aminomethyl resin produces functionalized methyl-PAM resin; loading of the Boc–amino acid is achieved by an esterification reaction (see Section 5.22). Regardless of the approach, there are no chloromethyl groups on PAM resin, so the side reaction of quaternization and its implications (see Section 5.16) are not an issue.[31,32]

31. AR Mitchell, SBH Kent, M Englehard, RB Merrifield. A new synthetic route to *tert*-butoxycarbonylaminoacyl-4-(oxymethyl)phenacetamidomethyl-resin, an improved support for solid-phase peptide synthesis. *J Org Chem* 43, 2845, 1978.
32. ML Valero, E Giralt, D Andreu. Solid phase-mediated cyclization of head-to-tail peptides. Problems associated with racemization. *Tetrahedron Lett* 37, 4229, 1996.

5.18 BENZHYDRYLAMINE RESIN FOR SYNTHESIS OF PEPTIDE AMIDES USING BOC/BZL CHEMISTRY

The first peptide amides prepared by solid-phase synthesis were obtained by ammonolysis of resin-bound benzyl esters of peptides in solvents containing methanol (Figure 5.16, A). The method was occasionally employed but was not popular because it was inefficient, producing some ester in addition to the amide. A new variant employing gaseous ammonia will likely rekindle this approach (see Section 8.3). During the early developments of solid-phase synthesis, it was known that the

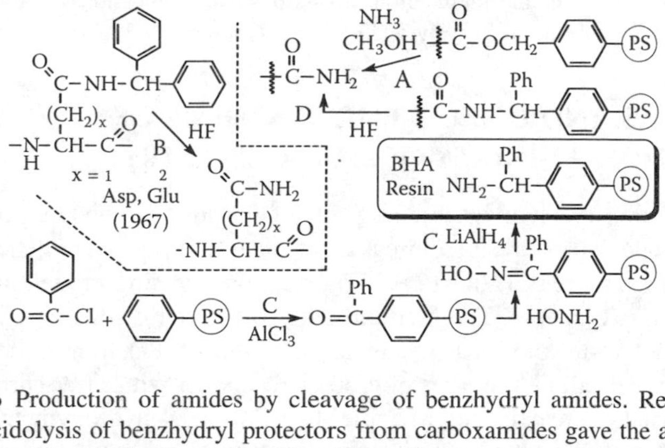

FIGURE 5.16 Production of amides by cleavage of benzhydryl amides. Recognition that removal by acidolysis of benzhydryl protectors from carboxamides gave the amides (B) led to development of benzhydrylamine (BHA) resin (C).[33] Treatment with HF of a peptide amide that has been assembled on a BHA resin using Boc/Bzl chemistry gives the peptide amide (D). Peptide amide is also obtainable by ammonolysis of the resin-bound benzyl ester (A), a reaction that is more efficient if gaseous NH_3 is employed (see Section 8.3).

carboxamido groups of asparagine and glutamine could be protected as the diphenylmethyl derivatives, with the protectors being removed by hydrogen fluoride (Figure 5.16, B). A resin for the synthesis of amides was developed on the basis of this information. The linker consists of benzylamine, with a phenyl ring attached to the methylene carbon. The ensuing diphenylmethyl moiety is known as benzhydryl, and hence the support is benzyhydrylamine, or BHA, resin. Note that the protector or linker is a benzyl whose sensitivity to acid has been increased by the adjacent phenyl ring. Analogous linkers are sensitized further by methoxylation or other (Figure 5.18). Benzhydrylamine resin is obtained by the aluminum chloride–catalyzed Friedel-Craft acylation reaction of benzoyl chloride with polystyrene resin (Figure 5.16, C). The resulting bound diphenyl ketone is converted by hydroxylamine to the oxime, which is then reduced to the substituted aminomethyl resin. This sequence of reactions has emerged as standard for the production of several modified benzhydryl resins. Preparation of amides on benzhydrylamine resin eliminates the side reactions associated with excess chloromethyl groups (see Section 5.16), as well as the obstacle of piperazine-2,5-dione formation after insertion of the second residue (see Section 6.19) that accompanies synthesis through the benzyl ester (Figure 5.16, A). A sensitized variant of BHA resin that gives higher yields of products for peptides with more bulky carboxy-terminal residues is 4-methylbenzhydrylamine, or MBHA, resin. The availability of benzhydrylamine resins provided a major advance in the synthesis of peptide amides.[33–35]

33. PG Pietta, PF Cavallo, K Takahashi, GR Marshall. Preparation and use of benzhydrylamine polymers in peptide synthesis. II. Syntheses of thyrotropin releasing hormone, thyrocalcitonin 26-32, and eledoisin. *J Org Chem* 39, 44, 1974.
34. GR Matsueda, J Stewart. A *p*-methylbenzhydrylamine resin for improved solid-phase synthesis of peptide amides. *Peptides* 2, 45, 1981.
35. JH Adams, M Cook, D Hudson, V Jammalamadaka, MH Lyttle, MS Songster. A reinvestigation of the preparation, properties and applications of aminomethyl and 4-methylbenzylamine polystyrene resins. *J Org Chem* 63, 3706, 1998.

5.19 RESINS AND LINKERS FOR SYNTHESIS OF PEPTIDES USING FMOC/TBU CHEMISTRY

Fmoc/tBu chemistry involves assembly of a chain that is anchored to the support by an ester bond that does not require a strong acid for its cleavage. The usual benzyl ester is rendered more sensitive to acid by electron-donating groups such as alkoxy, phenyl, and alkoxyphenyl. The first and simplest carrier of this type is the alkoxybenzyl alcohol resin known as Wang resin (Figure 5.17). It was developed as a support that would allow synthesis of protected peptides with a free carboxy terminus and is obtained by reaction of Merrifield resin with 4-hydroxymethylphenol and sodium methoxide in hot dimethylacetamide. It was employed for synthesis using Bpoc/Bzl chemistry (see Section 3.19), with the last residue being incorporated as the benzyloxycarbonyl derivative. Treatment with trifluoroacetic acid gave the peptide with all protectors intact. Fmoc/tBu chemistry was adapted for use in solid-phase synthesis employing Wang resin, as well as the equivalent stand-alone linker,

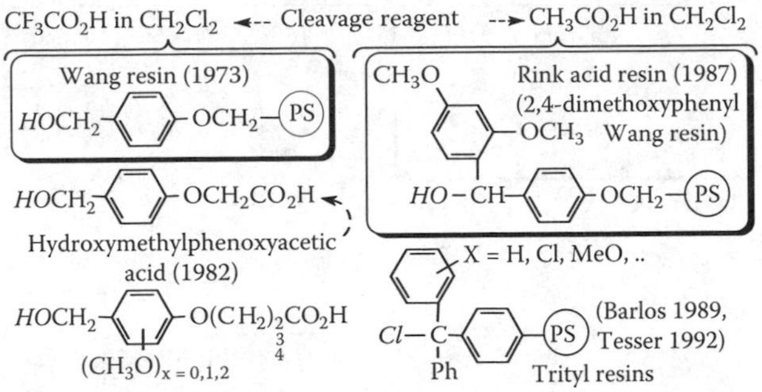

FIGURE 5.17 Resins and linkers for synthesis of peptides using Fmoc/tBu chemistry. The linkers are secured to supports by reaction with aminomethyl resins. A protected amino acid is anchored to the support as an ester by reaction with a hydroxyl or chloro group (italicized). The alkoxy and phenyl substituents render the benzyl esters sensitive to the cleavage reagents.

hydroxymethylphenoxyacetic acid (Figure 5.17), affixed to polyacrylamide resin (see Section 5.8). A variety of more-sensitized linkers, both stand-alone and fixed to resins (Figure 5.17) have emerged over the years. All incorporate substituents that modify the sensitivity of the ester bond to acid. When a linker such as that on Rink acid resin creates an ester bond that is too sensitive to the acid employed, one has the option of using the equivalent linker secured to the support through an amide bond, which stabilizes the ester, as is the case for PAM resin (see Section 5.17). An example is Rink acid aminomethyl resin. A reference amino acid can be inserted between the linker and the handle on the resin.[36–38]

36. S-S Wang. *p*-Alkoxybenzyl alcohol resin and *p*-alkoxybenzyloxycarbonyl-hydrazide resin for solid phase synthesis of protected peptide fragments. *J Am Chem Soc* 95, 1328, 1973.
37. RC Sheppard, B Williams. Acid-labile resin linkage agents in solid phase peptide synthesis. *Int J Pept Prot Res* 20, 451, 1982.
38. H Rink. Solid-phase synthesis of protected peptide fragments using trialkoxy-diphenyl-methylester resin. *Tetrahedron Lett* 28, 3787, 1987.

5.20 RESINS AND LINKERS FOR SYNTHESIS OF PEPTIDE AMIDES USING FMOC/TBU CHEMISTRY

Amides became accessible by solid-phase synthesis, using Fmoc/tBu chemistry, about 7 years after the latter was introduced. Presently in use for preparing amides are variants primarily of benzylhydrylamine resins and linkers (see Section 5.18), in which the carboxamido–methyl bonds anchoring the peptide chains have been rendered sensitive to trifluoroacetic acid by alkoxy groups, with the handles being attached to supports through an oxyalkyl group (Figure 5.18). All are presented as the Fmoc derivatives that are generated by reaction of the hydroxymethyl precursor with excess Fmoc-amide; the amino group is liberated for synthesis. Rink amide resin is 2,4-dimethoxy-substituted benzhydrylamine affixed to oxymethyl resin, whereas Sieber amide resin has the two phenyl rings of benzhydryl joined at the

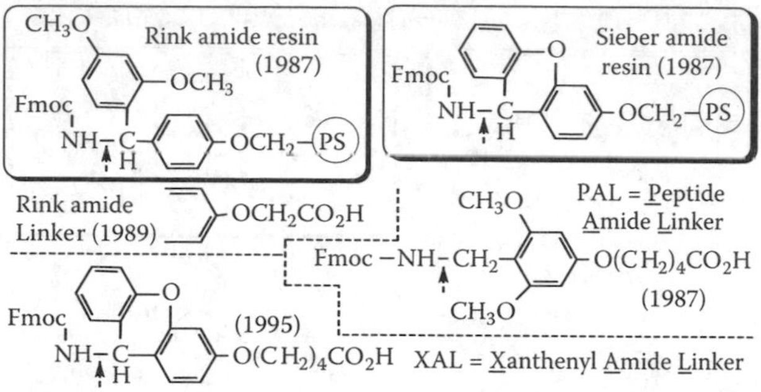

FIGURE 5.18 Resins and linkers for synthesis of peptide amides using Fmoc/tBu chemistry. Chain assembly is effected after removal of the Fmoc group. Treatment with CF_3CO_2H releases a peptide amide by cleavage at the NH–CH/CH_2 bond.

ortho positions through an oxygen atom, giving a xanthenyl moiety that is also affixed to oxymethyl resin. The corresponding stand-alone linkers secured to supports through amide bonds are also available. Use of the latter produces anchoring bonds of peptides that are slightly more stable to acid because the oxymethyl substituent is replaced by oxyacetamidomethyl, the same phenomenon observed for PAM resin, in comparison to Merrifield resin (see Section 5.17). The linker known as PAL is 2,6-dimethoxybenzylamine with oxypentanoic acid at position 4, and the linker known as XAL is 9*H*-xanthen-9-ylamine with oxypentanoic acid *meta* to the oxygen atom. The longer alkyl group makes the pertinent carboxyamido-methyl bond more acid-labile than it would be if the substituent were oxyacetic acid. The carboxyamido-methyl bond of the XAL moiety is much more sensitive to acid than that of the PAL moiety. An alternative to these more expensive linkers and resins is 4-methylbenzhydrylamine resin, from which a chain can be detached using trifluoromethanesulfonic acid. Use of a two-step deprotection procedure minimizes side reactions.[34,38–43]

34. GR Matsueda, J Stewart. A *p*-methylbenzhydrylamine resin for improved solid-phase synthesis of peptide amides. *Peptides* 2, 45, 1981.
38. H Rink. Solid-phase synthesis of protected peptide fragments using trialkoxy-diphenyl-methylester resin. *Tetrahedron Lett* 28, 3787, 1987.
39. P Sieber. A new acid-labile anchor group for the solid-phase synthesis of C-terminal peptide amides by the Fmoc method. *Tetrahedron Lett* 28, 2107, 1987.
40. MS Bernatowicz, SB Daniels, KH Köster. A comparison of acid labile linkage agents for the synthesis of peptide C-terminal amides. *Tetrahedron Lett* 30, 4645, 1989.
41. F Albericio, N Kneib-Cordonier, S Biancalana, L Gera, RI Masada, D Hudson, G Barany. Preparation and application of the 5-(4(9-fluorenylmethyloxycarbonyl)aminomethyl-3,5-dimethoxyphenoxy)-valeric acid (PAL) handle for solid-phase synthesis of C-terminal peptide amides under mild conditions. *J Org Chem* 55, 3730, 1990.
42. Y Han, SL Bontems, P Hegyes, MC Munson, CA Minor, SA Kates, F Albericio, G Barany. Preparation and application of xanthenylamide (XAL) handles for solid-phase synthesis of C-terminal peptide amides under particularly mild conditions. *J Org Chem* 61, 6326, 1996.

43. PE Thompson, HH Keah, PT Gomme, PG Stanton, MTW Hearn. Synthesis of peptide amides using Fmoc-based solid-phase procedures on 4-methylbenzhydrylamine resins. *Int J Pept Prot Res* 46, 174, 1995.

5.21 RESINS AND LINKERS FOR SYNTHESIS OF PROTECTED PEPTIDE ACIDS AND AMIDES

The synthesis of protected peptide acids and amides requires detachment of a chain from the resin without affecting the protectors. Chain assembly on a Wang resin using Bpoc/Bzl chemistry (see Section 3.22; Figure 5.19, A) with a Cbz-amino acid as the terminal residue followed by release of the chain with trifluoroacetic acid was the first approach available for preparing protected peptide acids. One unique approach allows synthesis of protected peptides using Boc/Bzl-chemistry, and that is use of a substituted oxime resin (Figure 5.19, B). The chain is constructed starting with the penultimate residue. The final step of the synthesis involves detachment of the chain by nucleophilic displacement by the amino group of the carboxy-terminal residue. The product is the target peptide. The displacing residue can be the amino acid anion, ester, or amide, thus rendering this approach highly versatile. The method is incompatible with Fmoc chemistry, which involves repeated use of a nucleophile, and it is prone to piperazine-2,5-dione formation (see Section 6.19) at the time of coupling to the dipeptide ester. It is compatible, however, with onium salt-based coupling reagents (see Section 2.16). An alternative for making protected peptide

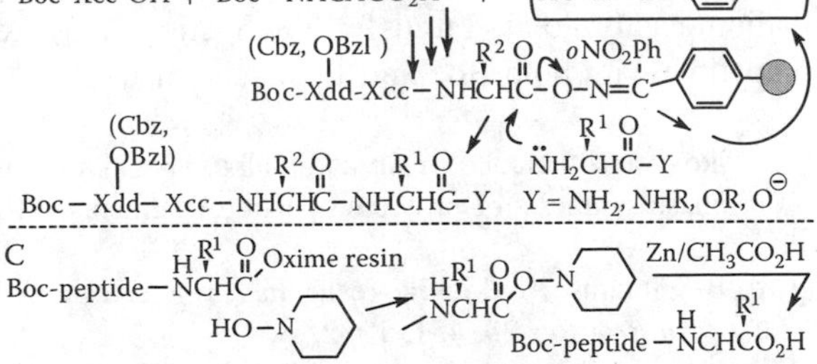

FIGURE 5.19 Approaches for synthesis of protected peptides using acid-sensitive protectors. For (A) the last residue is incorporated as the Cbz-derivative. (B) The chain is assembled on a substituted oxime resin,[44] starting at the penultimate residue. The chain is then detached from the resin by displacement by the carboxy-terminal residue of the target peptide, which may be a free (*tert*-butylammonium salt) or carboxy-substituted residue. (C) An alternative for producing acids by use of the oxime resin is transesterfication of the peptide with *N*-hydroxypiperidine, followed by reduction to remove the piperidino substituent.

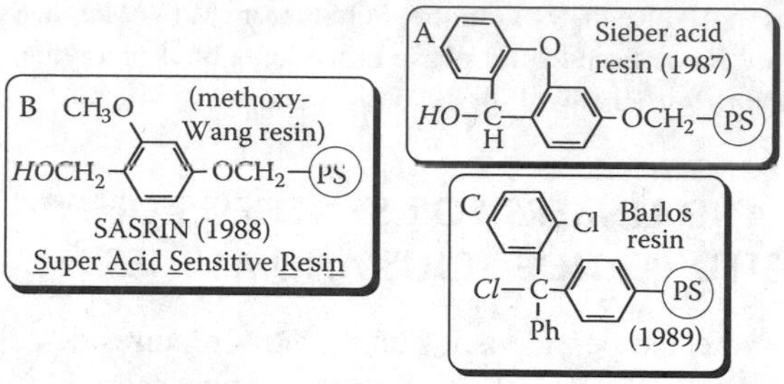

A Sieber acid resin (1987)
B (methoxy-Wang resin) SASRIN (1988) Super Acid Sensitive Resin
C Barlos resin (1989)

FIGURE 5.20 Resins for the synthesis of protected peptides using Fmoc/tBu chemistry. The first residue is esterified to the handle by reaction with the italicized functional group. The protected peptide is detached by cleavage of the ester bond with 1% CF_3CO_2H for (A) and (B) and 10% CF_3CO_2H for (C).

acids on the oxime resin is assembly of the whole chain followed by transesterification to the chirally stable piperidino ester by the addition of *N*-hydroxypiperidine (Figure 5.19, C). Cleavage of the ester by reduction with zinc dust in acetic acid gives the acid. The oxime resin is obtained from polystyrene (Figure 5.19, B), according to the general procedure (Figure 5.16), through the nitrobenzophenone (diphenylketone). A linker equivalent to that of the oxime resin is 4-hydroxymethylbenzamidomethyl. It provides a benzyl ester that is stable to acid but sensitive to nucleophiles, thus allowing preparation of amides, esters, and hydrazides of protected peptides.

There are several options for preparing protected peptide acids using Fmoc/tBu chemistry, including Sieber acid resin, the resin known as SASRIN, and Barlos resin (Figure 5.20). The highly sensitized ester linkages anchoring the first residues are cleavable by 1% trifluoroacetic acid. The linkages are also sensitive to 1-hydroxybenzotriazole if tertiary amine is not present. Protected peptide amides are accessible using the linker or resin, incorporating the *para*-alkoxyxanthenyl amide moieties (Figure 5.18). Linkers incorporating the allyl group in which the anchoring bond is cleavable by palladium-catalyzed allyl transfer (see Section 3.13) are also available. The latter are compatible with both Boc and Fmoc chemistry.[36,37,39,42,44–48]

36. S-S Wang. *p*-Alkoxybenzyl alcohol resin and *p*-alkoxybenzyloxycarbonyl-hydrazide resin for solid phase synthesis of protected peptide fragments. *J Am Chem Soc* 95, 1328, 1973.
37. RC Sheppard, B Williams. Acid-labile resin linkage agents in solid phase peptide synthesis. *Int J Pept Prot Res* 20, 451, 1982.
39. P Sieber. A new acid-labile anchor group for the solid-phase synthesis of C-terminal peptide amides by the Fmoc method. *Tetrahedron Lett* 28, 2107, 1987.
42. Y Han, SL Bontems, P Hegyes, MC Munson, CA Minor, SA Kates, F Albericio, G Barany. Preparation and application of xanthenylamide (XAL) handles for solid-phase synthesis of C-terminal peptide amides under particularly mild conditions. *J Org Chem* 61, 6326, 1996.
44. WF DeGrado, ET Kaiser. Polymer-bound oxime esters as supports for solid-phase peptide synthesis. Preparation of protected fragments. *J Org Chem* 45, 1295, 1980.

45. Nakagawa, ET Kaiser. Synthesis of protected peptide segments and their assembly on a polymer-bound oxime: application to the synthesis of a peptide model for plasma apolipoprotein A-I. *J Org Chem* 48, 678, 1983.
46. M. Megler, R Tanner, J Gosteli, P Grogg. Peptide synthesis by a combination of solid phase and solution methods. 1. A new very acid-labile anchor group for the solid phase synthesis of protected fragments on 2-methoxy-4-alkoxy-benzyl resins. *Tetrahedron Lett* 29, 4005, 1988.
47. H Kunz, B Dombo. Solid phase synthesis of peptides and glycopeptides on polymeric supports with allylic anchor groups. *Angew Chem Int Edn Engl* 27, 711, 1988.
48. RB Scarr, MA Findeis. Improved synthesis and aminoacylation of *p*-nitrobenzophenone oxime polystyrene resin for solid-phase synthesis of protected peptides. *Pept Res* 3, 238, 1990.

5.22 ESTERIFICATION OF FMOC-AMINO ACIDS TO HYDROXYMETHYL GROUPS OF SUPPORTS

During the first decade when solid-phase synthesis was executed using Fmoc/tBu chemistry, the first Fmoc–amino acid was anchored to the support by reaction of the symmetrical anhydride with the hydroxymethylphenyl group of the linker or support. Because this is an esterification reaction that does not occur readily, 4-dimethylaminopyridine was employed as catalyst. The basic catalyst caused up to 6% enantiomerization of the activated residue (see Section 4.19). Diminution of the amount of catalyst to one-tenth of an equivalent (Figure 5.21, A) reduced the isomerization substantially but did not suppress it completely. As a consequence, the products synthesized during that decade were usually contaminated with a small amount of the epimer. In addition, the basic catalyst was responsible for a second side reaction; namely, the premature removal of Fmoc protector, which led to loading of some dimer of the first residue. Nothing could be done about the situation,

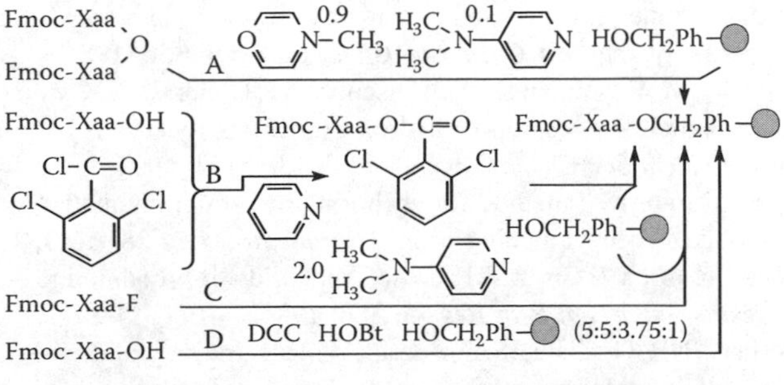

FIGURE 5.21 Methods for anchoring an Fmoc–amino acid to the hydroxymethyl group of a linker-resin. (A) 4-Dimethylaminopyridine-catalyzed acylation by the symmetrical anhydride.[19] (B) Acylation by a mixed anhydride obtained from 2,6-dichlorobenzoyl chloride.[39] (C) Acylation by the acid fluoride.[50] (D) Dicyclohexylcarbodiimide-mediated acylation in the presence of 1-hydroxybenzotriazole.[52]

however, because there was no better procedure available. Other methods of esterification are now in common use, including the reaction of the mixed anhydride formed from 2,6-dichlorobenzoyl chloride (Figure 5.21, B), the reaction of the Fmoc–amino acid fluoride (see Section 7.12) in the presence of two equivalents of 4-dimethylaminopyridine (Figure 5.21, C), and reaction of a three- to fourfold excess of the Fmoc–amino acid mediated by the same amount of dicyclohexylcarbodiimide in a weakly polar solvent in the presence of 0.75 equivalents of 1-hydroxybenzotriazole relative to the amount of substrate (Figure 5.19, D). The stoichiometries indicated in Figure 5.21 for C and D are crucial for success; rationalization for this is, however, unavailable. None of these methods produces any dimer. Preservation of chirality in the 4-dimethylaminopyridine-catalyzed reaction is attributed to the fact that esterification occurs in a few minutes. Loading of derivatives of cysteine and histidine, however, is not straightforward, as isomerization occasionally occurs.

An alternative to the above is esterification by reaction of the salt of the Fmoc–amino acid with the halomethylphenyl-support (see Section 3.17). It was established in the 1960s that this method of esterifying *N*-alkoxycarbonylamino acids, which does not involve electrophilic activation, is not accompanied by enantiomerization. Examples of supports with haloalkyl linkers are bromomethylphenoxymethyl-polystyrene and 2-chlorotrityl chloride resin (see Section 5.23).

The extent of loading of the first residue can be established by treating an aliquot of resin-bound Fmoc–amino acid with piperidine and measuring the absorbance (extinction coefficient = 7800) at 301 nm of a solution of the 9-methylfluorene-piperidine adduct that is liberated. Enantiomerization during anchoring can be established by analyzing for diastereomeric products produced by reacting the amino acid with a chiral reagent after its deprotection and detachment from the support, or after release of the dipeptide from the support if the reagent is an N^{α}-protected amino acid (see Section 4.24).[49–54]

49. E Atherton, NL Benoiton, E Brown, RC Sheppard, B Williams. Racemisation of activated, urethane-protected amino-acids by *p*-dimethylaminopyridine. Significance in solid-phase synthesis. *J Chem Soc Chem Commun* 336, 1981.
50. D Granitza, M Beyermann, H Wenschuh, H Haber, LA Carpino, GA Truran, M Bienert. Efficient acylation of hydroxy functions by means of Fmoc amino acid fluorides. *J Chem Soc Chem Commun* 2223, 1985.
51. P Sieber. An improved method for anchoring of 9-fluorenylmethoxycarbonyl-amino acids to 4-alkoxybenzyl alcohol resins. *Tetrahedron Lett* 28, 6147, 1987.
52. A Grandas, X Jorba, E Giralt, E Pedroso. Anchoring of Fmoc-amino acids to hydroxymethyl resins. *Int J Pept Prot Res* 33, 386, 1987.
53. JW Corbett, NR Graciani, SA Mousa, WF DeGrado. Solid-phase synthesis of a selective α,β integrin antagonist library. (bromomethylphenoxymethyl-polystyrene resin) *Bioorg Medicinal Chem* 7, 1371, 1997.
54. JG Adamson, T Hoang, A Crivici, GA Lajoie. Use of Marfey's reagent to quantitate racemization upon anchoring of amino acids to solid supports for peptide synthesis. *Anal Biochem* 202, 210, 1992.

5.23 2-CHLOROTRITYL CHLORIDE RESIN FOR SYNTHESIS USING FMOC/TBU CHEMISTRY

The availability of triphenylmethyl-based (see Section 3.19) protectors that are stable to base but sensitive to weak acid prompted investigation of trityl-based linkers for solid-phase synthesis. 2-Chlorotrityl chloride resin (Figure 5.22) emerged as the support with the most favorable properties. It is obtained by the aluminum chloride–catalyzed acylation of polystyrene by 2-chlorobenzoyl chloride (see Figure 5.16, C), followed by reaction of the chlorobenzophenone with phenylmagnesium bromide. The 2-chlorotrityl carbinol produced is converted to the chloride by acetyl chloride. 2-Chlorotrityl chloride resin reacts very efficiently with Fmoc–amino acid anions in weakly polar solvents, in which it exists in equilibium with the cation (Figure 5.22, A). Esterification occurs without enantiomerization (see Section 5.22). Unreacted chloromethyl groups are eliminated by conversion to methoxymethyl by reaction with methanol in the presence of diisopropylethylamine. The ester bond is sensitive to weak acid, the common cleavage reagent being 10% acetic acid in dichloromethane-trifluoroethanol (7:2), the alcohol serving to trap the cation released by acidolysis (Figure 5.22). *tert*-Butyl-based protectors are stable to this reagent, so the resin allows access to protected peptide acids as well as peptides using Fmoc/tBu chemistry (Figure 5.22, B). An interesting feature is that a peptide containing two *S*-trityl-cysteine residues that is detached from the resin is converted directly into the disulfide by the reagent if iodine (see Section 6.17) has been added (Figure 5.22, B). In addition, the 2-chlorotrityl linker possesses properties that are shared only with *tert*-butyl-based linkers. At the dipeptide stage of chain assembly, no piperazine-2,5-dione is formed, as is the case for other dipeptide esters, which have a tendency to cyclize (see Section 6.19). Similarly, peptide chains attached to a support through

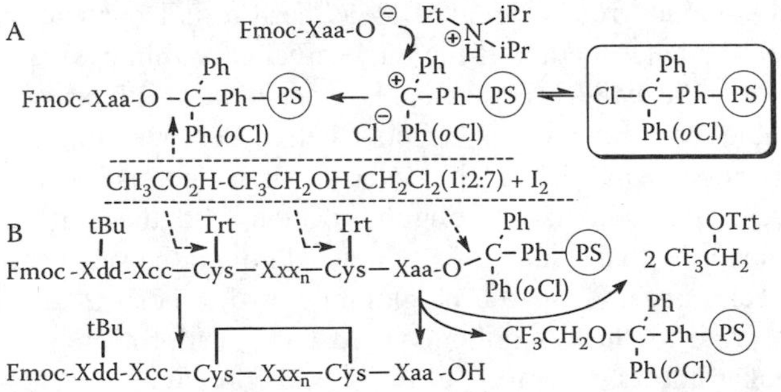

FIGURE 5.22 (A) Reaction of an Fmoc–amino acid with 2-chlorotrityl chloride resin.[56] The ester bond formed is cleavable by the mild acid, which does not affect *tert*-butyl-based protectors. (B) Generation of a protected peptide containing cystine by detachment of a chain, deprotection of cysteine residues, and oxidation of the sulfhydryls by the reagent containing iodine. The cations produced are trapped by CF_3CH_2OH.

the 2-chlorotrityl ester of cysteine are not subject to the epimerization that is caused by piperidine at cysteine residues attached to supports through other alkyl esters (see Section 8.1). 2-Chlorotrityl chloride resin is readily transformed into 2-chlorotrityl amine resin. Acylation of the latter with an amide linker (see Section 5.20) provides a support for the synthesis of amides.[55–58]

55. JMJ Fréchet, LJ Nuyens. Use of polymers as protecting groups in organic synthesis. III. Selective functionalization of polyhydroxy alcohols. *Can J Chem* 54, 926, 1976.
56. K Barlos, O Chatzi, D Gatos, G Stavropoulus. 2-Chlorotrityl chloride resin. Studies on anchoring of Fmoc-amino acids and peptide cleavage. *Int J Pept Prot Res* 37, 513, 1991.
57. K Barlos, D Gatos, S Kutsogianni, G Papaphotiou, C Poulus, T Tsegenidis. Solid phase synthesis of partially protected and free peptides containing disulphide bonds by simultaneous cysteine oxidation-release from 2-chlorotrityl resin. *Int J Pept Prot Res* 38, 562, 1991.
58. A van Vliet, RH Smulders, BH Rietman, GI Tesser. Protected peptide intermediates using a trityl linker on a solid support, in R Epton, ed. *Innovations and Perspectives in Solid Phase Synthesis. Proceedings of the 2nd Symposium.* Intercept, Andover, 1992, pp 475-477.

5.24 SYNTHESIS OF CYCLIC PEPTIDES ON SOLID SUPPORTS

Cyclic peptides have been synthesized in solution by activation of a linear peptide by the addition of diphenyl phosphorazidate [$(PhO)_2P(=O)N_3$); see Section 7.16] or by liberation of the amino group of a Boc-peptide pentafluorophenyl ester, followed by deprotonation of the amino group. Reaction between the activated carboxyl groups and the amino groups, referred to as head-to-tail cyclization, produces a cyclic peptide. The latter method is the more efficient. Aminolysis is carried out in dilute solution to try to avoid dimerization. Linear peptides with protected side chains are more soluble, so cyclization is effected before deprotection. Cyclization of a peptide that is bound to a resin is, however, a better approach because it is more efficient, there is minimal risk of oligomerization, and the reagents are readily eliminated. Anchoring to the support is achieved through the side chain of a trifunctional amino acid, usually aspartic or glutamic acid, whose α-carboxyl group is blocked by a protector that is orthogonal to the other protectors. Choice of the appropriate linker allows generation of either aspartic/glutamic acid–containing or asparagine/glutamine–containing peptides. Examples appear in Figure 5.23. For production of acids, the ω-carboxyl groups are best reacted as the cesium salts with bromomethyl linkers (see Section 3.17), as 4-dimethylaminopyridine-catalyzed esterification to hydroxymethyl linkers leads to enantiomerization of the residues (see Section 4.19). Both Boc/Bzl and Fmoc/tBu chemistries are applicable; the former has been established as superior for the synthesis of acids. Cleavage of the 9-fluorenylmethyl ester with piperidine leaves the carboxyl group as the piperidine salt. The secondary amine should be displaced by 1-hydroxybenzotriazole or a

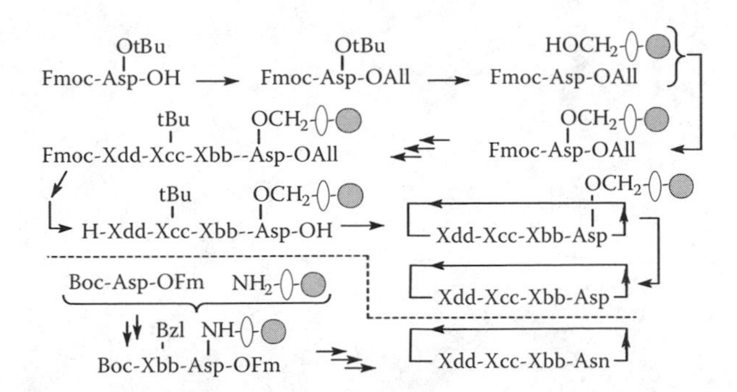

FIGURE 5.23 Synthesis of cyclic peptides by head-to-tail cyclization of resin-bound peptides using Boc/Bzl chemistry[59] and Fmoc/tBu chemistry.[60] The carboxy-terminal protectors are orthogonal to the other protectors. The nature of the linker determines the nature of the product. Both chemistries are compatible with the two types of linkers. All = allyl.

tertiary amine, so that no piperidide is formed during cyclization, or the non-nucleophilic base 1,8-diazabicyclo[5.4.0]undec-7-ene (see Section 8.12) should be used for the deprotection. Less oligomerization occurs when the support is polystyrene than when it is a polyethyleneglycol–polystyrene copolymer. Several onium salt-based reagents have been employed for activation. More than one report indicates that the tetrafluoroborate salt of *O*-benzotriazol-1-yl-*N,N,N′,N′*-tetramethyluronium (TBTU) gives better results than the hexafluorophosphate salt (HBTU). In any event, chiral preservation during bond formation is not guaranteed because cyclization involves activation of a segment (see Section 4.4).[31,59–62]

31. AR Mitchell, SBH Kent, M Englehard, RB Merrifield. A new synthetic route to *tert*-butoxycarbonylaminoacyl-4-(oxymethyl)phenacetamidomethyl-resin, an improved support for solid-phase peptide synthesis. *J Org Chem* 43, 2845, 1978.
59. P Rovero, L Quartara, G Fabbri. Synthesis of cyclic peptides on solid support. *Tetrahedron Lett* 23, 2639, 1991.
60. A Trzeciak, W Bannwarth. Synthesis of ‘head-to-tail’ cyclic peptides on solid support by FMOC chemistry. *Tetrahedron Lett* 33, 4557, 1992.
61. W Neugebauer, J-P Gratton, M Ihara, G Bkaily, P D’Orléans-Juste. Solid phase synthesis of head-to-tail cyclic peptide — ET_A antagonist, in R Ramage, R Epton, eds. *Peptides 1996. Proceedings of the 24th European Peptide Symposium.* Mayflower, Kingswinford, 1998, pp 677-678.
62. M-L Valero, E Giralt, D Andreu. A comparitive study of cyclization strategies applied to the synthesis of head-to-tail cyclic analogs of a viral epitope. *J Pept Res* 53, 56, 1999.

6 Reactivity, Protection, and Side Reactions

6.1 PROTECTION STRATEGIES AND THE IMPLICATIONS THEREOF

Peptide synthesis involves the fusion of two amino acids through an amide bond. Amino acids possess two or three functional groups capable of undergoing reactions. The objective is the formation of the bond that is desired while avoiding reaction at the other functional groups. The tendency for reaction depends on the electrophilicity or nucleophilicity of the groups. The reactivity of a functional group is suppressed by combining it with a protector that is removable preferably without side reactions. Reaction with functional groups of lesser reactivity is less likely if the activation required for peptide-bond formation is moderate. There are three strategies of protection for synthesis, with the adoption of any one having its own implications and consequences. Protection can be maximal, where all functional groups are blocked; minimal, where only functions that have to be blocked are protected; or intermediate, where some groups are blocked during the bond-forming reaction and then unblocked during the subsequent reaction. Functions that have to be blocked are the amino groups in the component to be activated and sulfhydryls.

Protecting or not protecting has implications for the accessibility of starting materials, the solubility of the reacting components and products, the applicability of coupling methods, the number of operational steps, the extent of side reactions, and the ease of purification of intermediates and products. A simpler starting material is less costly and more readily available. The effect on solubility varies. Side-chain protectors improve the solubility of small polar peptides in coupling solvents, but extensive protection on larger molecules can have the opposite effect. Charged or polar groups improve the solubility of larger molecules in the polar solvents used for coupling and in the partially aqueous solvents used for purification. Unprotected functions restrict the use of coupling methods to those involving moderately activated intermediates. Fewer side reactions can be expected during coupling when functional groups are protected. In contrast, unprotected groups eliminate the steps of deprotection as well as the side reactions that might accompany the deprotections. Purification is facilitated by the presence of polar or charged groups. Maximum protection is usually associated with solid-phase synthesis, and minimum protection with synthesis in solution. Examples of functional groups that are initially blocked and then liberated after amide-bond formation are the imidazole of histidine from *bis*-Fmoc-histidine and the phenolic hydroxyl of tyrosine from *bis*-Boc-tyrosine.

How the functional groups on the side chains of residues are protected and the implications of protecting or not protecting during synthesis form the subject of this chapter. Each functional group is addressed in sequence, along with the side reactions in which it can participate. Recall that the standard protectors for the so-called less difficult residues seryl, threonyl, tyrosyl, aspartyl, glutamyl, and lysyl (see Section 3.2) are *tert*-butyl based for Cbz and Fmoc chemistries and benzyl-based for Boc chemistry (see Sections 3.20 and 5.12).[1]

1. JK Inman, in E Gross, J Meienhofer, eds. *The Peptides Analysis, Synthesis, Biology.* Vol 3. *Protection of Functional Groups in Peptide Synthesis*, Academic Press, New York, 1981, pp 254-302.

6.2 CONSTITUTIONAL FACTORS AFFECTING THE REACTIVITY OF FUNCTIONAL GROUPS

A chemical reaction involves a rearrangement of atoms that issues from contact between electron-rich and electron-deficient centers of the two reactants. The reactants may be separate molecules, functions on different residues of a chain, or functions on the same residue. Contact is favored by the appropriate juxtaposition of the participating functional groups, and reaction is encouraged by the driving force to produce five- and six-membered rings, which are ever-present because of their high stability. Pertinent reactions include desired as well as undesired ones, occurring during activation, aminolysis, protection, and deprotection and after deprotection. The reactivity of a function depends first on its nature. An α-amino group is a different entity than an amino group on a side chain because the adjacent atoms are different. The same holds for α- and ω-carboxyl groups (see Section 6.24). Thus, their reactivities are not the same. Second, the reactivity of a particular functional group such as the carboxamido of asparagine is affected by the location of the residue in the peptide chain. Three different situations exist: The functional group can be on the side chain of the residue that is activated, on the side chain of the aminolyzing residue, or on the side chain of an intrachain residue (Figure 6.1, A). Examples of

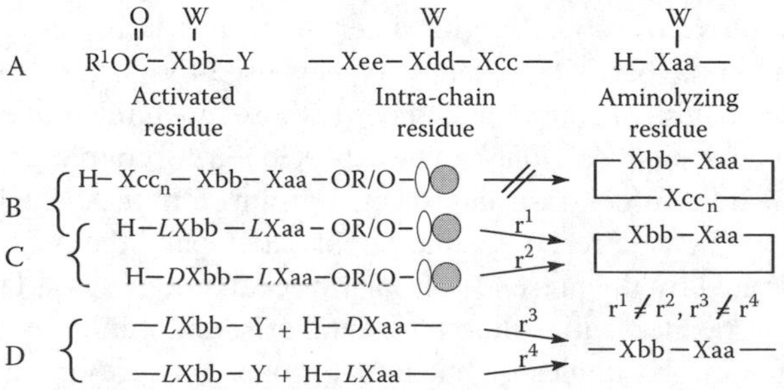

A: $R^1OC(=O)$–Xbb(W)–Y (Activated residue); —Xee–Xdd(W)–Xcc— (Intra-chain residue); H–Xaa(W)— (Aminolyzing residue)

B: H–Xcc_n–Xbb–Xaa–OR/O–resin ⇸ cyclo[Xbb–Xaa–Xcc_n]

C: H–*L*Xbb–*L*Xaa–OR/O–resin $\xrightarrow{r^1}$ cyclo[Xbb–Xaa]; H–*D*Xbb–*L*Xaa–OR/O–resin $\xrightarrow{r^2}$ cyclo[Xbb–Xaa]

$r^1 \neq r^2$, $r^3 \neq r^4$

D: —*L*Xbb–Y + H–*D*Xaa— $\xrightarrow{r^3}$ —Xbb–Xaa—; —*L*Xbb–Y + H–*L*Xaa— $\xrightarrow{r^4}$ —Xbb–Xaa—

FIGURE 6.1 Constitutional factors affecting the reactivity of functional groups. (A) The reactivity of W depends on the location of the residue. (B) The amino group of a dipeptide ester reacts with the ester carbonyl to form a cyclic dipeptide; amino groups of other peptide esters do not react in this manner. (C, D) Reactions between residues of identical configuration do not occur at the same rates as reactions between residues of opposite configuration.

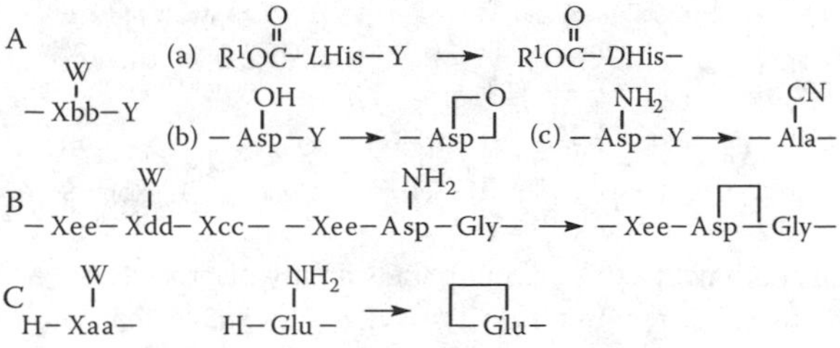

A
W
|
−Xbb−Y

(a) $R^1OC(=O)$−*L*His−Y → $R^1OC(=O)$−*D*His−

(b) −Asp(OH)−Y → −Asp(anhydride, O)−

(c) −Asp(NH_2)−Y → −Ala(CN)−

B
W
|
−Xee−Xdd−Xcc− −Xee−Asp(NH_2)−Gly− → −Xee−Asp−Gly− (imide ring)

C
W
|
H−Xaa− H−Glu(NH_2)− → Glu− (ring)

FIGURE 6.2 Constitutional factors affecting the reactivity of functional groups. Examples of reactions of functional group W that depend on the location of the residue. (A, a) Isomerization of activated histidinyl. (A, b) Activated aspartyl forming the anhydride. (A, c) Activated asparaginyl forming cyanoalaninyl. (B) Intrachain asparaginyl-glycyl forming the imide. (C) Terminal glutaminyl forming pyroglutamyl.

the dramatic differences in reactivity of functional group are illustrated by the reactions in Figure 6.2. Each reaction occurs only when W is on the residue indicated. Third, the reaction between two functional groups in the same molecule is affected by the distance that separates them. Of the compounds indicated in Figure 6.1, B, only the dipeptide ester has a tendency to produce a cyclic peptide (see Section 6.19). The terminal groups of an amino acid ester also do not react with each other. Fourth, the reactivity of functional groups on two different residues is affected by the stereochemistries of the residues, whether in the same molecule (Figure 6.1, C; see Section 6.19) or in different molecules (Figure 6.1, D). Allusion to this phenomenon has been encountered in the discussion on asymmetric induction (see Section 4.6). The generation of 5(4*H*)-oxazolones (see Section 1.7) is another example of a phenomenon occurring because of the tendency to produce rings.

6.3 CONSTITUTIONAL FACTORS AFFECTING THE STABILITY OF PROTECTORS

Protectors are employed to suppress the reactivity of functional groups. To achieve this objective, the protector must be stable to the chemistries to which the molecule is subjected. The stability of a protector depends obviously on the nature of the protector (see Sections 3.4, 3.5, and 3.19). In addition, it depends on the nature of the functional group and, in most instances, its dissociation constant. The more acidic the functional group that is protected, the more sensitive the protector is to acidolysis because it is easier to protonate. This is exemplified by the greater sensitivity of *O-tert*-butyl-tyrosyl relative to *O-tert*-butyl-seryl (Figure 6.3, A), as well as the greater sensitivity of a *tert*-butyl ester relative to a *tert*-butyl ether. Extending this, it follows that an α-*tert*-butyl ester is more sensitive to acid that an ω-*tert*-butyl ester. The same holds for amino groups. Protected α-amino is more sensitive to acid than the protected ε-amino of lysyl (Figure 6.3, C). This has been demonstrated by the preparation of N^{ε}-trityl- and N^{ε}-methyltrityl-lysine by acidolysis of the *bis*-substituted-lysines and by comparison of the rates of removal of N^{α}-biphenylisopropoxycarbonyl (see Section 3.19, Figure 3.23) relative to *tert*-butoxycarbonyl from the disubstituted lysines.

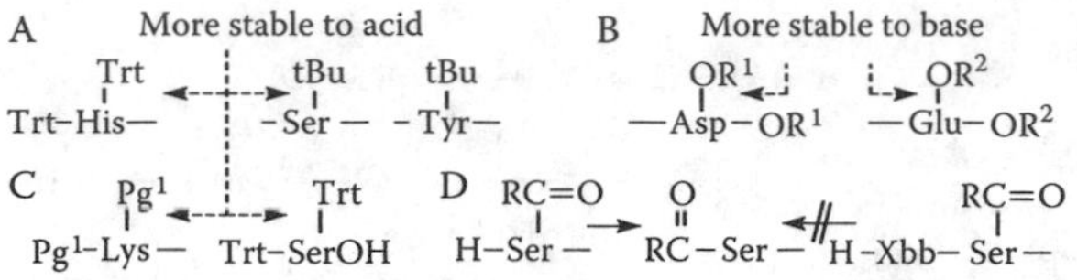

A More stable to acid
Trt tBu tBu
Trt–His– –Ser– –Tyr–
B More stable to base
OR^1 OR^2
–Asp–OR^1 –Glu–OR^2
C Pg^1 Trt
Pg^1–Lys– Trt–SerOH
D RC=O O RC=O
H–Ser– → RC–Ser– ↚ H-Xbb– Ser–

FIGURE 6.3 Constitutional factors affecting the stability of protectors. (A) The nature of the functional group protected. *tert*-Butyl on tyrosyl is less stable than on seryl; trityl on the imidazole of histidyl is more stable than on α-amino. The location of the protectors: (B) the terminal esters of aspartyl and glutamyl are easier to saponify than the side-chain esters. (C) Side-chain trityl and urethane substituents are more stable to acid than α-amino substituents. (D) The acyl group of seryl will migrate only to the amino group of the same residue. (E) As indicated in Figure 6.1, B, only the ester of a dipeptide reacts to form a cyclic peptide. Pg = protecting group.

N,O-di-Substituted serine and threonine can also be used to prepare the *O*-triphenylmethylamino acids. A similar phenomenon obtains for sensitivities to bases. A Dde protector (see Section 6.4) on the side chain of lysyl is more stable to piperidine than the protector on the α-amino function, and esters of the more acidic α-carboxylic groups are more readily attacked by hydroxide ions than esters of side-chain carboxylic groups (Figure 6.3, B). A third factor affecting the stability of a protector is the presence of a reactive site in the vicinity. An acyl group on the side chain of seryl or threonyl migrates to the unprotonated amino group of the same residue (Figure 6.3, D), through a ring intermediate (see Section 6.6); a dipeptide ester has a tendency to cyclize to the dilactam (Figure 6.1, B); and migration of a Dde protector (see Section 6.4) from one amino group of a residue to the other amino group occurs more readily when the residue is of shorter chain length. Other phenomena exist but are unexplainable. Some side-chain protectors are more stable when the peptide is larger, and some protectors such as methoxytrimethylbenzenesulfonyl on the guanidino group of arginyl (see Section 6.11) become resistant to cleavage when there are several of them on the same peptide. Reflection is in order when protecting arginine as a single protector on the guanidino group may be insufficient to suppress its reactivity completely (see Section 6.11).[2–4]

2. A Berger, E Katchalski. Poly-L-aspartic acid. (saponification). *J Am Chem Soc* 73, 4084, 1951.
3. P Sieber, B Iselin. Peptide synthesis using the 2-(*p*-diphenyl)-isopropoxycarbonyl (Dpoc) amino protecting group. *Helv Chim Acta* 51, 622, 1968.
4. A Aletras, K Barlos, D Gatos, S Koutsogianni, P Mamos. Preparation of very acid-sensitive Fmoc-Lys(Mtt)-OH. *Int J Pept Prot Res* 45, 488, 1995.

6.4 THE ε-AMINO GROUP OF LYSINE

The side-chain amino group of lysine is a strong nucleophile, the reactivity of which cannot be suppressed by protonation, so it must be protected at all times. Acyl groups such as formyl, which is stable to alkali, ammonia, and hydrogenation but sensitive to mild acid, and trifluoroacetyl (see Section 3.9), which is stable to piperidine and

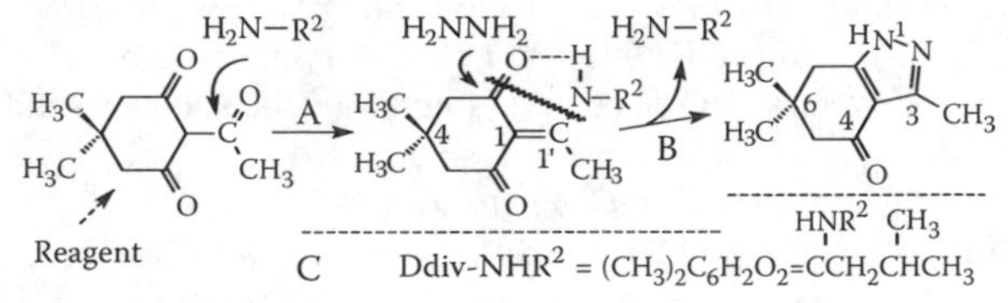

C Ddiv-NHR^2 = $(CH_3)_2C_6H_2O_2$=CCH_2CHCH_3

FIGURE 6.4 The (4,4-dimethyl-2,6-dioxocyclohex-1-ylidene)-1′-ethyl (Dde) protector for amino groups.[6] (A) Reaction of NH_2R^2 with 2-acetyldimedone giving Dde-NHR^2. (B) Release of NH_2R^2 by displacement of Dde by hydrazine facilitated by a hydrogen bond giving 3,6,6-trimethyl-4-oxo-4,5,6,7-tetrahydro-1*H*-indazole. (C) An amino group protected by the isovaleryl equivalent of Dde composed of Dde elongated with isopropyl.[9]

trifluoroacetic acid but sensitive to alkaline pH, have been employed, the latter to prevent the aggregation that occurred during chain assembly using Fmoc/tBu chemistry. *p*-Toluenesulfonyl is appropriate for syntheses terminating in deprotection by reduction with sodium in liquid ammonia. Phthaloyl (see Section 3.12) is an effective protector that is sensitive to hydrazine and alkali, but its removal is not straightforward (see Section 3.8). 2-Chlorobenzyloxycarbonyl or allyloxycarbonyl are the most suitable for Boc chemistry, and *tert*-butoxycarbonyl is suitable for Cbz and Fmoc chemistries. A significant advance occurred in 1993 with development of the (4,4-dimethyl-2,6-dioxocyclohexylidene)ethyl or Dde group that is bound to an amino group through C-1 of ethyl (Figure 6.4). The bound function is generated by the reaction of the amino group with 4,4-dimethyl-2,6-dioxocyclohexylmethylketone (2-acetyldimedone), which produces a Schiff's base (see Section 5.4 and Figure 6.6 therein), followed by migration of the double bond to give the conjugated system (Figure 6.4). The protector is sensitive to nucleophilic displacement by hydrazine, which is facilitated by the effect of a hydrogen bond between the NH of the residue and the carbonyl of the protector. Hydrazine in dimethylformamide (1:50) is the cleavage reagent. The Dde group is stable to trifluoroacetic acid and sufficiently stable to piperidine-dimethylformamide (1:4) (3% loss in two hours) and hence is compatible with Boc/Bzl and Fmoc/tBu chemistries. It provides a third level of orthogonality for synthesis. Before introduction of the Dde protector, multichain peptides joined through the amino groups of lysine could be prepared only using Boc/Bzl chemistry. The protector is compatible with all coupling methods; the cyclohexylmethyl equivalent developed earlier was not compatible with tertiary-amine requiring reagents.

In contrast, a small amount of migration of the protector to other side-chain amino groups occurs during piperidine-mediated cleavage of Fmoc groups. Migration of an N^{α}-Dde protector is greater than that of an N^{ε}-Dde protector, and migration of the protector is avoided by use of the nonnucleophilic base DBU (see Section 8.12) instead of piperidine or by replacing it with the cyclohexylidene-3-methylbutyl (isovaleryl) equivalent, Ddiv (Figure 6.4, C), strangely known as ivDde. The migration problem is worse for diamino acids of shorter chain length. An alternative protector is methyltrityl, which is sensitive to 1% trifluoroacetic acid and is thus compatible with *tert*-butyl-based protectors and linkers that are not highly sensitive to acid.[4,5–9]

4. A Aletras, K Barlos, D Gatos, S Koutsogianni, P Mamos. Preparation of very acid-sensitive Fmoc-Lys(Mtt)-OH. *Int J Pept Prot Res* 45, 488, 1995.
5. E Atherton, V Wooley, RC Sheppard. Internal association in solid phase peptide synthesis. Synthesis of cytochrome C residues 66-104 on polyamide supports. (trifluoroacetyl) *J Chem Soc Chem Commun* 970, 1980.
6. BW Bycroft, WC Chan, SR Chhabra, ND Hone. A novel lysine-protecting procedure for continous flow solid phase synthesis of branched peptides. (Dde group) *J Chem Soc Chem Commun* 778, 1993.
7. A Aletras, K Barlos, D Gatos, S Koutsogianni, P Mamos. Preparation of the very acid-sensitive Fmoc-Lys(Mtt)-OH. Application in the synthesis of side-chain to side-chain cyclic peptides and oligolysine cores suitable for solid-phase assembly of MAPs and TASPs. *Int J Pept Prot Res* 45, 488, 1995.
8. K Augustyns, W Kraas, G Jung. Investigation on the stability of the Dde protecting group used in peptide synthesis: migration to an unprotectd lysine. *J Pept Res* 51, 127, 1998.
9. SR Chhabra, B Hothi, DJ Evans, PD White, BW Bycroft, WC Chan. An appraisal of new variants of Dde amine protecting groups for solid phase synthesis. (Ddiv group) *Tetrahedron Lett* 39, 1603, 1998.

6.5 THE HYDROXYL GROUPS OF SERINE AND THREONINE

The hydroxyl groups of serine and threonine are not ionized (see Section 1.3) under usual operating conditions and are weak nucleophiles, and, therefore, do not have to be protected. However, they are still reactive enough to undergo changes, the primary hydroxyl of serine being more reactive than the secondary hydroxyl of threonine. A hydroxyl in the aminolyzing component of a coupling is nearly inert to acyl azides and anhydrides and mildly reactive to activated esters and onium salt-based reagents; however, it should be borne in mind that excess activated component or excess base could induce its acylation. *O*-Acylation during aminolysis of an activated ester is suppressed by moisture. Hydroxyamino-acid derivatives with unprotected side chains have traditionally been coupled as the acyl azides. The use of carbodiimides gives modest yields. The mixed-anhydride and activated-ester methods have been used occasionally; *N*-alkoxycarbonylthreonines are coupled efficiently as the activated esters obtained through the mixed anhydride. Success has been achieved using phosphonium and uronium salt-based reagents, and this is probably the best approach. In addition to restricting the applicability of coupling methods, hydroxyl groups are subject to acylation by carboxylic acids that are employed as reagents or solvents for deprotection of functional groups, and this may be followed by acyl transfer if the residue is amino terminal (see Section 6.6).

Common protectors of hydroxyls are benzyl and 2-bromobenzyl for Boc chemistry and *tert*-butyl for Fmoc chemistry. Trityl provides a third level of selectivity for both chemistries because it can be removed by mild acid (1% CF_3CO_2H in CH_2Cl_2), which does not affect *tert*-butyl based protectors. *O*-Allyl is not removable by palladium-catalyzed allyl transfer, so it is not appropriate. Protection by acyl such as benzyloxycarbonyl is possible, but O^{β}-acyl protectors can be problematic because of their tendency to shift to adjacent amino groups (see Section 6.6) and their vulnerability to base-catalyzed elimination (see Section 3.10), which dehydroxylates the residue.[10–13]

10. H Romovacek, SR Dowd, K Kawasaki, N Nishi, K Hofmann. Studies on polypeptides. 54. The synthesis of a peptide corresponding to positions 24-104 of the peptide chain of ribonuclease T_1. (acyl azides) *J Am Chem Soc* 101, 6081, 1979.
11. VM Titov, EA Meshcheryakova, TA Balashova, TM Andronova, VT Ivanov. Synthesis and immunological evaluation of the conjugates composed from muramyl peptide GMDP and tuftsin. (Z-Thr-ONSu) *Int J Pept Prot Res* 45, 348, 1995.
12. H Mostafavi, S Austermann, W-G Forssmann, K Adermann. Synthesis of phospho-urodilatin by combination of global phosphorylation with the segment coupling approach. (Fmoc-Ser-OH + TBTU + H-Xaa-OR) *Int J Pept Prot Res* 48, 200, 1996.
13. N Sin, L Meng, H Auth, CM Crews. Eponemycin analogues: synthesis and use as probes for angiogenesis. (RCO_2H + TBTU + H-Ser-OR) *Bioorg Med Chem* 6, 1209, 1998.

6.6 ACID-INDUCED *O*-ACYLATION OF SIDE-CHAIN HYDROXYLS AND THE *O*-TO-*N* ACYL SHIFT

The major side reaction associated with peptides containing unprotected seryl and threonyl residues is acid-induced acylation of the hydroxyl groups. Hydroxyl groups of peptides undergo acylation when the latter are dissolved in strong carboxylic acids such as formic and trifluoroacetic acids. In the weaker acetic acid, acylation occurs if the acid is saturated with hydrogen bromide or chloride. The reaction is referred to as a reverse esterification. It occurs even in trifluoroacetic acid that contains 10% of water. In synthesis, it is encountered during removal of benzyl-based protectors by hydrogen bromide in acetic acid (see Section 3.5), and *tert*-butyl-based protectors by trifluoroacetic acid. Acylation does not occur if *p*-toluenesulfoninc acid in acetic acid is employed for removal of the latter. The acyl groups are easy to dispose of with 5% hydrazine in dimethylformamide; however, they present a serious obstacle to synthesis because they migrate to amino groups as soon as the latter are deprotonated, and in particular to amino groups of the same residue (Figure 6.5, paths C, D). The transfer of acyl groups from oxygen to nitrogen atoms of residues is known as the *O*-to-*N* acyl shift. *O*-Acylation of peptides also occurs in strong mineral acids such as sulfuric and phosphoric acids, but in this case the acyl group originates by a shift of the acyl component of a peptide bond to the adjacent hydroxyl group (Figure 6.5, paths A, B). This *N*-to-*O* acyl shift occurs when protected peptides are dissolved in hydrogen fluoride at the end of chain assembly. It is readily reversed by aqueous ammonia. Acyl transfer is postulated to occur through the formation of oxazolidine and oxazoline rings (paths A and C, respectively). Transfer of acyl from oxygen to nitrogen occurs immediately after the amino group is deprotonated. Acyl transfer also occurs in neat trifluoroacetic acid, but not if the latter contains 6 *M* hydrogen chloride; the inference is that an α-amino group is not completely protonated by the carboxylic acid. The analogous reaction of trifluoroacetylation of Merrifield resin that is prematurely deacylated during acidolysis reactions has been alluded to (see Section 5.16).

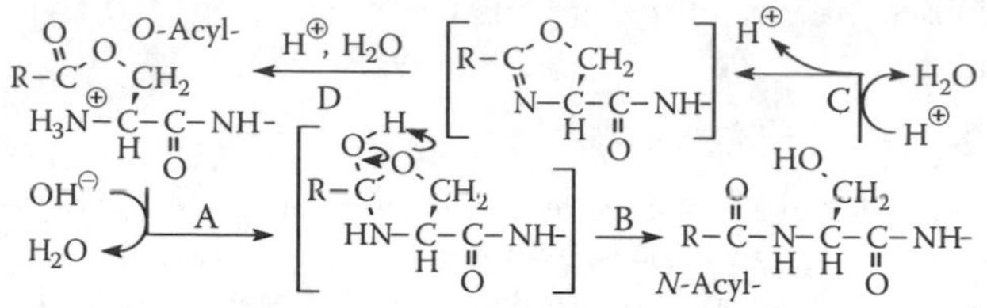

FIGURE 6.5 Transfer of acyl between the amino and hydroxyl groups of seryl. (A) Deprotonation of *O*-acylseryl– induces oxazolidine formation, which is followed by (B) rearrangement to *N*-acylseryl–. (C) Protonation of the carbonyl of *N*-acylseryl– by mineral acid results in dehydration to the oxazoline, which is followed by hydrolysis (D) at the double bond giving protonated *O*-acylseryl–.

The *O*-acylation of side-chain hydroxyl groups was first recognized in proteins dissolved in concentrated acids and encountered in peptide work when hydrogen bromide in acetic acid had been employed for deprotection. A major implication of the *O*-to-*N* acyl shift is that a peptide carrying an *O*-acyl group such as fatty acyl cannot be assembled by routine Fmoc/tBu chemistry because the base necessary for N^{α}-deprotection after incorporation of the β-acyloxy-Fmoc-amino acid triggers the shift. In contrast, the shift has been taken advantage of for several purposes. N^{α}-Trifluoroacetylserine and threonine peptides have been deacylated by treatment with 0.1 *M* hydrogen chloride in methanol. The acid promotes the *N*-to-*O* shift, which is followed by a transesterification, thus liberating the peptide. Racemic *N*-acetylserine, a substrate for resolution by enzymes (see Section 4.23), was obtained by reverse esterification of serine followed by *O*-to-*N* acyl shift (Figure 6.6). Water-insoluble acylamido-substituted enzyme inhibitors bearing a hydroxyl group on the carbon atom adjacent to the carbon bearing the derivatized amino group are rendered soluble for delivery by presentation as the protonated *N*-unsubstituted *O*-acyl derivatives (Figure 6.6). The acylation of *N*-(2-hydroxy-4-methoxybenzyl) amino acid residues is believed to proceed by *O*-acylation, followed by the shift of acyl to nitrogen (see Section 8.5).[14–19]

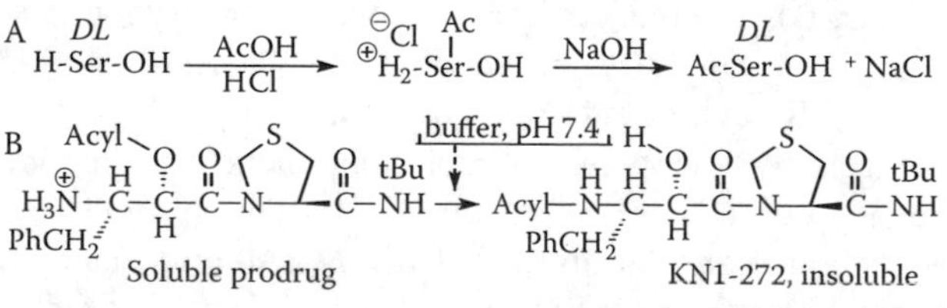

FIGURE 6.6 Practical uses of the *O*-to-*N* acyl shift. (A) Synthesis of *O*-acetyl-DL-serine by reverse esterification followed by migration of acetyl. Enzymatic resolution of the unisolated product provided the enantiomers of serine in good yield, in contrast to the classical method of resolving *N*-chloroacetyl-DL-serine, which was inefficient.[17] (B) HIV-protease inhibitor KN1-272 was too insoluble in aqueous media for delivery. Presentation as the soluble prodrug (half-life at pH 7.4 = 0.5 min) gave the desired activity, attributed to the drug generated by the acyl shift.[19]

14. Elliott DF. Specific chemical methods for the fission of peptide bonds. I. *N*-acyl to *O*-acyl transformation in the degradation of silk fibroin. *Biochem J* 542, 1952.
15. K Narita. Isolation of acetylseryltyrosine from the chymotryptic digests of five strains of tobacco mosaic virus. *Biochim Biophys Acta* 30, 352, 1958.
16. K Narita. Reaction of anhydrous formic acid with proteins. *J Am Chem Soc* 81, 1751, 1959.
17. L Benoiton. An enzymic resolution of serine. *J Chem Soc* 763, 1960.
18. G Hübener, W Göhring, H-J Musiol, L Moroder. N^{α}-Trifluoroacetylation of N-terminal hydroxyamino acids: a new side reaction in peptide synthesis. *Pept Res* 5, 287, 1992.
19. Y Hamada, J Ohtake, Y Sohma, T Kimura, Y Hayashi, Y Kiso. New water-soluble prodrugs of HIV protease inhibitors based on *O*→*N* intramolecular acyl migration. *Bioorg Med Chem* 10, 4155, 2002.

6.7 THE HYDROXYL GROUP OF TYROSINE

The side chain of tyrosine is moderately reactive, so it is usually protected during synthesis. The hydroxyl of tyrosine is a better nucleophile than the hydroxyls of serine and threonine, the phenolic moiety is a good acceptor of cations, and the phenolic anion is a better nucleophile than an α-amino group, so unprotected tyrosine invites side reactions during couplings as well as during acidolytic deprotections. An excess of activated component or tertiary amine in a coupling will readily induce acylation of a tyrosine hydroxyl in the aminolyzing component. The acylated product that is formed is an activated ester that must be destroyed before further action is taken; otherwise, it will react with amino groups. *N*-Alkoxycarbonyltyrosine can be incorporated by the usual coupling methods, probably most advantageously by use of onium salt-based reagents, with the exception of the acyl-azide method, in which the nitrous acid employed to generate the azide modifies the phenol ring (see Section 2.13). Peptide azides in which the activated residue is tyrosyl also have a tendency to decompose to the isocyanate.

The main considerations in protecting the hydroxyl of tyrosine are the stability to acid of the protector and the protector's tendency to alkylate the *ortho*-position of the ring when it is removed (Figure 6.7; see Section 3.7). The standard protector for Fmoc chemistry is *tert*-butyl, but the substituent is sensitive to acid. Preferable

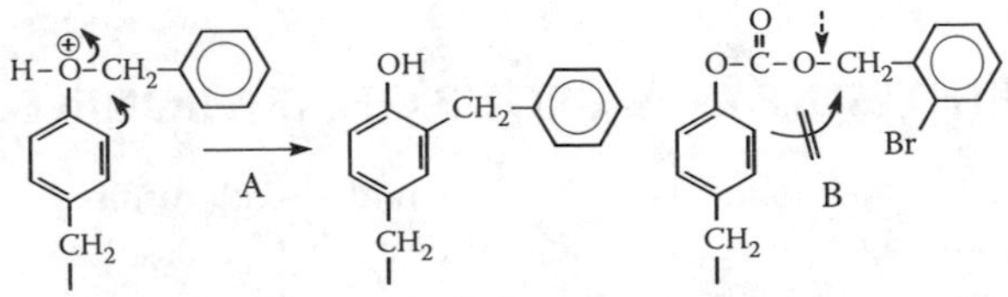

FIGURE 6.7 (A) Rearrangement to the substituted phenol during acidolytic debenzylation of *O*-benzyltyrosine. Alkylation also occurs by intermolecular reaction. (B) Alkylation does not occur during acidolysis of 2-bromobenzyloxycarbonyltyrosine. The oxycarbonylphenol produced is a weaker nucleophile than phenol, and the cation that is generated is farther away from the nucleophile.

is 2-chlorotrityl, which is more stable to acid and is completely trapped by scavenger when released. Benzyl is suitable for Boc chemistry in solution but is too sensitive to trifluoroacetic acid for use in solid-phase synthesis. In addition, its removal by acid generates a cation that is not well trapped by scavenger. 2,6-Dichlorobenzyl is 5000 times more stable than benzyl, but it also generates too much alkylated product. Cyclohexyl has the appropriate stability and minor tendency to alkylate, but the necessary derivative is not easy to prepare. The best alkyl protector for Boc chemistry seems to be 3-pentyl, which is the noncyclic lower homologue of cyclohexyl. It is effectively trapped by scavengers. Tyrosine hydroxyls are occasionally blocked by alkoxycarbonyl — in particular when the residue is at the amino terminus of a chain. Typical derivatives are *N,O-bis*-Boc– and *N,O-bis*-Fmoc-tyrosine. The protected functional group is a mixed carbonate (see Section 3.15) activated by virtue of the phenyl moiety and is thus sensitive to piperidine, and, consequently, incompatible with Fmoc chemistry. 2-Bromobenzyloxycarbonyl possesses the acid stability and alkylating property required for successful synthesis using Boc chemistry. The cation produced by acidolysis is completely trapped by scavenger in part because it is not near the nucleophilic site when released. 2,4-Dimethyl-3-pentyloxycarbonyl (Dmpoc) is a new protector that is superior because of its lesser sensitivity to piperidine. The classical scavenger for protecting tyrosine side chains from alkylation during acidolysis has been anisole ($PhOCH_3$); trialkyl silanes are emerging as the best for the purpose (see Section 6.23).[20–24]

20. BW Erickson, RB Merrifield. Acid stability of benzylic protecting groups used in solid-phase peptide synthesis. Rearrangement of *O*-benzyltyrosine to 3-benzyltyrosine. *J Am Chem Soc* 95, 3750, 1970.
21. D Yamashiro, CH Li. Protection of tyrosine in solid-phase synthesis. (BrZ) *J Org Chem* 38, 591, 1973.
22. M Englehard, RB Merrifield. Tyrosine protecting groups. Minimization of rearrangement to 3-alkyltyrosine during acidolysis. (*O*-cyclohexyltyrosine) *J Am Chem Soc* 100, 3559, 1978.
23. K Rosenthal, A Karlström, A Undén. The 2,4-dimethylpent-3-yloxycarbonyl (Doc) group as a new nucleophile-resistant protecting group for tyrosine in solid phase peptide synthesis. *Tetrahedron Lett* 38, 1075, 1997.
24. J Bódi, Y Nishiuchi, H Nishio, T Inui, T Kimura. 3-Pentyl (Pen) group as a new base resistant side chain protector for tyrosine. *Tetrahedron Lett* 39, 7117, 1998.

6.8 THE METHYLSULFANYL GROUP OF METHIONINE

The side chain of methionine is inert to peptide-bond forming reactions but is sensitive to atmospheric oxygen, which converts it to the sulfoxide (Figure 6.8). Oxidation occurs more quickly in acidic medium and less rapidly in alcohols. It is suppressed by the presence of a thio ether alcohol such as methylsulfanylethanol ($CH_3SCH_2CH_2OH$), which consumes the oxygen. It is avoided by removing the air. The sulfoxide exists as two stereoisomers that emerge separately after HPLC; this interferes with the monitoring of reactions. Unfortunately, there is no simple way to protect the thio ether from oxidation, so it is common practice to oxidize it completely at the beginning of a synthesis and reduce it back to the thio ether at the

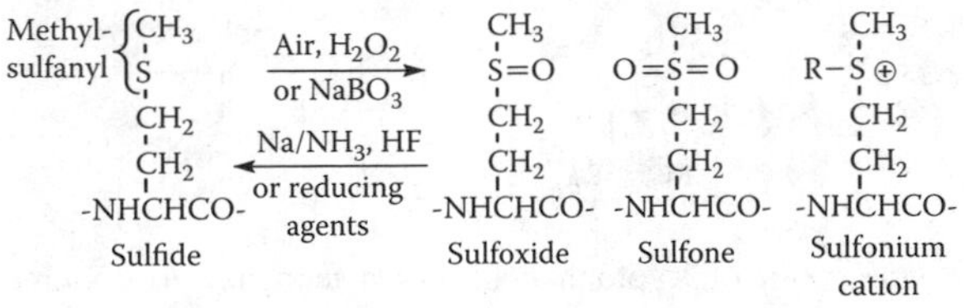

Methyl-sulfanyl CH_3–S–CH_2–CH_2–NHCHCO- (Sulfide) → Air, H_2O_2 or $NaBO_3$ → CH_3–S=O–CH_2–CH_2–NHCHCO- (Sulfoxide); ← Na/NH_3, HF or reducing agents

CH_3–O=S=O–CH_2–CH_2–NHCHCO- (Sulfone)

CH_3–R–S⊕–CH_2–CH_2–NHCHCO- (Sulfonium cation)

FIGURE 6.8 The functional group of methionine is oxidized to the sulfoxide by atmospheric oxygen or by reagents employed to prevent its alkylation to the sulfonium cation.[25] Overoxidation produces the sulfone. The sulfide is regenerated by cleavage reagents or reagents such as *N*-methylsulfamylacetamide [$CH_3(C{=}S)NHCH_3$][27] or iodide in CF_3CO_2H.[26]

end. Oxidation is effected using an equivalent of hydrogen peroxide or sodium perborate. Very little sulfone (Figure 6.8) is produced by overoxidation. The reagents employed for final deprotection after chain assembly using Boc/Bzl chemistry, namely, hydrogen fluoride and sodium in liquid ammonia, reduce the sulfoxide to the thio ether. Trifluoroacetic acid does not reduce the sulfoxide. Mercaptoacetic acid serves the purpose for reduction after assembly, using Fmoc/tBu chemistry, but it has a tendency to acylate amino groups. The equally effective *N*-methylsulfamylacetamide (Figure 6.8) is preferred. Other reducing reagents are ammonium iodide in mild excess in trifluoroacetic acid in the presence of dimethyl sulfide, or trimethylsilyl bromide in the presence of ethanedithiol $(CH_2SH)_2$ to reduce the liberated bromine. The sulfide of methionine is a good acceptor of cations. Alkylation produces sulfonium ions (Figure 6.8), so deprotections are carried out before the reduction when possible. Good scavengers for minimizing alkylation of methionine side chains are anisole ($PhOCH_3$) and dimethyl sulfide (CH_3SCH_3). An implication of the presence of methionine in a peptide is that deprotection by catalytic hydrogenation fails because the sulfide poisons the catalyst. However, there are alternative methods for hydrogenolytic deprotection of methionine-containing peptides (see Section 6.21). An intriguing difference between methionine and methionine sulfoxide is that a benzyloxycarbonyl substituent on the latter is removed by mild acid, whereas *N*-benzyloxycarbonylmethionine is stable to mild acid, as expected.[25–28]

25. BM Iselin. Derivatives of L-methionine sulfoxide and their use in peptide syntheses. *Helv Chim Acta* 44, 61, 1961.
26. E Izeboud, HC Beyerman. Synthesis of substance P via its sulfoxide by the repetitive excess mixed anhydride method. (iodide for reduction) *Rec Trav Chim Pays-Bas* 97, 1, 1978.
27. RA Houghten, CH Li. Reduction of sulfoxides in peptides and proteins. (*N*-methylsulfamylacetamide) *Anal Biochem* 98, 36, 1979.
28. W Beck, G Jung. Convenient reduction of S-oxides in synthetic peptides, lipopeptides and peptide libraries. (trimethylsilyl bromide) *Lett Pept Sci* 1, 31, 1994.

6.9 THE INDOLE GROUP OF TRYPTOPHAN

The indole group of tryptophan is inert to activation and aminolysis reactions, so it does not have to be protected for constructing a peptide. It is sensitive, however, to

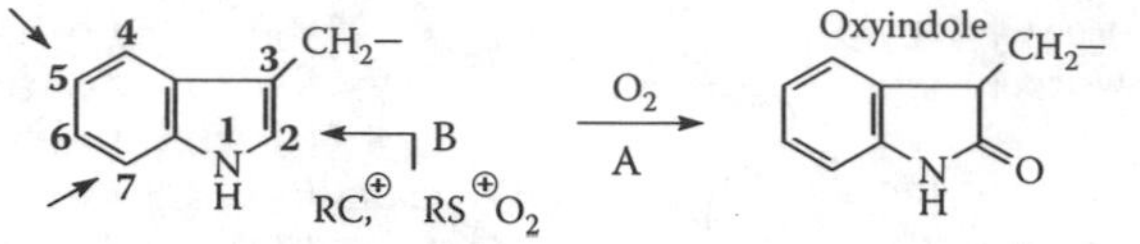

FIGURE 6.9 The side chain of tryptophan has as a tendency to undergo reaction at the electron-rich centers indicated: (A) oxidation in acidic medium,[29] and (B) electrophilic addition by one or more carbo[30] or arylsulfonyl cations.

nitrous acid employed to obtain acyl azides. Many peptides containing tryptophan have been prepared using Boc/Bzl chemistry without protecting the indole rings. However, there are two serious obstacles to success in synthesizing tryptophan-containing peptides because of the nucleophilic character of the rings, primarily at C-2 but also at C-5 and C-7 (Figure 6.9). There is a strong tendency for oxidation at C-2, particularly in an acidic medium. Oxidation is averted by replacing air with nitrogen, minimizing exposure to acidic conditions, and ensuring that reagents do not contain oxidants. Inclusion of a scavenger such as N^{α}-acetyltryptophan or pre-treatment of trifluoroacetic acid solution with indole removes oxidants as well as aldehyde that might be the sources of side reactions. Oxidation is also prevented by substitution at the nitrogen atom of the ring. The second major obstacle is electrophilic addition of carbo and arylsulfonyl cations (Figure 6.9) that are generated during removal of protectors from other residues. In addition, there occurs dimerization during acidolysis, between the C-2 atoms of two side chains. Both obstacles are best overcome by employing a protector for tryptophan that is removed separately after removal of other protectors. Allyloxycarbonyl satisfies this requirement and is applicable to both Boc and Fmoc chemistry if the non-nucleophilic base DBU (see Section 8.12) is employed for Fmoc cleavage. *tert*-Butoxycarbonyl is the standard protector for Fmoc chemistry. Protectors for Boc chemistry have been formyl that is stable to hydrogen fluoride and removable by alkali, piperidine or hydrogen fluoride with a soft nucleophile (see Section 6.22), and 2,4-dichlorobenzyloxycarbonyl that was introduced at the dipeptide ester stage by acylation of the nitrogen anion produced by unsolvated fluoride ion generated by a crown ether. The dihalo protector is removable by hydrogen fluoride, hydrazine, or catalytic hydrogenation, though the latter can partially reduce the heterocyclic ring. When removed by acidolysis, these protectors of the indole ring generate cations that must be trapped by scavengers. Dissatisfaction with the protectors available for synthesis using Boc chemistry led to development of the cyclohexyloxycarbonyl (Hoc) protector, which is cleavable by hydrogen fluoride with cresol (see Section 6.22) without the need for the usual thiol scavengers to prevent alkylation. Very little alkylation occurs because of the nature of the cyclohexyl cation (see Section 6.23). The exclusion of thiols precludes side reactions that are associated with their use. An analogous protector derived from a secondary alcohol is 2,4-dimethylpentyloxycarbonyl (see Section 6.7). The alkoxycarbonyl substituents based on secondary alcohols are stable to nucleophiles and acids weaker than hydrogen fluoride and, hence, are suitable for Fmoc and Boc chemistries.[29–33]

29. WE Savige, A Fontana. New procedure for the oxidation of 3-substituted indoles to oxyindoles. (preparative oxidation) *J Chem Soc Chem Commun* 599, 1976.
30. M Löw, L Kisfaludy, P Sohár. *tert*-Butylation of the indole ring of tryptophan during removal of the *tert*-butyloxycarbonyl group in peptide synthesis. *Hoppe Seyler's Z Physiol Chem* 359, 1643, 1978.
31. Y Nishiuchi, H Nishio, T Inui, T Kimura, S Sakakibara. N^{in}-Cyclohexyloxycarbonyl group as a new protecting group for tryptophan. *Tetrahedron Lett* 37, 7529, 1996.
32. A Karlström, A Undén. Protection of the indole ring of tryptophan by the nucleophile-stable, acid-cleavable N^{in}-2,4-dimethylpent-3-yloxycarbonyl (Doc) protecting group. *J Chem Soc Chem Commun* 1471, 1996.
33. T Vorherr, A Trzeciak, W Bannwart. Application of the allyloxycarbonyl protecting group for the indole of Trp in solid-phase peptide synthesis. *Int J Pept Prot Res* 48, 553, 1996.

6.10 THE IMIDAZOLE GROUP OF HISTIDINE

The imidazole group of a histidine residue that is activated reacts with activating reagents, so it has been traditional to activate N^{im}-unprotected histidyl as the azide that is obtained from N^{α}-alkoxycarbonylhistidine methyl ester. However, regardless of the method of activation, histidine derivatives with an unprotected side chain undergo enantiomerization during coupling (see Section 4.3). In addition, the imidazolyl of histidine in the aminolyzing component of a coupling is a nucleophile that competes with the amino group for the activated component (Figure 6.10). This reduces efficiency even if it is not an obstacle; the unsubstituted imidazolyl is readily regenerated. In contrast, formation of the imidazolide indeed is a problem, because the latter is an activated form of the acyl group, which it may transfer to other nucleophiles such as amino and hydroxyl groups. Thus, the side chain of histidine is usually protected to prevent isomerization and subsequent acyl transfer. Isomerization results from abstraction of the α-proton by the basic π-nitrogen of the ring (Figure 6.10; see Section 4.3). Substitution at the π-nitrogen prevents isomerization. Unfortunately, derivatization of the ring occurs primarily at the τ-nitrogen, so protection of the π-nitrogen must be achieved indirectly (see Section 6.24), and this makes the derivatives expensive.

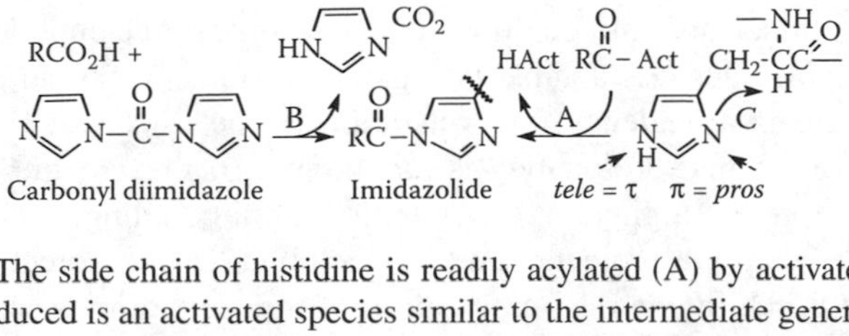

FIGURE 6.10 The side chain of histidine is readily acylated (A) by activated residues. The imidazolide produced is an activated species similar to the intermediate generated by reaction (B) of a carboxylic acid with coupling reagent carbonyldiimidazole. (Staab, 1956). Imidazolides acylate amino and hydroxyl groups. Isomerization of histidyl during activation results from abstraction (C) of the α-proton by the π-nitrogen.

Substitution at the τ-nitrogen alters the basicity of the π-nitrogen and may diminish its ability to abstract the α-proton. Efficient protectors are N^{π}-benzyloxymethyl (Bom) for Boc chemistry and N^{π}-1-adamantyloxymethyl (1-Adom) for Fmoc chemistry. The former is cleavable by hydrogenolysis as well as acidolysis. Formaldehyde that is liberated must be trapped by scavengers. The N^{π}-allyl and N^{π}-allyloxymethyl (Alom) derivatives are also available for Boc chemistry. Despite the above considerations, the most popular derivative for synthesis using Fmoc chemistry is N^{α}-Fmoc-N^{τ}-trityl-L-histidine. It is very simple to prepare, and the electron-withdrawing effect of the substituent reduces the basicity of the π-nitrogen so that activation and coupling occur with minimal enantiomerization. Analogous τ-substituted derivatives are N^{α}-Fmoc, N^{τ}-Boc-histidine and N^{α}-Boc-N^{τ}-dinitrophenyl-histidine. The protector of the latter is removed by thiolysis (see Section 3.12); it is, however incompatible with 1-hydroxybenzotriazole. N^{α},N^{τ}-*bis*-Boc-histidine and N^{α},N^{τ}-*bis*-Fmoc-histidine are available for special purposes.[34–39]

34. T Brown, JH Jones, JD Richards. Further studies on the protection of histidine side chains in peptide synthesis: the use of the π-benzyloxylmethyl group. *J Chem Soc Perkin Trans 1* 1553, 1982.
35. P Sieber, B Riniker. Protection of histidine in peptide synthesis: a reassessment of the trityl group. *Tetrahedron Lett* 48, 6031, 1987.
36. SJ Harding, I Heslop, JH Jones, ME Wood. The racemization of histidine in peptide synthesis. Further studies, in HLS Maia, ed. *Peptides 1994. Proceedings of the 23rd European Peptide Symposium*, Escom, Leiden, 1995, pp 189-190.
37. Y Okada, J Wang, T Yamamoto, Y Mu, T Yokoi. Amino acids and peptides. Part 45. Development of a new N-protecting group of histidine, N^{π}-(1-adamantyloxymethyl)-histidine, and its evaluation for peptide synthesis. *J Chem Soc Perkin Trans 1* 2139. 1996.
38. AM Kimbonguila, S Boucida, F Guibé, A Loffet. On the allyl protection of the imidazole ring of histidine. *Tetrahedron* 53, 12525, 1997.
39. SJ Harding, JH Jones. π-Allyloxymethyl protection of histidine. *J Pept Sci* 5, 399, 1999.

6.11 THE GUANIDINO GROUP OF ARGININE

The guanidino group of arginine is a strong base, p*K* 12.5, which is protonated (as HCl, HBr, etc.) under normal conditions and is a strong nucleophile. The charged group renders molecules less soluble in organic solvents. When an arginyl residue is activated, there is a tendency for cyclization to the δ-lactam, Figure 6.11; the tendency is so great that protonation is insufficient to prevent the intramolecular reaction. So unprotected arginyl is rarely employed for coupling. Cyclization occurs even for residues that are protected on the side chain if the protector is not at the δ-atom (Figure 6.11). In contrast, protonation of the guanidino group of an aminolyzing residue or segment suppresses its nucleophilicity sufficiently to allow peptide-bond formation to occur without significant acylation of the side chain. Any N^{g}-acylated product that might be formed is sensitive to base; piperidine ruptures the δN-εC bond of the adduct (Figure 6.11), generating an ornithine residue. Protection against these side reactions is provided by a substituent on the δ-nitrogen

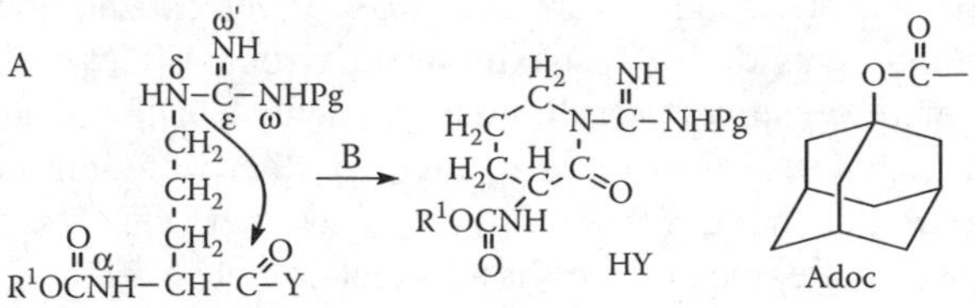

FIGURE 6.11 (A) An N^g (guanidino)-protected arginyl residue with the nitrogen atoms is identified. A side reaction of intramolecular aminolysis giving the lactam (B) can occur if the δ-nitrogen is unprotected. N^{δ},N^{ω}-*bis*-(Adoc = 1-adamantyloxycarbonyl)– substituted Boc- and Cbz-arginine[42] are derivatives that can be coupled without this side-reaction occurring.

atom or one or more substituents on the terminal atoms that are bulky or electron withdrawing enough to reduce the nucleophilicity of the δ-nitrogen. The protection provided is not always complete. The original protector was nitro, giving conjugated structure $-(H_2N)C{=}N\text{-}NO_2$; the guanidino group is best restored by catalytic hydrogenolysis though acidolysis by hydrogen fluoride-anisole (see Section 6.22) is an alternative. Z-Nitroarginine and Boc-nitroarginine are still valuable for synthesizing short peptides. Useful peracylated derivatives with substitution at the δ-nitrogen, accessible by acylation of N^{α}-protected arginine or by guanidinylation of ornithine derivatives, are *tris*-benzyloxycarbonylarginine and Fmoc-arginine with two 1-adamantyloxycarbonyl moieties (Figure 6.11) on the side chain. The latter is a modified Boc group that is more stable to acid as a result of the bulkiness of the three interconnected cyclohexyl rings. Acylation of guanidino may not be completely prevented by these moieties. The more common protectors are arylsulfonyl substituents that are synthesized from Z-arginine and removed by acidolysis, with N^{α}-Boc-N^{ω}-*p*-toluenesulfonylarginine being the classical derivative. For Fmoc-chemistry, the arylsulfonyl moiety has been sensitized to acidolysis by changing the ring substituent to methoxy and trimethyl and then by incorporating the ether into a second contiguous ring. Thus, emerged the methoxytrimethylbenzenesulfonyl, Pmc and Pbf groups (Figure 6.12), in that order. The heterocyclic rings fix the positions of the oxygen

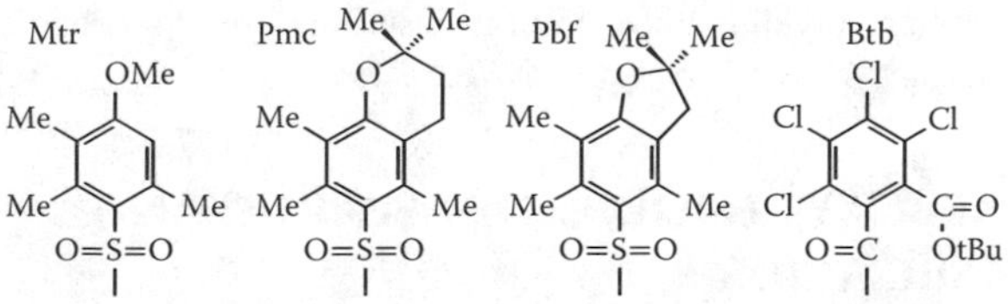

FIGURE 6.12 Protectors for the guanidino group of arginine:

Mtr = 4-methoxy-2,3,6-trimethylbenzenesulfonyl, [Atherton et al., 1983],
Pmc = 2,2,5,7,8-pentamethylchroman-6-sulfonyl,[44]
Pbf = 2,2,4,6,7-pentamethyldihydrobenzofuran-5-sulfonyl,[46]
Btb = *bis*(*o*-*tert*-butoxycarbonyltetrachlorobenzoyl).[45]

The ease of removal by trifluoroacetic acid is in the order Pbf > Pmc > Mtr.

atoms so that the lone electron pairs are available for delocalization with the phenyl ring π-systems, thus providing the maximum electronic effect. The Pbf group is slightly (1.3×) more sensitive to trifluoroacetic acid than the Pmc group. It had previously been established that a five-membered ether ring annealed to a phenyl ring is more electron donating than a six-membered ether ring. The dual-ring structures were developed after experience showed that complete removal of 4-methoxy-2,3-6-trimethylsulfonyl protectors could not be achieved when there was more than one methoxytrimethylbenzenesulfonyl group in the molecule. Arylsulfonyl protectors are released by acidolysis as sulfonylium ions (see Section 6.22) and, hence, must be trapped by scavengers. Pbf gives rise to less arylsulfonylation of tryptophan than Pmc. A different type of protection for guanidino that is suitable for Fmoc chemistry comprises two *o-tert*-butoxytetrachlorobenzoyl groups (Figure 6.12) on different terminal (ω, ω′) atoms. The protectors are removed by acidolysis by a two-stage process involving assistance by the carboxyl groups after they have been deesterified. No acylation of tryptophan accompanies the cleavage.[40–46]

40. M Bergmann, L Zervas, H Rinke. New process for the synthesis of arginine peptides. (nitroarginine). *Hoppe-Seyler's Z Physiol Chem* 224, 40, 1932.
41. L Zervas, M Winitz, JP Greenstein. Studies on arginine peptides. I. Intermediates in the synthesis of N-terminal and C-terminal arginine peptides. *J Org Chem* 22, 1515, 1957.
42. G Jäger, R Geiger. The adamantyl-(1)-oxycarbonyl group as protecting group for the guanidino function of arginine. *Chem Ber* 103, 1727. 1970.
43. R Schwyzer, CH Li. A new synthesis of the pentapeptide L-histidinyl-L-phenylalanyl-L-arginyl-L-tryptophyl-glycine and its melanocyte-stimulating activity. (*p*-toluene-sulfonyl) *Nature (London)* 182, 1669, 1958.
44. R Ramage, J Green. N_G-2,2,5,7,8-Pentamethylchroman-6-sulphonyl-L-arginine: a new acid labile derivative for peptide synthesis. (Pmc) *Tetrahedron Lett* 28, 2287, 1987.
45. T Johnson, RC Sheppard. A new t-butyl-based acid-labile protecting group for the guanidine function of N^{α}-fluorenylmethyoxycarbonyl-arginine. (Btb) *J Chem Soc Chem Commun* 1605, 1990.
46. LA Carpino, H Shroff, SA Triolo, EME Mansour, H Wenschuh, F Albericio. The 2,2,4,6,7-pentamethyldihydrobenzofuran-5-sulfonyl group (Pbf) as arginine side chain protectant. *Tetrahedron Lett* 34, 7829, 1993.

6.12 THE CARBOXYL GROUPS OF ASPARTIC AND GLUTAMIC ACIDS

The carboxyl groups on the side chains of dicarboxylic-acid residues are nearly always protected because unblocked carboxyl groups react with activated acyl moieties. If the activated function is on a residue different from that carrying the free carboxyl group, their combination generates an unsymmetrical anhydride that can undergo aminolysis at either of the carbonyls to produce undesired peptides. If the activated function is on the same residue as the free carboxyl group, their combination generates an internal anhydride that undergoes aminolysis at either of the carbonyls, producing a mixture of side-chain-linked peptide and normal peptide

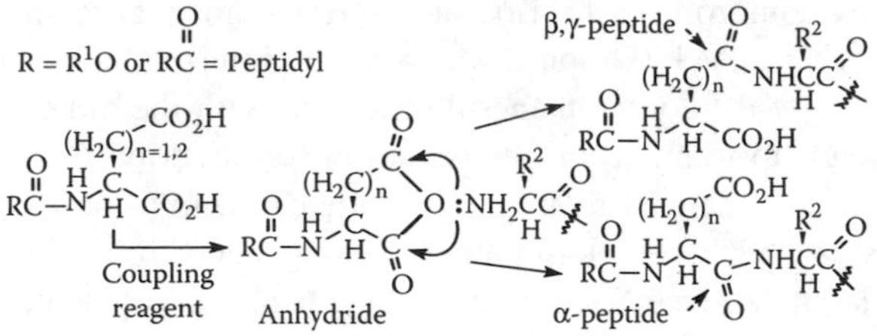

R = R^1O or RC(=O) = Peptidyl

Coupling reagent

Anhydride

β,γ-peptide

α-peptide

FIGURE 6.13 Activation of N^{α}-substituted aspartic or glutamic acid in the presence or absence of amine nucleophile gives the anhydride. Aminolysis of the anhydride gives a mixture of two peptides. [Melville 1935; Bergmann et al., 1936].

(Figure 6.13). Attempts to avoid anhydride formation during activation have been fruitless. There are, however, three exceptions to the above. A carboxyl group on an aminolyzing segment does not react with an acyl azide. Activation of *p*-toluenesulfonylglutamic acid gives the corresponding pyroglutamate (see Section 6.17) and not the anhydride. There is one way to react an activated dicarboxylic acid without protecting the side chain — by activation as the *N*-carboxyanhydride (Figure 6.14; see Section 7.13). The two side reactions described above are eliminated by protection of the carboxyl group as the ester. A third possible side reaction is combination of the carboxyl group of the side chain of a residue with the nitrogen atom of the residue to which it is linked to produce an imide (see Section 6.13). The reaction is of minor significance; however, it becomes a major consideration once the carboxyl group is in the form of an ester and the ester is subjected to the action of base or strong acid (see Section 6.13). *tert*-Butyl protection is ideal for Cbz chemistry, in which hydrogenolysis is employed for cleaving the urethane. 4-Nitrobenzyl is ideal for Boc chemistry because the ester is stable to acid and cleavable by hydrogenolysis. Allyl is appropriate for Cbz and Boc chemistries because the protector is orthogonal to other protectors and does not require acid or base for its removal. *tert*-Butyl and allyl are appropriate for Fmoc chemistry, provided the secondary amine employed for Fmoc removal is neutralized by an acid (see Section 6.13). As for protectors removed by acid, benzyl is not stable enough to withstand the repeated acidolysis associated with chain assembly using Boc chemistry. In contrast, a protector that is more stable to acid, and hence more suitable, becomes less suitable at the stage of its removal because the stronger acid required to remove it produces more imide. 2-Chlorobenzyl is a case in point.

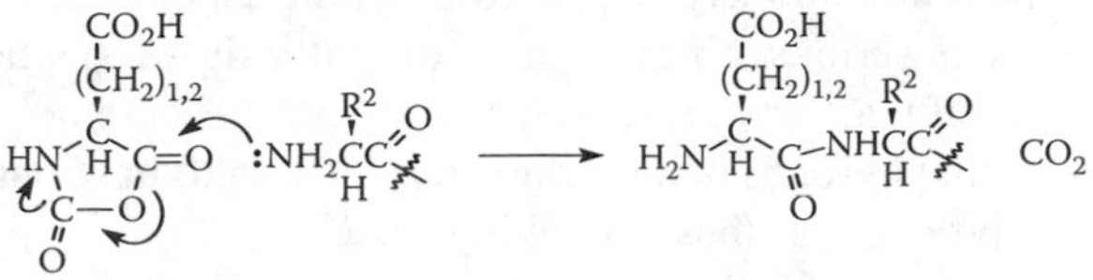

CO_2

FIGURE 6.14 Peptide-bond formation by aminolysis of the *N*-carboxyanhydride of aspartic or glutamic acid, followed by release of carbon dioxide.[48]

Thus, the best compromises for Boc and Fmoc chemistries seem to be cyclohexyl and 2,4-dimethylpent-3-yl (Dmpn), which is of intermediate stability, and the removal of which by trifluoromethanesulfonic acid with the aid of thioanisole (see Section 6.22) leads to minimal imide formation (see Section 6.13). Points to note are that acidolysis of esters by hydrogen fluoride can lead to fission at the oxy–carbonyl bond instead of the alkyl–oxy bond, thus generating acylium ions that can react with nucleophiles (see Sections 6.16 and 6.22), and that benzyl esters may undergo transesterification if left in methanol. The side reactions of cyclization (see Section 6.16) and acylation of anisole (see Section 6.22) caused by acylium ion formation do not occur at the side chain of aspartic acid.[47–51]

47. WJ Le Quesne, GT Young. Amino acids and peptides. Part I. An examination of the use of carbobenzoxyglutamic anhydride in the synthesis of glutamyl peptides. *J Chem Soc* 1954, 1950.
48. RG Denkewalter, H Schwam, RG Strachan, TE Beesley, DF Veber, EF Schoenewaldt, H Barkemeyer, WJ Paleveda, TA Jacob, R Hirschmann. The controlled synthesis of peptides in aqueous solution. I. The use of α-amino acid *N*-carboxyanhydrides. *J Am Chem Soc* 88, 3164, 1966.
49. JP Tam, TW Wong, MW Riemen, FS Tjoenj, RB Merrifield. Cyclohexyl ester as a new protecting group for aspartyl peptides to minimize aspartimide formation in acidic and basic treatments. *Tetrahedron Lett* 42, 4033, 1979.
50. H Kunz, H Waldmann, C Unverzagt. Allyl ester as temporary protecting group for the β-carboxy function of aspartic acid. *Int J Pept Prot Res* 26, 493, 1985.
51. A Karsltröm, A Undén. Design of protecting groups for the β-carboxylic group of aspartic acid that minimize base-catalyzed aspartimide formation. (dimethylpentyl) *Int J Pept Prot Res* 48, 305, 1996.

6.13 IMIDE FORMATION FROM SUBSTITUTED DICARBOXYLIC ACID RESIDUES

It was established in the 1950s that the action of alkali on aspartyl and glutamyl peptides causes a rearrangement that produces a mixture of two peptides (Figure 6.15). The rearrangement occurs through a cyclic imide intermediate that is stable enough to be isolated. An aspartyl residue is affected more than a glutamyl residue, the aspartimide is more stable to hydrolysis than the glutamimide, and more ω-peptide is formed than α-peptide. Subsequent studies showed that the phenomenon was also caused by hydrazine, strong acids, and piperidine. Because strong acids and secondary amines are routinely employed in synthesis, the side reaction of imide formation presents a serious obstacle in the synthesis of peptides containing glutamyl, and especially aspartyl, residues. The reaction occurs only when both the α-amino and α-carboxyl groups of a residue are substituted. In strong acid, the side-chain carbonyl is protonated, thus provoking attack by the nitrogen atom of the adjacent residue (Figure 6.15). Base deprotonates the nitrogen atom of the adjacent residue, promoting a nucleophilic attack at the side-chain carbonyl. The tendency for attack by the nitrogen atom is influenced by the nature of the substituent on the side-chain carbonyl and the nature of the residue located at the carboxylic function of the susceptible residue. The reaction occurs much more readily when the

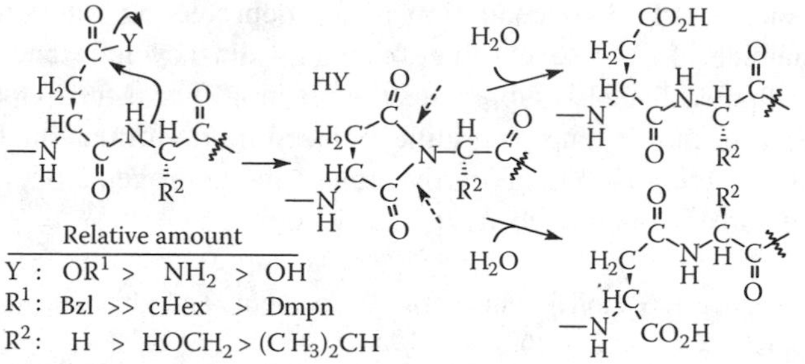

Relative amount

Y :	OR^1 > NH_2 > OH
R^1:	Bzl >> cHex > Dmpn
R^2:	H > $HOCH_2$ > $(CH_3)_2CH$

FIGURE 6.15 Imide formation from a dipeptide sequence containing an aspartyl residue with side-chain functional group in various states followed by generation of two peptide chains resulting from cleavage at the bonds indicated by the dashed arrows. The reaction is catalyzed by base[52] or acid. [Merrifield, 1967]. The table shows the effect of the nature of the substituent on the extent of this side reaction. Dmpn = 2,4-dimethylpent-3-yl.

ω-function is esterified than when it is unsubstituted, is favored by ester groups that are electron withdrawing, and is impeded by ester groups that are severely hindered. Carboxamido functions have a moderate tendency to undergo the cyclization reaction. Hindrance by the side chain of the adjacent residue also disfavors the reaction, which is more prevalent in polar solvents.

The amount of imide formed for an –Asp(OR)-Xbb– sequence is in the order OBzl >> OcHex > ODpm (2,4-dimethylpent-3-yl) and Gly/Asn > Ser/Thr > Val. For unknown reasons, an –Asp(OR)-Asn(Trt)– sequence is also very sensitive to imide formation. In addition to being sequence dependent, the reaction is configuration dependent, with the amounts being different when D-residues are included in the chain. An additional implication of imide formation is partial isomerization at the cyclized residue. Formation of imide is not a problem during chain assembly using Boc chemistry because it is not caused by trifluoroacetic acid, but formation of aspartimide is an obstacle at the stage of final deprotection with hydrogen fluoride, or even hydrogen bromide in trifluoroacetic acid. Glutamimide formation does not occur under the same conditions. An approach that minimizes the reaction is use of a phenacetyl (OCH_2COPh) ester instead of a benzyl ester, which is removed immediately before the treatment with hydrogen fluoride. The unprotected aspartyl leads to much less of a side reaction than the esterified aspartyl. One option is to avoid strong acid by use of Bpoc/tBu chemistry (see Section 3.20). If precautions are not taken, imide is formed at every cycle of the synthesis of a peptide containing esterified dicarboxylic-acid residues, employing Fmoc chemistry. *tert*-Butyl, phenacyl, and allyl esters all produce imide in the presence of triethylamine or piperidine. A phenacyl ester even produces imide when triethylamine is added to neutralize the trifluoroacctatc salt of an amino terminus. With piperidine as base, the product is a mixture of two peptide piperidides, –Asp(NC_5H_{10})-Xbb– and –Asp(Xbb-)-NC_5H_{10}, resulting from an opening of the pyrrolidine ring by the nucleophile. The β-piperidide predominates.

Efficient, though not total, suppression of the reaction is achieved in a synthesis using Fmoc chemistry by including an acid such as 1-hydroxybenzotriazole or

2,4-dinitrophenol at 0.1 *M* concentration in the deprotecting solution (piperidine-dimethylformamide, 1:4). Use of piperazine (1,4-diazacyclohexane; see Section 8.12) instead of piperidine also suppresses the reaction. The 3-methylpent-3-yl ester may be the best for minimizing piperidine-induced imide formation. The only way to eliminate the reaction is to temporarily replace the hydrogen atom of the peptide bond by 2-Fmoc-oxy-4-methoxybenzyl (see Section 8.5).[52–62]

52. E Sondheimer, RW Holley. Imides from asparagine and glutamine. (effect of alkali on ester) *J Am Chem Soc* 76, 2467, 1954.
53. A Battersby, JC Robinson. Studies on the specific fission of peptide links. Part 1. The rearrangement of aspartyl and glutamyl peptides. *J Chem Soc* 259, 1955.
54. BM Iselin, R Schwyzer. Synthese of peptide intermediates for the construction of β-melanophore-stimulating hormone (β-MSH) of beef. I. Protected peptide sequences 1-6 and 1-7. [imide by saponification of ROCO-Asp(OMe)-] *Helv Chim Acta* 45, 1499, 1962.
55. MA Ondetti, A Deer, JT Sheehan, J Plušček. Side reactions in the synthesis of peptides containing the aspartylglycyl sequence. *Biochemistry* 7, 4069, 1968.
56. CC Yang, RB Merrifield. The β-phenacyl ester as a temporary protecting group to minimize cyclic amide formation during subsequent treatment of aspartyl peptides with HF. *J Org Chem* 41, 1032, 1976.
57. M Bodanszky, JZ Kwei. Side reactions in peptide synthesis. V11. Sequence dependence in the formation of aminosuccinyl derivatives from β-benzyl-aspartyl peptides. *Int J Pept Prot Res* 12, 69, 1978.
58. M Bodanszky, J Martinez. Side reactions in peptide synthesis. 8. On the phenacyl group in the protection of the β-carboxyl function of aspartyl peptides. *J Org Chem* 43, 3071, 1978.
59. J Martinez, M Bodanszky. Side reactions in peptide synthesis. IX. Suppression of the formation of aminosuccinyl peptides with additives. *Int J Pept Prot Res* 12, 277, 1978.
60. JP Tam, MW Rieman, RB Merrifield. Mechanisms of aspartimide formation: the effects of protecting groups, acid, base, temperature and time. *Pept Res* 1, 6, 1988.
61. R Dölling, M Beyermann, J Haenel, F Kernchen, E Krause, P Franke, M Brudel, M Bienert. Piperidine-mediated side product formation for Asp(OtBu)-containing peptides. *J Chem Soc Chem Commun* 853, 1994.
62. A Karsltröm, A Undén. A new protecting group for aspartic acid that minimizes piperidine-catalyzed aspartimide formation in Fmoc solid phase peptide synthesis. (3-methylpent-3-yl) *Tetrahedron Lett* 37, 4234, 1996.

6.14 THE CARBOXAMIDE GROUPS OF ASPARAGINE AND GLUTAMINE

The carboxamido groups of asparagine and glutamine must be blocked during activation of the residues to prevent cyclization to the imides (see Section 6.13) and dehydration to the cyano function (see Section 6.15). In contrast to the acid- or base-induced reactions (Figure 6.15), cyclization occurs by attack of the nitrogen atom of the carboxamide on the activated carbonyl. Activated esters of the N^{α}-protected residues undergo the same reaction when stored. In addition, derivatives of N^{α}-protected glutamine are sparingly soluble in the usual coupling solvents. Substitution

β,γ- $-CH_2C(=O)-NH_2 + HOR^1$ or $-CH_2CO_2H + H_2NR^2$

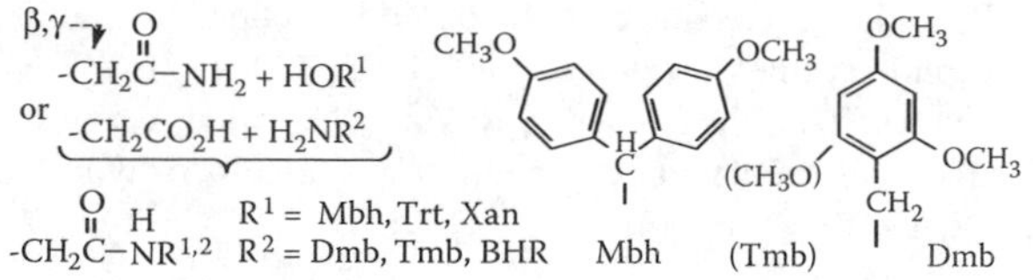

$-CH_2C(=O)-NHR^{1,2}$ R^1 = Mbh, Trt, Xan; R^2 = Dmb, Tmb, BHR

CH_3O, OCH_3, Mbh; (CH_3O), (Tmb); OCH_3, OCH_3, CH_2, Dmb

FIGURE 6.16 Protecting groups for carboxamides. Derivatives are obtained by reaction of the carboxamide with an alcohol or the acid with an amine. Mbh = 4,4′-dimethoxybenzhydryl[64]; Trt = trityl; [Sieber & Iselin, 1991]; Xan = 9-xanthenyl[63]; Dmb = 2,4-dimethoxybenzyl[66]; Tmb = 2,4,6-trimethoxybenzyl[65]; BHR = benzhydryl-resin.[68]

on the carboxamido group increases the solubility of derivatives in organic solvents and eliminates the side reactions. In contrast, the carboxamide group does not undergo acylation or alkylation during chain assembly. Thus, protection is not essential after the residue has been incorporated into the chain, though some imide might be formed at an –Asn-Gly– sequence if a base is present. A benzyl amide is not cleaved by acid, so protectors consist of benzyl with phenyl and methoxy substituents to render the moiety sensitive to acid (Figure 6.16). The traditional protector for Cbz chemistry has been 4,4′-dimethoxybenzyhydryl which is stable to hydrogenolysis. The same protector was initially employed for Fmoc chemistry; it has now been replaced by trityl. Di- and trimethoxybenzyl (Figure 6.16), as well as the substituted benzhydryl, have served as protectors for Boc chemistry; the methoxybenzyl aspartamides are too sensitive to base for use with Fmoc chemistry. One tactic involves use of 9-xanthenyl (see Section 5.20) protection; the protector is removed by acid at the same time as the Boc group for chain assembly using Boc chemistry and at the end of a synthesis employing Fmoc chemistry. Acidolysis of all carboxamido protectors produces stable cations; hence, scavengers are necessary to trap them and to force the cleavage reactions to go to completion. Acidolysis proceeds more slowly when it is the carboxamide of an amino-terminal residue that is being liberated because the amino group is protonated. A methyl on the 4-position of trityl speeds up detritylation. The two approaches for the preparation of derivatives are presented in (Figure 6.16). The solid-phase synthesis of a peptide that has asparagine or glutamine at the carboxy terminus is best achieved by carrying out chain assembly on a benzhydrylamine or equivalent resin (see Sections 5.18 and 5.20), with an aspartyl or glutamyl residue attached through its side chain (Figure 6.16).[63–68]

63. S Akabori, S Sakakibara, Y Shimonishi. Protection of amide nitrogen for peptide synthesis. A novel synthesis of peptides containing C-terminal glutamine. (xanthenyl) *Bull Chem Soc Jpn* 34, 739, 1961.

64. W König, R Geiger. A new amide protecting group. (methoxybenzhydryl) *Chem Ber* 103, 2041, 1970.

65. F Weygand, W Steglich, J Bjarnason. Easily cleavable protective groups for acid amide groups. III. Derivatives of asparagine and glutamine with 2,4-dimethoxybenzl and 2,4,6-trimethoxybenzyl protected amide groups. *Chem Ber* 101, 3642, 1968.

66. PG Pietta, P Cavallo, GR Marshall. 2,4-Dimethoxybenzyl as a protecting group for glutamine and asparagine in peptide synthesis. *J Org Chem* 36, 3966, 1971.

67. P Sieber, B Riniker. Protection of carboxamido functions by the trityl residue. Application to peptide synthesis. *Tetrahedron Lett* 32, 739, 1991.
68. NA Abraham, G Fazal, JM Ferland, S Rakhit, J Gauthier. A new solid phase strategy for the synthesis of mammalian glucagon. (asparagine-benzhydryl resin) *Tetrahedron Lett* 32, 577, 1991.

6.15 DEHYDRATION OF CARBOXAMIDE GROUPS TO CYANO GROUPS DURING ACTIVATION

A major side reaction occurs when a derivative of asparagine or glutamine is activated; namely, dehydration of the carboxamide function to a cyano group (Figure 6.17). The side reaction was first encountered when a synthetic peptide containing an asparaginyl residue showed the presence of 2,4-diaminobutyric acid on analysis. It transpired that the cyano group of the peptide had undergone reduction to the alkylamine during the final deprotection by sodium in liquid ammonia. The peptide had been assembled using dicyclohexylcarbodiimide as coupling reagent. The reaction was later shown to occur during activation of a derivative as the symmetrical anhydride and during BOP-, PyBOP- and HBTU-mediated reactions (see Sections 2.17–2.19), and more recently, during the attempt to form a peptide bond between the carboxyl group of a terminal isoasparaginyl residue and a side-chain amino group (Figure 6.17, B). More β-cyanoalanine is formed from asparagine than γ-cyanobutyrine is formed from glutamine. The intermediate involved is the isoimide formed by nucleophilic attack at the activated carbonyl by the oxygen atom of the carboxamide function (Figure 6.17, A). The cyano group does not interfere with couplings, so dehydrated residues are incorporated into the peptide that is synthesized. The dehydration is completely reversed by hydrogen fluoride, partly reversed by trifluoroacetic acid, and very effectively suppressed for a carbodiimide- or HBTU-mediated reaction by the presence of one equivalent of 1-hydroxybenzotriazole. The additive is a better nucleophile than the carbonyl of the carboxamide. The classical option for circumventing the reaction is the use of activated esters for the couplings. Any cyano contaminant that is generated during preparation of the ester is removed

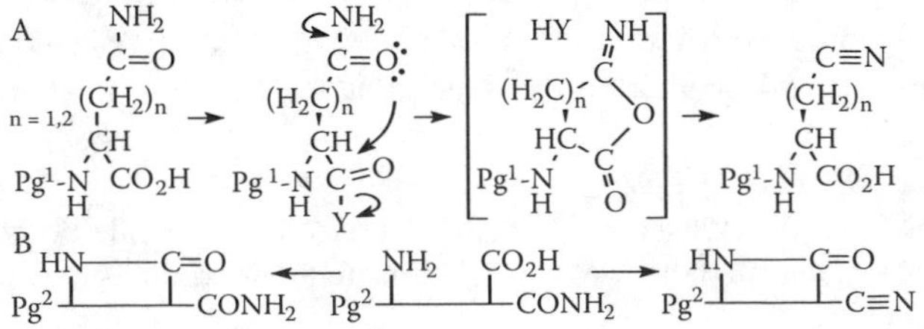

FIGURE 6.17 (A) Dehydration of the carboxamide of N^{α}-protected asparagine or glutamine during activation, producing an ω-cyanoamino acid.[69] The isoimide (brackets) is the postulated intermediate for the reaction.[70,71] [Stammer, 1961]. Cyclization of an octapeptide with a terminal isoasparagine residue (B) gave the desired peptide and also the dehydrated product.[74] Pg^2 = Fmoc on α,γ-diaminobutyroyl.

by recrystallization of the derivative, which, however, must be done with care, so as not to produce more contaminant. The side reaction is avoided by use of derivatives with protected carboxamide functions. Cyano groups can be converted quantitatively into carboxamide groups by the action of hydrogen peroxide in the presence of aqueous sodium carbonate. ω-Cyano-α-amino acids can be prepared from the α-amino-protected ω-carboxamido derivatives by the action of dicyclohexylcarbodiimide or cyanuric chloride [*c*(-N=CHCl-)$_3$; see Section 7.11].[69–75]

69. C Ressler. Formation of α,γ-diaminobutyric acid from asparagine-containing peptides. *J Am Chem Soc* 78, 5956, 1956.
70. B Liberek. Tertiary butyl esters of protected β-cyano-L-alanine peptides as possible intermediates in the preparation of L-asparaginyl peptides. (peroxide for oxidation) *Chem Ind* 987, 1961.
71. DV Kashelikar, C Ressler. An oxygen-18 study of the dehydration of asparagine amide by N,N′-dicyclohexylcarbodiimide and *p*-toluenesulfonyl chloride. *J Am Chem Soc* 86, 2467, 1964.
72. S Mojsov, AR Mitchell, RB Merrifield. A quantitative evaluation of methods for coupling asparagine. *J Org Chem* 45, 555, 1980.
73. H Gausepohl, M Kraft, RW Frank. Asparagine coupling in Fmoc solid phase peptide synthesis. *Int J Pept Prot Res* 34, 287, 1989.
74. P Rovero, S Pegoraro, F Bonelli, A Triolo. Side reactions in peptide synthesis: dehydration of C-terminal aspartylamide peptides during side chain to side chain cyclization. *Tetrahedron Lett* 34, 2199, 1993.
75. P Maetz, M Rodriguez. A simple preparation of N-protected chiral α-aminonitriles from N-protected α-amino acids. *Tetrahedron Lett* 38, 4221, 1997.

6.16 PYROGLUTAMYL FORMATION FROM GLUTAMYL AND GLUTAMINYL RESIDUES

When glutamic acid is heated, an amide bond is formed between the γ-carboxyl group and the amino group to give 2-oxo-pyrrolidine-5-carboxylic acid (Figure 6.18), which is known by the trivial name pyroglutamic acid (pGlu). The same cyclization reaction occurs in a peptide containing a glutamine residue at the amino terminus of a chain, which is left at pH 2–3, and during chain assembly of a peptide, using

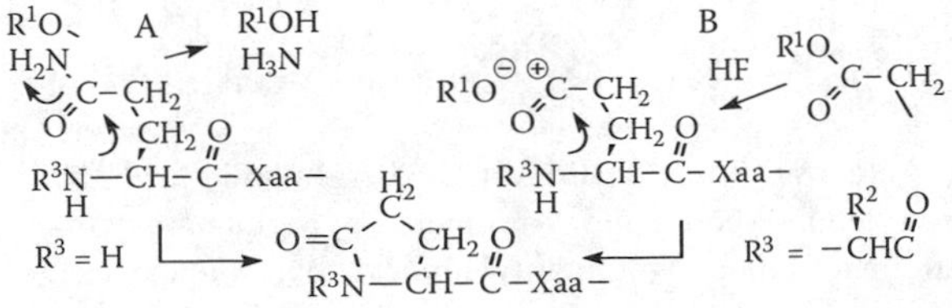

FIGURE 6.18 Formation of pyroglutamyl by cyclization (A) of a terminal glutaminyl[78] or esterified glutamyl residue that is catalyzed by weak acid, and (B) of an intrachain esterified glutamyl residue that is acidolyzed by hydrogen fluoride.[80] The strong acid generates the acylium ion.

Boc/Bzl chemistry after deprotection of a Boc-glutaminyl peptide. The cyclization is catalyzed by weak carboxylic acids such as acetic acid, trifluoroacetic acid, and Boc-amino acids, as well as by 1-hydroxybenzotriazole. Hence, amino-terminal glutaminyl residues should be kept away from these acids as much as possible. Tactics to accomplish this are the use of symmetrical anhydrides (see Section 2.5) instead of carbodiimides for couplings and replacement of trifluoroacetic acid by methanesulfonic acid (see Section 6.22) or hydrogen chloride in dioxane for deprotection or at least removal of trifluorocetic acid as quickly as possible. Pyroglutamyl is not formed when the carboxamide groups are protected (see Section 6.14). The equivalent reaction does not occur for asparaginyl residues, but it occurs when a peptide containing an amino-terminal glutamic acid γ-benzyl ester is in the presence of base and when *N*-protected glutamic acid α-ester is activated. Secondary γ-alkyl esters are less susceptible. It follows that onium salt-mediated couplings to γ-alkyl glutamyl residues should be effected as quickly as possible because they are carried out in the presence of tertiary amine. The reaction also occurs at intrachain esterified glutamyl residues when peptides are treated with hydrogen fluoride and anisole (see Section 6.22) at temperatures above 0°C for more than 30 minutes (Figure 6.18). The reagent dehydrates the protonated carboxyl group that is liberated, giving the acylium ion (see Section 6.23), which reacts with the nitrogen atom of the peptide bond.

Pyroglutamic acid can be coupled directly to an amino group. However, *N*-alkoxycarbonylpyroglutamic acids are sometimes employed instead because the derivatives are more soluble in organic solvents. Acid-sensitive derivatives are obtained by leaving the protected anhydride in the presence of dicyclohexylamine (Figure 6.19). Fmoc-pyroglutamic acid chloride, which can be converted to the acid or succinimido ester, is obtained by slow spontaneous cyclization of the dichloride (Figure 6.19). Residual dichloride is destroyed by addition of water, which decomposes the 2-alkoxy-5(4*H*)-oxazolone that is produced. There have been reports that specimens of pyroglutamic and Boc-pyroglutamic acids were not chirally pure.[76–82]

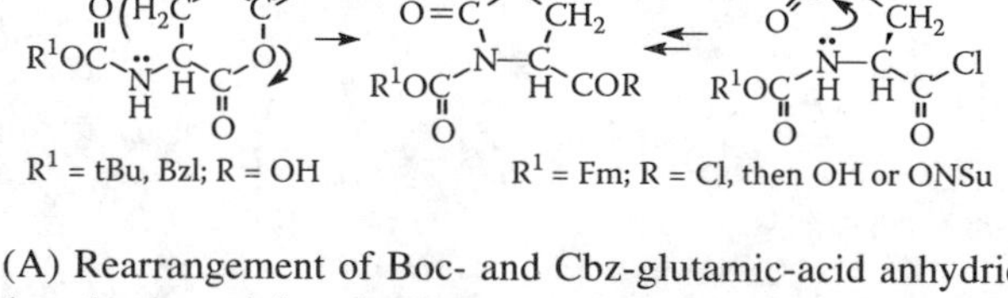

A

R^1 = tBu, Bzl; R = OH

B

R^1 = Fm; R = Cl, then OH or ONSu

FIGURE 6.19 (A) Rearrangement of Boc- and Cbz-glutamic-acid anhydrides in the presence of dicyclohexylamine, giving the *N*-protected pyroglutamic acids. (Gibian & Kliger, 1961). (B) Generation of Fmoc-pyroglutamyl chloride by spontaneous cyclization of Fmoc-glutamyl dichloride.[82] The monochloride can be transformed into the acid and the succinimido ester.

76. J Rudinger. Amino-acids and peptides. X. Some derivatives and reactions of 1-*p*-toluenesulphonyl-L-pyrrolid-5-one-2-carboxylic acid. *Coll Czech Chem Commun* 19, 365. 1954.
77. HE Klieger. On peptide synthesis I. Synthesis of glutamic acid peptides using carbobenzoxy-L-pyroglutamic acid. *Justus Liebig's Ann Chem* 640, 145, 1961.
78. HC Beyerman, TS Lie, CJ van Veldhiuzen. On the formation of pyrogutamyl peptides in solid phase peptide synthesis, in H Nesvadba, ed. *Peptides 1971. Proceedings of the 11th European Peptide Symposium*, North Holland, Amsterdam, 1973, pp 162-164.
79. G Jäger. Preparation of sequence 1-8 of human-proinsulin-peptide-C. (pGlu- from Z-Glu- hydrogenated in acetic acid). *Chem Ber* 106, 206, 1973.
80. RS Feinberg, RB Merrifield. Modification of peptides containing glutamic acid by hydrogen fluoride-anisole mixtures. γ-Acylation of anisole or the glutamyl nitrogen. *J Am Chem Soc* 97, 3485, 1975.
81. RD Dimarchi, JP Tam, SBH Kent, RB Merrifield. Weak-acid catalyzed pyrrolidone carboxylic acid from glutamine during solid phase peptide synthesis. Minimization by rapid coupling. *J Pept Prot Res* 19, 88, 1982.
82. NL Benoiton, FMF Chen. *N*-9-Fluorenylmethoxycarbonylpyroglutamate. Preparation of the acid, chloride and succinimidyl ester. *Int J Pept Prot Res* 43, 321, 1994.

6.17 THE SULFHYDRYL GROUP OF CYSTEINE AND THE SYNTHESIS OF PEPTIDES CONTAINING CYSTINE

The sulfhydryl of cysteine is a good nucleophile that competes with amino groups for activated residues. The result is production of activated esters, which leads to other side reactions. The sulfhydryl also readily undergoes oxidation in the presence of air, so it is essential that it be protected during synthesis. Because most cysteine-containing peptides consist of one or more pairs of cysteine residues joined together through their side chains as disulfides, one option for synthesis is to begin with derivatives of the disulfide-linked pair cystine. Unfortunately, the disulfide bond between two cysteine residues is unstable under several of the operating conditions of peptide synthesis (see Section 6.18), so this approach is rarely employed. Instead, synthesis is effected by assembly of the chain containing the two cysteine residues that are identically protected, the protectors on the functional groups on the side chains of the other residues and at the termini are removed, and the sulfhydryls are liberated and oxidized to the disulfide, as the last step, by bubbling air through the solution at pH 7.0 or adding ferric ion, which gives a faster reaction (Figure 6.20, A). A second option is to protect one of the sulfhydryls by a moiety that also activates the sulfur atom, to remove the protectors on the other side chains and at the termini, to deprotect the other sulfhydryl group, and to then let it react with the activated sulfhydryl (Figure 6.20, B). Nitrogen atmosphere during oxidation favors monomer formation over dimerization. The first peptides containing cysteine were synthesized with benzyl removed by sodium in liquid ammonia (see Section 3.4) for sulfhydryl protection, and with benzyloxycarbonyl for amino protection. These achievements by V. du Vigneaud were deemed significant enough to merit the Nobel Prize for

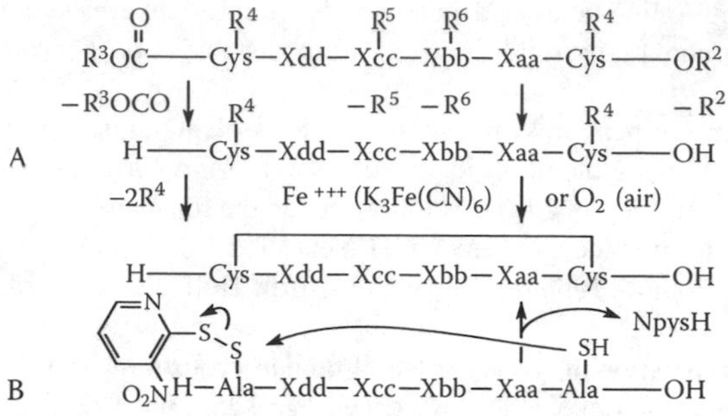

A

$R^3OC(=O)$—Cys(R^4)—Xdd—Xcc(R^5)—Xbb(R^6)—Xaa—Cys(R^4)—OR^2

$-R^3OCO$ ↓ $-R^5$ $-R^6$ ↓ $-R^2$

H—Cys(R^4)—Xdd—Xcc—Xbb—Xaa—Cys(R^4)—OH

$-2R^4$ ↓ Fe^{+++} ($K_3Fe(CN)_6$) ↓ or O_2 (air)

H—Cys—Xdd—Xcc—Xbb—Xaa—Cys—OH

NpysH

SH

B

O_2N H—Ala—Xdd—Xcc—Xbb—Xaa—Ala—OH

FIGURE 6.20 Synthesis of cystine-containing peptides from cysteine-containing peptides by removal of other protectors followed by (A) deprotection of the sulfhydryls and their oxidation to the disulfide, and (B) formation of the disulfide bond by reaction of a liberated sulfhydryl with a sulfhydryl that is protected and activated by 3-nitro-2-pyridylsulfanyl (Npys).[89]

chemistry in 1955. Because some dibenzyl $(PhCH_2)_2$ is produced during the deprotection, the mechanism is believed to involve free radicals. Dethiobenzylation by β-elimination (see Section 3.11), giving the dehydroalanine residue, is a second side reaction accompanying benzyl removal by sodium in liquid ammonia.

Catalytic hydrogenation cannot be employed to remove the benzyl group because sulfur poisons the catalyst, though hydrogenolysis of benzyloxycarbonyl in the presence of sulfur can still be achieved if the reaction is effected in liquid ammonia. However, the presence of several sulfhydryls in a molecule has favorable implications. There are various protectors that can be removed, with the exception of the arylalkyls, by a variety of mechanisms other than acidolysis, with a selectivity that allows pairs of sulfhydryls to be deprotected in sequence. These include reaction with iodine in the presence of an alcohol and displacement by heavy metal ions and thiols. Benzyl is too stable to hydrogen fluoride for routine use as a protector. Commonly employed derivatives with protectors presented in the order of increasing sensitivity to acid are Boc-Cys(Acm, MeBzl, MeOBzl, StBu, tBu, 1-Ada)-OH and Fmoc-Cys(Acm, StBu, tBu, 1-Ada, Trt, Xan)-OH, and Boc-Cys(Trt)-OH for synthesis in solution. Acetamidomethyl (Acm) is removable by mercury(II) acetate or by reaction with iodine in methanol but not in trifluoroethanol (Figure 6.21, A). *tert*-butyl and 1-adamantyl (1-Ada, see Section 6.10) are stable to iodine and removable by mercury(II) acetate in acetic or trifluoroacetic acid (Figure 6.21, B). Trityl is removable by mercury(II) acetate, iodine in the presence of trifluoroethanol, or acidolysis in the presence of triethylsilane. Thus, trityl is removable by iodine without removing acetamidomethyl. The metal ion binds the sulfur irreversibly, thus forcing the reactions to completion, after which the metal is displaced by hydrogen sulfide or an excess of thiol. The alkylsilane reduces the trityl cation to triphenylmethane. The alcohols trap the carbenium ions and thus shift the reactions to completion. Unreacted iodine is advantageously eliminated by adsorption on charcoal. *tert*-Butylsulfanyl, which gives a mixed disulfide ($tBuSSCH_2$-), is sensitive to sodium in liquid ammonia and is removable by thiolysis (Figure 6.22; see Section 3.12). The mixed disulfide does not undergo disulfide interchange (see Section 6.17) because a tertiary carbon atom is linked to the sulfur atom. Acetamidomethyls can be removed

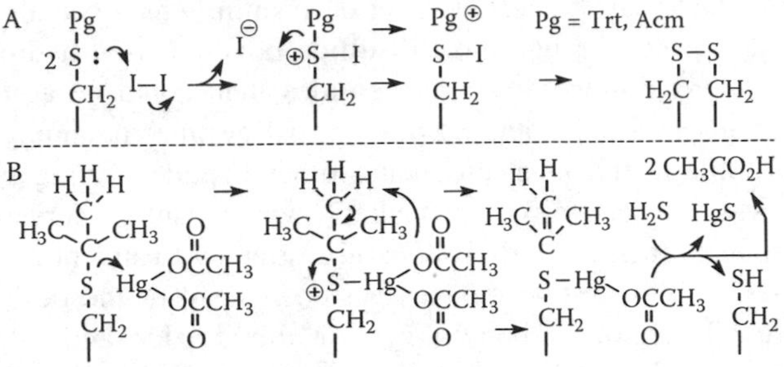

FIGURE 6.21 (A) Removal of trityl and acetamidomethyl from sulfhydryl by oxidative cleavage by iodine. (B) Cleavage of *tert*-butylsulfanyl by mercury(II) acetate,[88] followed by displacement of the metal ion by hydrogen sulfide.

with concomitant formation of disulfide bonds, using thallium(III) trifluoroacetate, which is a mild oxidant with soft-acid (see Section 6.22) character.[83–91]

83. L Zervas, DM Theodoropoulus. N-Tritylamino acids and peptides. A new method of peptide synthesis. (*S*-trityl) *J Am Chem Soc* 78, 1359, 1959.
84. E Wünsch, R Spangenburg. A new S-protecting group for cysteine. (in German) (*S-tert*-butyl) in E Scoffone, ed. *Peptides 1969. Proceedings of the 10th European Peptide Symposium*, North-Holland, Amsterdam, 1971, pp 30-34.
85. DF Veber, JD Milkouski, RG Denkewalter, R Hirschmann. The synthesis of peptides in aqueous solution. IV. A novel protecting group for cysteine. (*S*-acetamidomethyl) *Tetrahedron Lett* 3057, 1968.
86. DF Veber, JD Milkouski, SL Varga, RG Denkewalter, R Hirschmann. Acetamidomethyl. A novel thiol protecting group for cysteine. *J Am Chem Soc* 94, 5456, 1972.
87. P Sieber, B Kamber, A Hartmann, A Jöhl, B Riniker, W Rittel. Total synthesis of human insulin. IV. Description of the final steps. *Helv Chim Acta* 60, 27, 1977.
88. AM Felix, MH Jimenez, T Mowles, J Meienhofer. Catalytic hydrogenolysis in liquid ammonia. Cleavage of N^{α}-benzyloxycarbonyl groups from cysteine-containing peptides with *tert*-butyl side chain protection. *Int J Pept Prot Res* 11, 329, 1978.
89. R Matsueda, K Aiba. A stable pyridinesulfenyl halide. *Chem Lett (Jpn)* 951, 1978.
90. N Fujii, A Otaka, S Funakoshi, K Bessho, T Watanabe, K Akaji, H Yajima. Studies on peptides. CL1. Syntheses of cystine-peptides by oxidation of S-protected cysteine peptides with thalliumIII trifluoroácetate. *Chem Pharm Bull (Jpn)* 35, 2339, 1987.
91. D Sahal. Removal of iodine by solid phase adsorption to charcoal following iodine oxidation of acetamidomethyl-protected peptide precursors to their disulfide bonded products: oxytocin and a Pre-S_1 peptide of hepatitis B illustrate the method. *Int J Pept Res* 53, 91, 1999.

6.18 DISULFIDE INTERCHANGE AND ITS AVOIDANCE DURING THE SYNTHESIS OF PEPTIDES CONTAINING CYSTINE

The synthesis of peptides containing cystine is rarely performed starting with derivatives of cystine because of the obstacle of disulfide interchange. Disulfide interchange was discovered by F. Sanger during his attempts to determine the primary structure of the dual-chain 52-residue peptide insulin, which contains six cysteine residues linked together through three disulfide bonds. Partial hydrolysis by mild acid produced more cystine-containing peptides than could be accounted for by hydrolysis. It transpired that some unsymmetrical cystine-containing peptides had undergone disulfide interchange, producing other peptides during treatment. The phenomenon was first recognized in synthetic work when L. Zervas attempted to prepare the monohydrazide of a derivative of cystine by treatment of the N^{α}-benzyloxycarbonyl-$N^{\alpha'}$-trityl-L-cystine dimethyl ester. Instead of the desired result, the products were dibenzyloxycarbonyl-L-cystine dihydrazide and ditrityl-L-cystine dimethyl ester, with the trityl group having prevented hydrazinolysis as expected because of hindrance (Figure 6.22, A). Further work established that unsymmetrical derivatives of cystine undergo conversion to two symmetrical derivatives of cystine in the presence of strong acid such as hydrogen bromide and weak base such as pH 7.5. Acid generates the alkylthio cation (Figure 6.22, B), whereas alkaline pH produces the alkylthio anion (Figure 6.22, C), both of which initiate a chain reaction. Thiols suppress the acid-catalyzed reaction and promote the base-catalyzed reaction by shifting the equilibrium of the reactions. No interchange occurs at pH 6.5.

Interchange occurs less readily when one of the sulfur atoms is linked to a secondary carbon atom and does not occur when it is linked to a tertiary carbon atom (see Section 6.17). The effect is unrelated to steric factors. It is, however, possible to avoid disulfide interchange by use of mild operating conditions. This was elegantly demonstrated by the work of Swiss scientists in their synthesis of insulin (Figure 6.23). A cysteinyl dipeptide *tert*-butyl ester was reacted with a trityl-protected peptide containing a side-chain-activated cysteine residue. The chains were extended by 1-hydroxybenzotriazole-assisted carbodiimide-mediated couplings with a Bpoc-protected peptide and a peptide *tert*-butyl ester, respectively. The trityl group was removed with mild acid, and the chain was extended by coupling with a Boc-protected peptide containing an acetamidomethyl-protected cysteine residue, the Bpoc group was removed by warm trifluoroethanol, and the chain was extended by coupling with a Boc-peptide containing an acetamidomethyl-protected cysteine

A
Cbz-NH O
SCH_2CHC-OMe
SCH_2CHC-OMe
Trt-NH O
$\xrightarrow{N_2H_4}$
Cbz-NH O
SCH_2CHC-N_2H_3
SCH_2CHC-N_2H_3
Cbz-NH O
+
Trt-NH O
SCH_2CHC-OMe
SCH_2CHC-OMe
Trt-NH O

B Initiation: $R^1S\text{-}SR^2 + H^{\oplus} \rightleftharpoons R^1\overset{\oplus}{S}(H)\text{-}SR^2 \rightleftharpoons R^1SH + {}^{\oplus}SR^2$

Propagation: ${}^{\oplus}SR^2 + R^1S\text{-}SR^2 \longrightarrow R^2S\text{-}SR^2 + R^1S^{\oplus}$

C Initiation: $R^1S\text{-}SR^2 + B: \rightleftharpoons R^1SB + {}^{\ominus}SR^2$

Propagation: ${}^{\ominus}SR^2 + R^1S\text{-}SR^2 \longrightarrow R^2S\text{-}SR^2 + R^1S^{\ominus}$

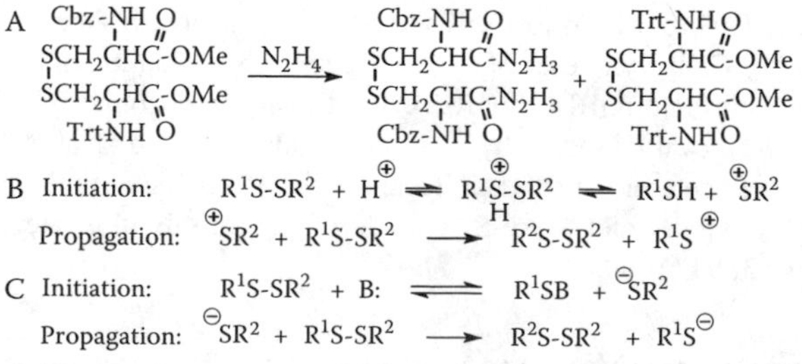

FIGURE 6.22 Disulfide interchange.[92] (A) Discovered in synthesis when hydrazinolysis of an unsymmetrical derivative of cystine gave two symmetrical products instead of the expected monohydrazide at the urethane-protected cysteine moiety of the derivative.[95] (B) Mechanism for interchange catalyzed by strong acid,[94] which is suppressed by thiols. (C) Mechanism for interchange catalyzed by weak alkali, which is enhanced by thiols.

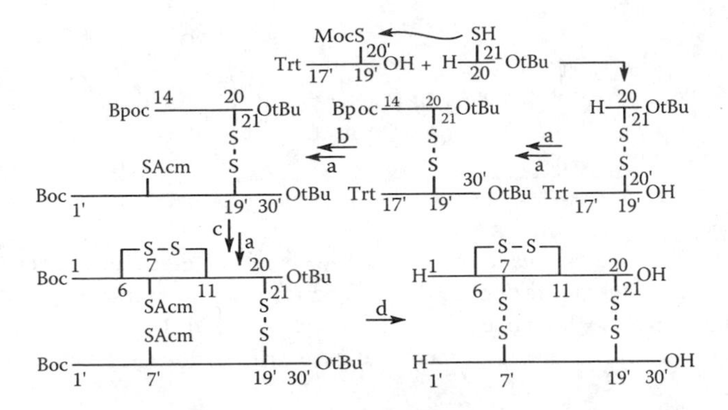

FIGURE 6.23 The synthesis of insulin, starting with a cystine-containing peptide. [Kamber et al., 1977]. Moc = methoxycarbonyl, Bpoc = biphenylisopropoxycarbonyl, Trt = trityl, Acm = acetamidomethyl. (a) HOBt-assisted carbodiimide-mediated coupling; (b) removal of Trt by HCl in CF_3CH_2OH-CH_2Cl_2 (9:1) at pH 3.5; (c) removal of Bpoc by CF_3CH_2OH-CH_2Cl_2 (9:1) at 60°C; (d) removal of Acm and oxidation by iodine.

residue as well as two cysteine residues linked through their side chains. The third disulfide bond was formed by removal of the acetamidomethyl groups and oxidation by iodine in acetic acid, and the *tert*-butyl protectors were removed by 95% trifluoroacetic acid. Thus, choice of appropriate conditions for deprotection and coupling allows the synthesis of peptides containing cystine in the absence of significant disulfide interchange.[92–97]

92. AP Ryle, F Sanger. Disulfide interchange reactions. *Biochem J* 535, 1955.
93. AP Ryle, F Sanger, LF Smith, R Kitai. Disulfide bonds of insulin. *Biochem J* 541, 1955.
94. RE Benesch, R Benesch. The mechanism of disulfide interchange in acid solution; role of sulfenium ions. *J Am Chem Soc* 80, 1666, 1958.
95. L Zervas, L Benoiton, E Weiss, M Winitz, JP Greenstein. Preparation and disulfide interchange reactions of unsymmetrical open-chain derivatives of cystine. *J Am Chem Soc* 81, 1729, 1959.
96. P Sieber, B Kamber, A Hartmann, A Jöhl, B Riniker, W Rittel. Total synthesis of human insulin under directed formation of disulfide bonds. *Helv Chim Acta* 57, 2617, 1974.
97. P Sieber, B Kamber, A Hartmann, A Jöhl, B Riniker, W Rittel. Total synthesis of human insulin. IV. Description of the final product. *Helv Chim Acta* 60, 27, 1977.

6.19 PIPERAZINE-2,5-DIONE FORMATION FROM ESTERS OF DIPEPTIDES

A peptide chain consists of a succession of –C(=O)-N-C^{α}– atoms that are coplanar, with the C-N bond being shorter than that of a normal amide. There is a partial sharing of the π-electrons between the C=O and the C-N bond, giving the latter double-bond character (~40%), so that it is unable to rotate freely. The NH-proton and the oxygen atom are in the same plane but in a *trans* relationship (Figure 6.24).

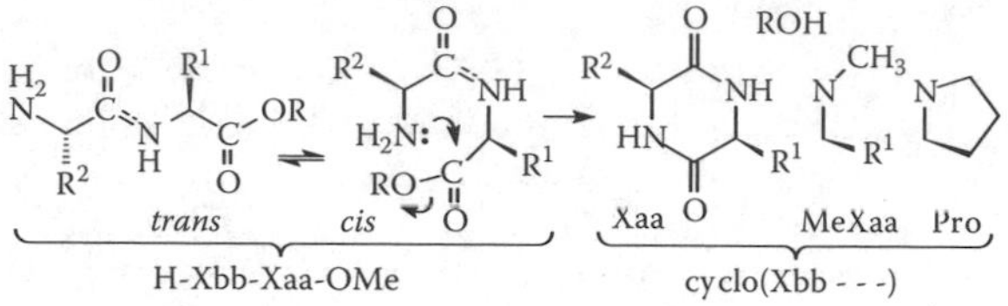

FIGURE 6.24 The *cis* and *trans* forms of the amide bond of a dipeptide ester and cyclization of the compound to the piperazine-2,5-dione. The tendency to cyclize is greater when the carboxy-terminal residue is proline or an *N*-methylamino acid. In these cases the predominating form is *cis*, which places the amino and ester groups closer together.

In the case of a peptide bond incorporating the nitrogen of a secondary amino acid such as proline or an *N*-methylamino acid, these atoms are in a *cis* relationship (Figure 6.24). When a peptide chain is assembled, at the dipeptide ester stage there is a tendency for the terminal functional groups to react together to form a ring. The product is a piperazine-2,5-dione (2,5-dioxopiperazine, diketopiperazine), which is a cyclic dipeptide. Cyclization occurs more readily when the alkoxy or aryloxy leaving group is electron withdrawing, such as in an activated ester, and when the aforementioned atoms are in a *cis* relationship that places the reacting groups closer together (Figure 6.24). The tendencies for compounds H-Phe-Xxx-OMe to cyclize are in the order Xxx = MeXaa > Pro > Gly > Val. A dramatic example of the ease of cyclization is provided by the fact that it is impossible to determine the p*K* of glycyl-*N*-methylglycine methyl ester by titration because the latter cyclizes too quickly when the pH is raised above neutral. Cyclization is not limited to dipeptide esters. Peptides containing the tyrosyl-tetrahydroquinoline-3-carboxylic-acid sequence at the amino terminus spontaneously release the corresponding piperazinedione after dissolution in dimethylsulfoxide. Dipeptides with the secondary amino acid at the amino terminus cyclize slightly more readily than a normal dipeptide ester, but this is because of the greater basicity and nucleophilicity of the amino groups. A dipeptide ester with residues of opposite configuration cyclizes faster than one with residues of identical configuration. The cyclization is catalyzed by mild acid as well as base.

Once it is part of a cyclic dipeptide, the prolyl residue becomes susceptible to enantiomerization by base (see Section 7.22). The implication of the tendency of dipeptide esters to form piperazine-2,5-diones is that their amino groups cannot be left unprotonated for any length of time. The problem arises during neutralization after acidolysis of a Boc-dipeptide ester and after removal of an Fmoc group from an Fmoc-dipeptide ester by piperidine or other secondary amine. The problem is so severe with proline that a synthesis involving deprotection of Fmoc-Lys(Z)-Pro-OBzl produced only the cyclic dipeptide and no linear tripeptide. The problem surfaces in solid-phase synthesis after incorporation of the second residue of a chain that is bound to the support by a benzyl-ester type linkage. There is also the added difficulty that hydroxymethyl groups are liberated, and they can be the source of other side reactions.

Dioxopiperazine formation is avoided by esterifying the carboxy terminus of a chain with a tertiary alcohol, which provides a poor leaving group. *tert*-Butyl and trityl esters (see Section 5.23) of dipeptides are resistant to the cyclization reaction. The reaction is circumvented in a synthesis by incorporating the second and third residues together as the protected dipeptide. The reaction is minimized by liberating the amino group of a protected dipeptide ester in the presence of the derivative of the third residue that is already activated, such as the activated ester, or that is being activated by a quick-acting reagent, such as onium salt-based reagents. Dipeptide esters are sometimes employed as aminolyzing components of model systems employed to obtain information on enantiomerization during couplings. It follows from the above that only esters that are resistant to the cyclization reaction will provide data that are reliable.[98–101]

98. HN Rydon, PWG Smith. Self-condensation of the esters of peptides of glycine and proline. *J Chem Soc* 3642, 1956.
99. BF Gisin, RB Merrifield. Carboxyl-catalyzed intramolecular aminolysis. A side reaction in solid-phase peptide synthesis. *J Am Chem Soc* 94, 3102, 1972.
100. JC Purdie, NL Benoiton. Piperazinedione formation from esters of dipeptides containing glycine, alanine and sarcosine: the kinetics in aqueous solution. *J Chem Soc Perk Trans* 2, 1845, 1973.
101. BJ Marsden, TM Nguyen, PW Schiller. Spontaneous degradation via diketopiperazine formation of peptides containing a tetrahydroquinoline-3-carboxylic acid residue in the 2-position of the peptide sequence. *Int J Pept Prot Res* 41, 313, 1993.

6.20 N-ALKYLATION DURING PALLADIUM-CATALYZED HYDROGENOLYTIC DEPROTECTION AND ITS SYNTHETIC APPLICATION

Two reports in the literature of the early 1990s described the side reaction of *N*-methylation that occurred during the palladium-catalyzed hydrogenolytic cleavage of *N*-benzyloxycarbonyl groups in anhydrous methanol. The same reaction had been described more than a decade earlier. Alkylation occurs during hydrogenolytic deprotection if the system is not completely freed of oxygen. In the presence of oxygen, the palladium catalyst dehydrogenates methanol and ethanol to the corresponding aldehydes (Figure 6.25). Amino groups that are liberated by reduction react with the aldehydes to produce Schiff's bases, which readily undergo hydrogenation to

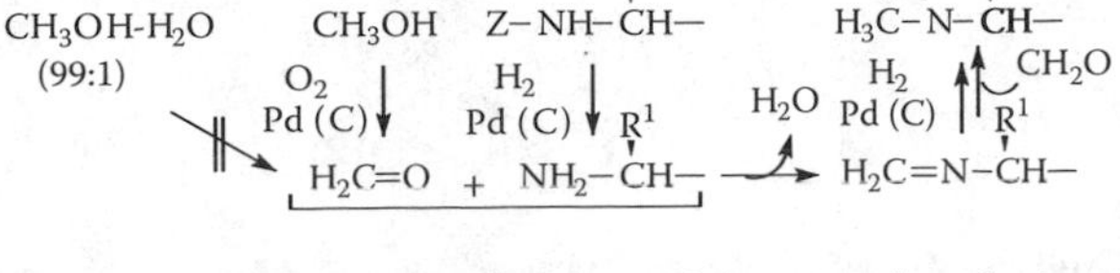

CH_3OH-H_2O (99:1) ⇸ ; CH_3OH $\xrightarrow{O_2,\ Pd(C)}$ $H_2C{=}O$; Z–NH–CH(R^1)– $\xrightarrow{H_2,\ Pd(C)}$ NH_2–CH(R^1)–

$H_2C{=}O$ + NH_2–CH(R^1)– $\xrightarrow{-H_2O}$ $H_2C{=}N$–CH(R^1)– $\xrightarrow{H_2,\ Pd(C)}$ … $\xrightarrow{CH_2O}$ H_3C–N(CH_3)–CH(R^1)–

FIGURE 6.25 Catalytic hydrogenolysis of *N*-protecting groups in anhydrous methanol in the presence of oxygen produces *N,N*-dimethylated products, (Chen & Benoiton, 1976) which originate by reductive alkylation of the Schiff's base formed by reaction of the amino group with formaldehyde,[102] generated by the palladium-catalyzed dehydrogenation of methanol. (Wieland, 1912).

alkyl groups — a well-known process referred to as reductive alkylation. Dialkylation occurs because secondary amines are more readily alkylated than primary amines. The reaction has been employed for synthetic purposes such as the preparation of N^{α}-Boc-N^{ε},N^{ε}-dimethyl-L-lysine from N^{α}-Boc-N^{ε}-Cbz-L-lysine. 2-Propanol is not oxidized by palladium catalyst. *N*-Isopropylamino acids are accessible by reduction of amino groups in the presence of acetone; the reaction stops at the monoalkyl stage because the reactant is a ketone. Palladium-catalyzed oxidation of alcohols does not occur if a trace of water is present. Thus, palladium-catalyzed hydrogenolytic deprotections should not be carried out in anhydrous methanol or ethanol unless the system is completely freed of oxygen or contains a small amount (5%) of water. 2-Propanol is also a suitable solvent.[102–104]

102. RE Bowman, HH Stroud. N-Substituted amino acids. I. A new method of preparation of dimethylamino acids. *J Chem Soc* 1342, 1950.
103. FMF Chen, NL Benoiton. Reductive *N,N*-dimethylation of amino acid and peptide derivatives using methanol as the carbonyl source. *Can J Biochem* 56, 150, 1978.
104. NL Benoiton. On the side reaction of *N*-alkylation of amino groups during hydrogenolytic deprotection in alcohol-containing solvents. *Int J Pept Prot Res* 41, 611, 1993.

6.21 CATALYTIC TRANSFER HYDROGENATION AND THE HYDROGENOLYTIC DEPROTECTION OF SULFUR-CONTAINING PEPTIDES

It was demonstrated in 1952 that ethylene or acetylene bonds of compounds undergo reduction if they are in the presence of cyclohexene and palladium. From this observation has emerged a process called catalytic transfer hydrogenation, in which hydrogen atoms are transferred from one organic compound to another by a palladium catalyst. No hydrogen or protons are generated — the dehydrogenation and hydrogenation reactions occur simultaneously. The hydrogen donors are not restricted to cyclohexene, but cyclohexene in boiling ethanol is the donor of choice for practical purposes (Figure 6.26). Equally popular now are formic acid or ammonium formate in methanol at ambient temperature, with the reacting species being formate anion. The reaction is employed for removal of protecting groups that are susceptible to reduction. The method obviates the need for a cylinder of hydrogen

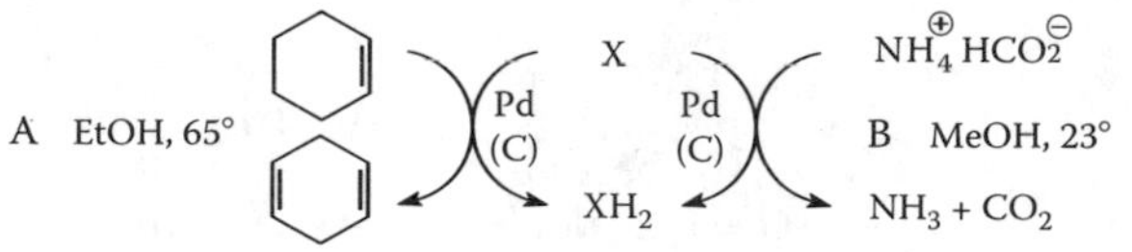

A EtOH, 65° — X — Pd (C) — XH_2 — $NH_4^{\oplus}HCO_2^{\ominus}$ — B MeOH, 23° — $NH_3 + CO_2$

FIGURE 6.26 Catalytic transfer hydrogenation,[105] with (A) cyclohexene as hydrogen donor,[108] and (B) ammonium formate as hydrogen donor.[110] X, cleavable at the indicated bond, can be $PhCH_2$–CO_2C–, $PhCH_2$–CO_2NH–, $PhCH_2$–OCH_2–, –Arg–(NO_2)–, and –His–(Bzl or Trt)–.

as a source of the reductant, or a closed or pressurized reaction assembly for effecting the reaction. The reductions at ambient temperature can be carried out in batches or on a column containing the catalyst. The amount of catalyst required, which is usually 10% palladium on charcoal, is relatively large, but the same catalyst can be used many times. *N*-Benzyloxycarbonyl and benzyl esters are cleaved in minutes, benzyl ethers in an hour or so, and *N*-9-fluorenylmethoxycarbonyl after several hours. Nitroarginine is reduced successfully. The reagents obviously do not affect carbonyls, but they dehalogenate chloroaromatics, thus allowing replacement of chloro in chlorophenylalanine-containing peptides by isotopes of hydrogen. The use of ammonium formate, acetic acid, and palladium hydroxide, which leads to the deposition of palladium black, allows for deprotection and detachment of peptides that are bound to resins as the benzyl esters. The use of cyclohexadiene as hydrogen donor allows removal of benzyl esters, which includes benzyloxycarbonyl, without affecting benzyl ethers.

Catalytic hydrogenolysis with hydrogen gas as the reactant (see Section 3.3) cannot be employed for removing protectors from peptides containing cysteine or methionine residues because the sulfur poisons the catalyst. There are, however, several alternatives for hydrogenolytic deprotection of sulfur-containing peptides. Boron trifluoride etherate suppresses the poisoning effect in some cases. Catalytic hydrogenation in liquid ammonia with palladium black as catalyst removes benzyl esters, which includes benzyloxycarbonyl, from sulfur-containing peptides, with the S-benzyl groups remaining intact. Catalytic transfer hydrogenation is applicable to methionine-containing peptides, and catalytic hydrogenolysis of *N*-benzyloxycarbonyl methionine-containing peptides is successful if tertiary amine is present. In fact, catalytic hydrogenolysis occurs more quickly than in the presence of acid because the imino tautomer [$PhCH_2OC(OH)=N-$] is reduced faster than the urethane [$PhCH_2OC(=O)NH-$].[105–115]

105. RP Linstead, EA Braude, PWD Mitchell, KRH Wooldridge, LM Jackman. Transfer of hydrogen in organic systems. *Nature (London)* 169, 100, 1952.

106. H Medzihradsky-Schweiger. Promoted hydrogenolysis of carbobenzoxyamino acids in the presence of organic bases. *Acta Chim (Budapest)* 76, 437, 1973.

107. K Kuromizu, J Meienhofer. Removal of the N^{α}-benzyloxycarbonyl group from cysteine-containing peptides by catalytic hydrogenolysis in liquid ammonia, exemplified by a synthesis of oxytocin. *J Am Chem Soc* 96, 4978, 1974.

108. AE Jackson, RAW Johnstone. Rapid, selective removal of benzyloxycarbonyl groups from peptides by catalytic transfer hydrogenation. *Synthesis* 685, 1976.

109. GM Anantharamaiah, KM Sivanandaiah. Transfer hydrogenation. A convenient method for the removal of commonly used protecting groups in peptide synthesis. (formic acid) *J Chem Soc Perk Trans 1* 490, 1977.

110. MK Anwer, AF Spatola. An advantageous method for the rapid removal of hydrogenolysable protecting groups under ambient conditions; synthesis of leucine-enkephalin. (ammmonium formate) *Synthesis* 929, 1980.

111. G Losse, H-U Stiehl, B Schwenger. Hydrogenolytic debenzylation of sulfur-containing peptides. *Int J Pept Prot Res* 19, 114, 1982.

112. Y Okada, N Ohta. Amino acids and peptides. VII. Synthesis of methionine-enkephalin using transfer hydrogenation. (cyclohexene) *Chem Pharm Bull (Jpn)* 30, 581, 1982.

113. MK Anwer, AF Spatola. Quantitative removal of a pentadecapeptide ACTH fragment analogue from a Merrifield resin using ammonium formate catalytic transfer hydrogenation: synthesis of [Asp^{25},Ala^{26},Gly^{27},Gln^{30}]-ACTH-(25-30)-OH. *J Org Chem* 48, 3503, 1983.
114. MK Anwer, RA Porter, AF Spatola. Applications of ammonium formate-catalytic transfer hydrogenation. Part V (1). Transfer hydrogenation of peptides containing *p*-chlorophenylalanine as a convenient method for preparation of deuterium labeled peptides. *Int J Pept Prot Res* 30, 489, 1987.
115. JS Bajwa. Benzyl esters in the presence of benzyl ethers. *Tetrahedron* 33, 2299, 1992.

6.22 MECHANISMS OF ACIDOLYSIS AND THE ROLE OF NUCLEOPHILES

Removal of protectors by acidolysis involves protonation, followed by spontaneous rupture of the cation, a unimolecular or S_N1 reaction (see Section 3.5), or rupture of the cation by attack by a nucleophile, by a bimolecular or S_N2 reaction, or by rupture of the protonated intermediate by both mechanisms (Figure 6.27). In the first case, the protector or part thereof is released as a cation; in the second case, the same moiety is displaced and secured by the nucleophile (Figure 6.27). The latter mechanism is preferred because an expelled cation has a tendency to attack the nucleophilic centers on the side chains of the peptide (see Section 3.7). The nucleophile can be the conjugate base of the acid or another compound that either facilitates the cleavage or is required to effect it. In addition to actively participating in the cleavage reaction, the nucleophile may also serve as a scavenger for the cation that is liberated. The cations can be carbenium ($R^1R^2R^3C^+$), acylium ($RC^+{=}O$), or arylsulfonylium (RS^+O_2) ions, including resin-bound species. Whether side-chain functional groups are targets for the electrophilic cations depends on whether they are protonated or not, which in turn depends on the strength of the acid employed. In any acidolysis, one or more nucleophiles are added to the acid to trap the cations that are generated. These scavengers are bases that are weak enough not to be protonated by the acid.

The popular acids for deprotection by acidolysis are hydrogen fluoride for benzyl-based protectors and trifluoroacetic acid for *tert*-butyl-based protectors. The use of hydrogen fluoride for deprotection emerged from the observation that it is a good solvent for dissolving enzymes (because of the *N*-to-*O* acyl shift; see Section 6.6), and that the enzymatic activity is recovered (*O*-to-*N* acyl shift) in saline solution. Two different approaches are employed for removal of benzyl-based

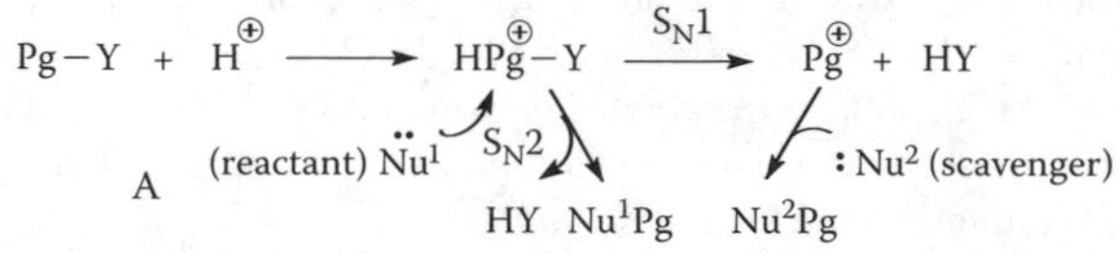

$$Pg{-}Y + H^{\oplus} \longrightarrow HPg^{\oplus}{-}Y \xrightarrow{S_N1} Pg^{\oplus} + HY$$

A (reactant) $\ddot{N}u^1$ S_N2 → HY Nu^1Pg; $Pg^{\oplus}$ + $:Nu^2$ (scavenger) → Nu^2Pg

FIGURE 6.27 Acidolysis removes a protector by one or both of the depicted mechanisms. Pg = protector, Y = residue, Nu = nucleophile. Nu^1 and Nu^2 may be identical or different. CO_2 is liberated when Pg = alkoxycarbonyl.

protectors using hydrogen fluoride. Acid at high concentration (>55%), which cleaves by the S_N1 mechanism with anisole ($PhOCH_3$), as scavenger, and acid at lower concentration with the assistance of a second nucleophile, which cleaves by the S_N2 mechanism. The fluoride ion, in contrast to the bromide anion (see Section 3.5), is too weak a nucleophile to participate in the reaction, but the hydrogen halide is strong enough to protonate most side-chain functional groups, so they are not targets for the cations. However, one nucleophilic center that is subject to alkylation, the *ortho* carbon to the hydroxyl group of tyrosine, remains. This presents a problem because the more stable ethers require a high concentration of acid to be cleaved, and the S_N1 removal of this side-chain substituent leads to transfer of the moiety to the adjacent position in amounts as high as 30%.

The recommended protocol for acidolysis by hydrogen fluoride is thus use of the acid diluted with dimethylsulfide, which acts as nucleophile to remove protecting moieties by the S_N2 mechanism, followed by evaporation of the diluent to give the high acid concentration required for removal of the more stable protectors by the S_N1 mechanism. The inclusion of dimethylsulfide has the added feature that at low acid concentration, the reagent reduces any methionine sulfoxide (see Section 6.8) back to methionine. A second side reaction associated with acidolysis by hydrogen fluoride assisted by anisole is cleavage of a benzyl ester at oxycarbonyl ($PhCH_2O–C=O$) instead of at benzyloxy ($PhCH_2–OC=O$), followed by acylation of anisole by the acylium ion (Figure 6.28; see Section 6.12). The acylation is avoided by effecting the cleavage at 0°C for less than 30 minutes. Experience has shown that thioanisole ($PhSCH_3$) is superior to anisole for minimizing alkyl transfer giving 3-alkyl-tyrosine. The thio ether has two nucleophilic sites (sulfur atom and *para*-carbon atom) and had been introduced as nucleophile for acidolysis, using boron trifluoride. There is a greater propensity for the cations (soft acids) to go to the sulfur atom (soft base), and the ether (hard base) is more readily protonated by H^+ (hard acid) than sulfur. However, the alkylated thioanisole acts as a reagent that methylates methionine, so it is common to use anisole as nucleophile for peptides that do not contain tyrosine, and thioanisole for tyrosine-containing peptides.

Trifluoroacetic acid removes *tert*-butyl-based protectors by the S_N1 mechanism, with the cation being trapped by the trifluoroacetate anion; however, the *tert*-butyl trifluoroacetate produced is an alkylating agent, and the acid is not strong enough to protonate the side chains of methionine, tryptophan, and cysteine, so these are acceptors of *tert*-butyl. A scavenger is required to prevent their alkylation. Anisole

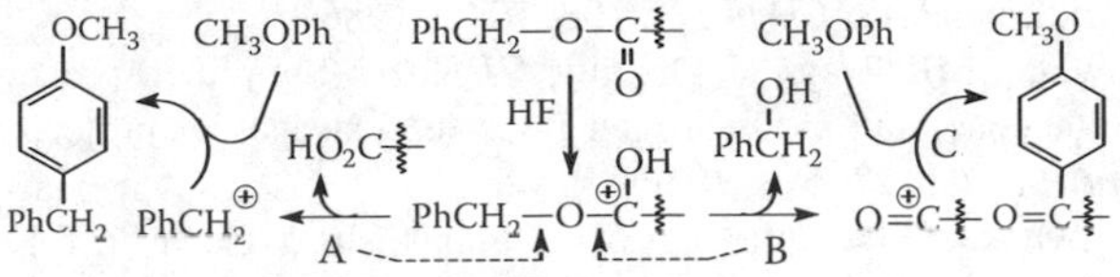

FIGURE 6.28 Protonation of a benzyl ester by hydrogen fluoride,[116] followed by S_N1 cleavage, (A) giving the benzyl cation and (B) giving the acylium ion,[118] and their reactions with anisole ($PhOCH_3$). (C) = Friedel-Crafts acylation. Cresol as the nucleophile would react with the acylium ion, generating an ester ($–CO_2C_6H_4CH_3$) that is saponifiable.

was the original scavenger; water is also an effective scavenger for *tert*-butyl. No migration of *tert*-butyl from the *para*-position of tyrosine to the adjacent nucleophilic carbon occurs because of the tertiary nature of the alkyl group. Deprotection by trifluoroacetic acid occurs by the S_N2 mechanism if thioanisole is present; benzyloxycarbonyl is cleaved by this mixture. Two other acids mixed with trifluoroacetic acid are employed for acidolysis; namely, trifluoromethanesulfonic acid and methanesulfonic acid. The former is a viscous liquid stronger than hydrogen fluoride but not requiring special equipment. The weaker methanesulfonic acid has provided efficient acidolysis for smaller peptides assembled by solid-phase synthesis. Trifluoromethanesulfonic acid trimethylsilyl ester in trifluoroacetic acid with cresol or anisole provides a simple alternative to hydrogen fluoride.[80,116–125]

80. RS Feinberg, RB Merrifield. Modification of peptides containing glutamic acid by hydrogen fluoride-anisole mixtures. γ-Acylation of anisole or the glutamyl nitrogen. *J Am Chem Soc* 97, 3485, 1975.
116. S Sakakibara, Y Shimonishi, Y Kishada, M Okada, H Sugiraha. Use of anhydrous hydrogen fluoride in peptide synthesis. I. Behavior of various protective groups in anhydrous hydrogen fluoride. *Bull Chem Soc Jpn* 40, 2164, 1967.
117. W Bauer, J Pless. The use of boron tristrifluoroacetate (BTFA) in the synthesis of biologically active peptides, in R Walter, J Meienhofer, eds. *Peptides: Chemistry, Structure, Biology. Proceedings of the 4th American Peptide Symposium.* Ann Arbor Science, Ann Arbor, MI, 1975, pp 341-345.
118. S Sano, S Kawashini. Hydrogen fluoride-anisole catalyzed reaction with glutamic acid containing peptides. *J Am Chem Soc* 97, 3480, 1975.
119. H Yajima, Y Kiso, H Ogawa, N Fujii, H Irie. Studies on peptides. L. Acidolysis of protecting groups in peptide synthesis by fluorosulphonic acid and methanesulfonic acid. *Chem Pharm Bull (Jpn)* 23, 1164, 1975.
120. Y Kiso, K Ukawa, T Akita. Efficient removal of *N*-benzyloxycarbonyl group by a "push-pull" mechanism using thioanisole-trifluoroacetic acid, exemplified by a synthesis of Met-enkephalin. *J Chem Soc Chem Commun* 101, 1980.
121. JW van Nispen, JP Polderdijk, WPA Janssen, HM Greven. Replacement of hydrogen fluoride in solid phase peptide synthesis by methanesulfonic acid. *Rec Trav Chim Pays-Bas* 100, 435, 1981.
122. JP Tam, WF Heath, RB Merrifield. S_N2 Deprotection of synthetic peptides with a low concentration of HF in dimethylsulfide: evidence and application in peptide synthesis. *J Am Chem Soc* 105, 6442, 1983.
123. JP Tam, WF Heath, RB Merrifield. Mechanism for the removal of benzyl protecting groups in synthetic peptides by trifluoromethanesulfonic acid-trifluoroacetic acid-dimethyl sulfide. *J Am Chem Soc* 108, 5242, 1986.
124. RA Houghten, MK Bray, ST Degraw, CJ Kirby. Simplified procedure for carrying out simultaneous multiple hydrogen fluoride cleavages of protected peptide resins. *Int J Pept Prot Res* 27, 673, 1986.
125. N Fujii, A Otaka, O Ikemura, K Akiji, S Funakoshi, Y Hayashi, Y Kuroda, H Yajima. Trimethylsilyl trifluoromethylsulfonate as a useful deprotecting reagent in both solution and solid phase peptide syntheses. *J Chem Soc Chem Commun* 274, 1987.

6.23 MINIMIZATION OF SIDE REACTIONS DURING ACIDOLYSIS

The main side reaction associated with removal of protectors by acidolysis is the reaction of the liberated moiety with the nucleophilic centers of side-chain functional groups. No moiety is liberated when bond fission is achieved by the S_N2 mechanism (see Section 6.22). Hence, cleavage should be effected by this mechanism if possible. Cleavage by the S_N1 mechanism liberates a cation, whose reactivity depends on its nature. Rearrangement to a neutral species, such as *tert*-butyl$^+$ to isobutene and cyclohexyl$^+$ to methylenecyclopentane, eliminates its reactivity. The more stable the cation, which rests on the ability of the charge to be delocalized, the more reactive it is. Reactivity is less if the charged atom is hindered and if the cation can tautomerize to a more hindered form (Figure 6.29). The more reactive cation theoretically will lead to the most side reactions. The order of reactivity for cations is roughly xanthenyl, methylbenzhydryl > acetamidomethyl, trityl, *tert*-butyl > benzyl > cyclohexyl > dichlorobenzyl >> allyl, dimethylpentyl. In contrast, the more reactive cation is more readily trapped by a competing nucleophile or scavenger. Scavengers are nucleophilic bases that are too weak to be protonated by the acid. Their constitution may resemble the functional group they are intended to protect. Their efficacy may be restricted to trapping one cation or protecting only one functional group from alkylation. The efficiency of the soft-base nucleophiles (see Section 6.22) varies with the nature of the protector. Some facilitate the cleavage reaction by forcing it to go to completion. As a consequence, cleavage mixtures often contain more than one scavenger if there are a variety of cations generated, and if there are several nucleophilic centers to protect from attack by the cations. Performance as much as rationalization has established the suitability of scavenges for various purposes.

Detailed analysis of the efficacy of various cleavage "cocktails" is beyond the scope of this treatise. Traditional scavengers have been anisole ($PhOCH_3$), thioanisole, and *o*/*m*-cresol ($CH_3C_4H_4OH$) for benzyls; 1,2-ethanedithiol or dithiothreitol $(HSCH_2CHOH)_2$, with its less offensive odor, for *tert*-butyl trifluoroacetate; methanol and trifluoroethanol for trityl and trimethoxybenzyl; water for *tert*-butyl; and ethyl- or phenylmethylsulfide or cresol for arylsulfonyls. Cresol is better than anisole as additive for debenzylation of carboxyl groups because the adduct formed with the acylium ion (Figure 28, path B) is an ester ($-CO_2C_4H_4CH_3$) that is saponifiable.

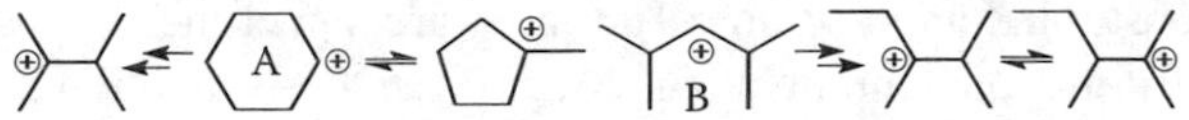

FIGURE 6.29 Carbenium ions are less reactive if the charged atom is hindered and the molecule can tautomerize to a more hindered form. (A) cyclohexyl, (B) dimethylpentyl (see ref. 132).

Thiocresol-thioanisole (1:1) prevents cleavage of arylsulfonyls on the wrong side of sulfonyl (R–SO_2NH–), which produces sulfonated arginine. Dialkylsulfides also suppress the oxidation of methionine. Acetyltryptophan and ethanedithiol prevent the alkylation of tryptophan. Thiols and thioethers should be avoided as scavengers if the peptide contains *S*-acetamidomethyl or *S*-*tert*-butylsulfanylcysteine because of the danger of disulfide interchange (see Section 6.18). Recent developments indicate that trialkylsilanes may be the best scavengers. Triethyl- or triisopropylsilane effectively eliminate *tert*-butyl, trityl, trimethoxybenzyl, and arylsulfonyl cations. Triethylsilane destroys triphenylmethyl$^+$ by reducing it to triphenylmethane, but it is incompatible with tryptophan, whose ring it reduces.[126–132]

126. DA Pearson, M Blanchette, ML Baker, CA Guindon. Trialkylsilanes as scavengers for the trifluoroacetic acid deblocking of protecting groups in peptide synthesis. *Tetrahedron Lett* 30, 2739, 1989.
127. DS King, CG Fields, GB Fields. A cleavage method which minimizes side reactions following Fmoc solid phase peptide synthesis. *Int J Pept Prot Res* 36, 255, 1990.
128. AG Beck-Sickinger, G Schnorrenberg, J Metger, G Jung. Sulfonation of arginine residues as side reaction in Fmoc-peptide synthesis. *Int J Pept Prot Res* 38, 25, 1991.
129. A Mehta, R Jaouhari, TJ Benson, KT Douglas. Improved efficiency and selectivity in peptide synthesis: use of triethylsilane as a carbocation scavenger in deprotection of t-butyl and t-butoxycarbonyl-protected sites. *Tetrahedron Lett* 33, 5441, 1992.
130. NA Solé, G Barany. Optimization of solid-phase synthesis of [Ala^8]-dynorphin A$^{1-3}$. (scavenger Reagents B, K, R) *J Org Chem* 57, 5399, 1992.
131. A Surovoy, JW Metger, G Jung. Optimized deprotection procedure for peptides containing multiple Arg(Mtr), Cys(Acm), Trp and Met residues, in CH Schneider, AN Eberle, eds. *Peptides 1992. Proceedings of the 22nd European Peptide Symposium*, Escom, Leiden, 1993, pp 241-242.
132. A Karlström, K Rosenthal, A Undén. Study of the alkylation propensity of cations generated by acidolytic cleavage of protecting groups in Boc chemistry. *J Pept Res* 55, 36, 2000.

6.24 TRIFUNCTIONAL AMINO ACIDS WITH TWO DIFFERENT PROTECTORS

When trifunctional amino acids are incorporated into a peptide, the side-chain function and either the α-amino or α-carboxy functions are protected by substituents that are not identical. These disubstituted derivatives are obtained by derivatization of the mono-substituted amino acids. The latter are sometimes obtained by derivatization of the trifunctional amino acids. More often, derivatization produces disubstituted amino acids, and these are converted to monosubstituted derivatives by removal of one of the protectors. With a few exceptions, the reagents employed are chloroformates, mixed carbonates, arylsulfonyl chlorides, or aryl halides. β/γ-Esters of dicarboxylic acids are accessible by acid-catalyzed esterification (Figure 6.30, A; see Section 3.17); the side-chain carboxyls have p*K* values about two units higher than those of the α-carboxyl groups, and, hence, they are easier to protonate. Alkoxycarbonylation (see Sections 3.14 and 3.15) of arginine can be effected exclusively at the α-amino group by keeping the guanidino group protonated by not

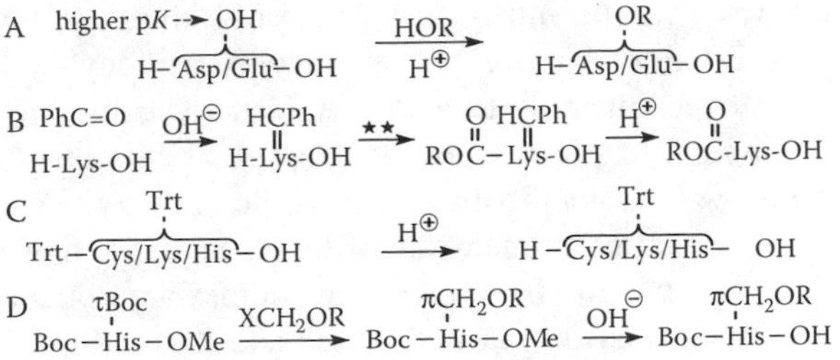

A higher pK--> OH; H–Asp/Glu–OH $\xrightarrow[H^{\oplus}]{HOR}$ H–Asp/Glu(OR)–OH

B PhC=O, H-Lys-OH $\xrightarrow{OH^{\ominus}}$ H-Lys(=HCPh)-OH $\xrightarrow{**}$ ROC(=O)–Lys(=HCPh)-OH $\xrightarrow{H^{\oplus}}$ ROC(=O)-Lys-OH

C Trt–Cys/Lys/His(Trt)–OH $\xrightarrow{H^{\oplus}}$ H–Cys/Lys/His(Trt)–OH

D Boc–His(τBoc)–OMe $\xrightarrow{XCH_2OR}$ Boc–His(πCH_2OR)–OMe $\xrightarrow{OH^{\ominus}}$ Boc–His(πCH_2OR)–OH

FIGURE 6.30 Approaches for the synthesis of monosubstituted trifunctional amino acids. (A) Monoesterification of dicarboxylic acids. (B) N^{α}-Alkoxycarbonylation of lysine through the ε-benzylidene derivative [Bezas & Zervas, 1963]. (C) Selective N^{α}-detritylation of ditrityl derivatives.[138] (D) N-Alkoxymethylation of histidine by displacement of N^{τ}-substituents.[137] Cbz-His(CH_2OR)-OMe are obtained from Cbz-His(τAc)-OMe. ** = Acylating reagent.

allowing the pH of the solution to exceed 10. N^{α}-Fmoc-arginine derivatives are obtained by N^{α}-deprotection and acylation of the corresponding N^{α}-Cbz-arginine derivatives. Formylation of tryptophan with formic acid in the presence of hydrogen chloride gives the side-chain formyl derivative. *S*-Trityl-cysteine is obtainable by acid-catalyzed etherification with trityl carbinol. Acylation of lysine by nitrophenyl esters but not acyl chlorides is selective for the ε-amino group at pH 11 because of the latter's much higher nucleophilicity. A general approach for reaction at side-chain functional groups involves binding of the α-amino and α-carboxyl groups as the copper(II) complexes (Figure 6.31, A). The method is applicable for acylations of lysine and tyrosine and benzylations with aromatic halides of tyrosine and aspartic and glutamic acids, though the esterifications are not very efficient. The copper(II) is conveniently removed with the aid of a chelator such ethylenediamine tetraacetate, 8-hydroxyquinoline, or Chelex resin. Organic equivalents of cupric complexes are the boroxazolidones obtained with trisubstituted boranes (Figure 6.31, B) and the dimethylsilyl derivative obtained with dichlorodimethylsilane (Figure 6.31, C). The boroxazolidinones allowed the synthesis of benzyl, 9-fluorenylmethyl, and *p*-nitrophenyl glutamates and aspartates, with the active esters being obtained by carbodiimide-mediated reactions with *p*-nitrophenol. Detritylation by mild acid of

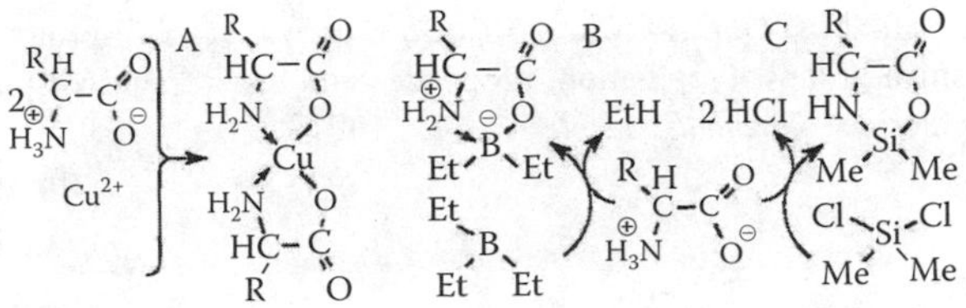

FIGURE 6.31 Simultaneous protection of the amino and carboxyl groups of an amino acid by reaction (A) with copper(II) basic carbonate[133] or acetate,[134] giving the copper complex, (B) with triethyl borane giving the boroxazolidone,[139] and (C) with dichlorodimethylsilane giving the dimethylsilyl derivative.[138,141]

α,ω-ditrityl derivatives gives the ω-trityl derivatives of lysine, histidine, and cysteine (Figure 6.30, C). N^{α}-Deprotection of N^{α},N^{ε}-dibenzyloxycarbonyl-lysine through the *N*-carboxyanhydride opens the way to a variety of derivatives (see Section 7.13). Saponification of diesters of *N*-protected dicarboxylic acids gives the side-chain substituted derivatives, the ester of the more acidic carboxyl group being more sensitive to base. Reactions at the imidazole of histidine give τ-substituted products (see Section 4.3). π-Substituted alkoxymethyl-histidines are obtained by displacement of the τ-substituents of Boc-His(Boc)-OMe and Z-His(Ac)-OMe by reaction with the aryloxymethyl halides (Figure 6.30, D). N^{α}-Substituted lysines are obtained by acylation of the benzylidene derivative, which is stable to base but destroyed by acid (Figure 6.30, B). Fmoc-lysine is similarly obtained by acylating the trimethylsilylated benzylidene derivative with $FmOCO_2NSu$ (A. Hong, personal communication). Boc-Trp(Aloc)-OH is obtained from Boc-Trp-OtBu, after which the terminal protectors are removed and the amino protector is restored. Boc-Trp(cHoc)-OH is obtained from the Boc-Trp-OBzl/OPac, followed by desterification.[133–141]

133. A Neuberger, F Sanger. The availability of acetyl derivatives of lysine for growth. (H-Lys(Z)-OH from the copper(II) salt) *Biochem J* 37, 515, 1943.
134. R Ledger, FHC Stewart. The preparation of substituted γ-benzyl L-glutamates and β-benzyl L-aspartates. (copper(II)) *Aust J Chem* 18, 1477, 1965.
135. J Leclerc, L Benoiton. On the selectivity of acylation of unprotected diamino acids. *Can J Chem* 46, 1047, 1968.
136. JW Scott, D Parker, DR Parrish. Improved syntheses of Nε-(*tert*-butyloxycarbonyl)-L-lysine and Nα-(benzyloxycarbonyl)-Nε-(*tert*-butyloycarbonyl)-L-lysine. *Synthetic Commun* 11, 303, 1981.
137. T Brown, JH Jones, JD Richards. Further studies on the protection of histidine side chains in peptide synthesis: Use of the π-benzyloxymethyl group. *J Chem Soc Perkin Trans 1* 1553, 1982.
138. K Barlos, D Papaioannou, D Theodoropoulos. Efficient "one-pot" synthesis of N-trityl amino acids. *J Org Chem* 47, 1324, 1982.
139. GHL Nefkens, B Zwanenburg. Boroxazolidones as simultaneous protection of the amino and carboxy group in α-amino acids. *Tetrahedron* 39, 2995, 1983.
140. F Albericio, E Nicolás, J Rizo, M Ruiz-Gayo, E Pedroso, E Giralt. Convenient syntheses of fluorenylmethyl-based side chain derivatives of glutamic, aspartic, lysine and cysteine. *Synthesis* 119, 1990.
141. K Barlos, O Chatzi, D Gatos, G Stavropoulos, T Tsegenidis. Fmoc-His(Mmt)-OH and Fmoc-His(Mtt)-OH. Two new histidine derivatives N^{im}-protected with highly acid-sensitive groups. Preparation, properties and use in peptide synthesis. (dimethylsilyldichloride) *Tetrahedron Lett* 32, 475, 1991.

7 Ventilation of Activated Forms and Coupling Methods

7.1 NOTES ON CARBODIIMIDES AND THEIR USE

Dialkylcarbodiimides are efficient peptide-bond-forming reagents (see Section 1.12) that react with carboxyl groups to give the *O*-acylisourea (Figure 7.1, path A). The acyl (RC=O) of the intermediate is immediately transferred to an amino group, to a second molecule of the acid, or to its own basic nitrogen atom (=*N*R^4) to give the *N*-acylurea (path B; see Sections 2.2–2.4). The net result is dehydration. They also react with amino groups to form guanidines (path C) if no carboxylic acid is available, so if amino groups are present, the carbodiimide should always be added to a reaction mixture after the acid. There are three popular carbodiimides, dicyclohexylcarbodiimide (DCC), ethyl(3-dimethylaminopropyl)carbodiimide hydrochloride (EDC), and diisopropylcarbodiimide (DIC). Each has its unique characteristics, but all are skin irritants and allergenic, so protective glasses and gloves should be worn when they are handled. They are not sensitive to water, so with the exception of EDC, they cannot be removed from an organic solvent by washing the solution with aqueous acid or base. A judicious approach is to add acetic acid after completion of a reaction with a few minutes delay before work-up to destroy any carbodiimide that has not been consumed. 1-Hydroxybenzotriazole, the common additive for carbodiimides (see Sections 2.12 and 2.25), is usually obtained as the hydrate; the water has no deleterious effect.

DCC is an inexpensive brittle solid, melting point 34–35°C, that must be crushed or warmed to a viscous liquid to withdraw it from a container. It generates a very bulky *N,N'*-dialkylurea that is insoluble in organic solvents, and, hence, it may interfere with mixing. Filtration removes most of the urea, but it is soluble enough in the usual solvents that its complete removal can be problematic. It is least soluble in water, hexane, or acetone; cooling a solution of a product in acetone for a few

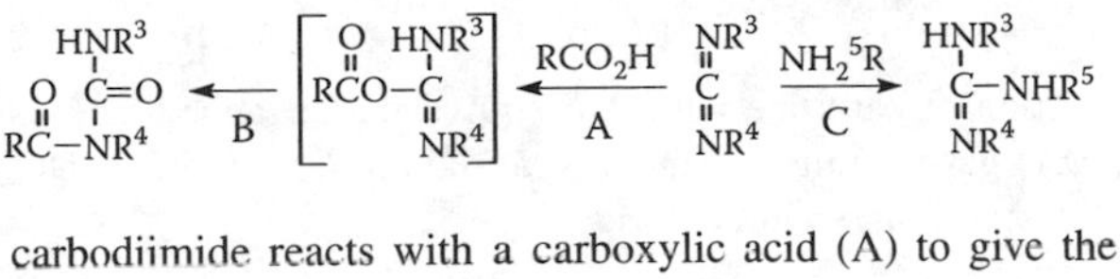

FIGURE 7.1 A carbodiimide reacts with a carboxylic acid (A) to give the *O*-acylisourea, which may rearrange (B) to the *N*-acylurea. It also reacts with amino groups (C) to produce guanidines if no carboxyl group is available.

hours allows removal of final traces of *N,N′*-dicyclohexylurea by filtration. The urea is soluble in trifluroacetic acid-dichloromethane (1:1), which is the deprotecting solution employed for solid-phase synthesis using Boc/Bzl chemistry. Operators who have developed sensitivities to DCC often have replaced it with 1-ethoxycarbonyl-2-ethoxy-1,2-dihydroquinoline (EEDQ; see Section 2.15).

DIC is a moderately expensive liquid that is employed in solid-phase synthesis to avoid the obstacles presented by the use of DCC. Both the reagent and the corresponding urea are soluble in organic solvents, and hence there is no bulky precipitate to contend with. The urea cannot be removed from an organic solution by aqueous extraction; however, it is soluble enough in water that final traces can be removed from a precipitated peptide by washing the latter with a water–ether mixture. Clean up of spills of DIC can cause temporary blindness if utmost care is not exercised.

EDC is an expensive crystalline material known as soluble carbodiimide (see Section 1.16) because both the reagent and the corresponding ureas are soluble in water. The commercial material may not be 100% pure, which could be a result of traces of oxidized product. The common form is the hydrochloride, though the free base has been used with the rationale that it ensures that the amino groups involved in the coupling reaction are not protonated. EDC can be employed for amide-bond-forming reactions in partially aqueous mixtures. An example is the preparation of carnitine 4-methoxyanilide from anisidine and carnitine chloride that is insoluble in organic solvents (Figure 7.2). It is used for modifying, and, hence, quantitating, exposed carboxyl groups in proteins. EDC is mildly acidic and thus can cause slight decomposition of Boc-amino acids that are activated in nonalkaline solution (see Section 7.15). Several methods for assaying the reagent are available.[1–3]

1. BS Jacobson, KF Fairman. A colorimetric assay for carbodiimides commonly usd in peptide synthesis and carboxyl group modification. *Anal Biochem* 106, 114, 1980.
2. NL Benoiton, FMF Chen, C Williams. Determination of carnitine by HPLC as carnitinyl-anisidine. *Proc Fed Am Soc Exp Biol*, Abstract 50775, 1987.
3. MA Gilles, AQ Hudson, CL Borders. Stability of water-soluble carbodiimides in aqueous solution. *Anal Biochem* 184, 244, 1990.

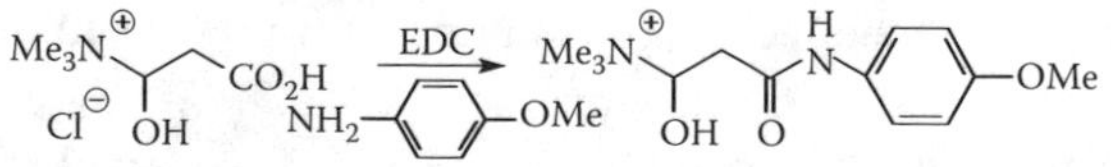

FIGURE 7.2 Reaction of carnitine with two equivalents of anisidine mediated by EDC in water-acetone (20:1) gave carnitine 4-methoxyanilide in 60% yield.[2] EDC = ethyl-(3-dimethylaminopropyl)carbodiimide hydrochloride.

7.2 CUPRIC ION AS AN ADDITIVE THAT ELIMINATES EPIMERIZATION IN CARBODIIMIDE-MEDIATED REACTIONS

The beneficial effects of Lewis acids, and zinc ion in particular, on the preservation of chirality during couplings have been known since the late 1970s, but only recently have the effects of one particular substance, namely, copper(II) chloride, been singled out as remarkable. This salt not only diminishes but eliminates the isomerization that occurs when small peptides are coupled using carbodiimides in dimethylformamide. The yields of product are modest; however, if 1-hydroxybenzotriazole is also present, the yields are as desired. The amounts of additive required are 0.1 equivalent of copper(II) chloride and 2 equivalents of HOBt. It is known that HOBt suppresses epimerization by minimizing formation of the 5(4*H*)-oxazolone (see Section 2.25), the tautomerization of which is the source of the isomerization (see Section 4.4). HOBt is not completely effective; copper(II) chloride prevents isomerization of the 5(4*H*)-oxazolone. The latter was established by experiments with the 5(4*H*)-oxazolone from *Z*-glycyl-L-valine. When dissolved in dimethylformamide at 25°C, the solution of oxazolone lost 27% of its optical activity within 3 hours. In the presence of copper(II) chloride, the solution lost none of its optical activity over the same period. The effects of the cupric salt and the combination of the two additives were dramatically illustrated by examination of the carbodiimide-mediated coupling of *N*-benzoyl-L-valine with L-valine methyl ester (Figure 7.3). More than 50% of -D-L- isomer was produced in the absence of additives, 2.5% in the presence of copper(II) chloride, and none in the presence of both additives. In the absence of cupric salt, *N*-hydroxysuccinimide was much superior as an additive than HOBt for this coupling, which involved an *N*-acylamino acid. The beneficial effect of copper(II) chloride can be attributed to coordination of the cupric ion with the ring of the oxazolone, the effect of which is analogous to protonation of the ring by noncarboxylic acids (see Section 2.25). A variant of the above approach for eliminating isomerization in carbodiimide-mediated couplings is the use of the salt formed from the two additives; namely, $Cu(OBt)_2$. It has the same effect as a mixture of the

Additive	
–	64
HOBt	48
HONSu	8.0
$CuCl_2$	2.5
$CuCl_2$ + HOBt	<0.1

FIGURE 7.3 The protective effects of additives on the EDC-mediated coupling of Bz-L-Val-OH with *H*-L-Val-OMe.TosOH/Et_3N.[5] Percentage –D-L– epimer formed in dimethylformamide at 0°C. EDC = ethyl-(3-dimethylaminopropyl)carbodiimide hydrochloride.

additives, and it prevented epimerization in various couplings of segments, except when the aminolyzing residue of the peptide ester was proline (see Section 7.22). An interesting point to note is that tertiary amine is not required for neutralizing the acidic moiety (HCl or CF_3CO_2H) of a peptide-ester salt when $Cu(OBt)_2$ is employed. The latter neutralizes the acid generating HOBt and the cupric salt of the acid. In contrast, $Cu(OBt)_2$ decreased the yields to unacceptable levels in the coupling of *N*-protected dipeptides to a resin-bound amino acid. Copper(II) chloride was also effective for eliminating epimerization in mixed-anhydride and EEDQ-mediated (see Section 2.15) reactions, but the yields of the products were unsatisfactory.[4–7]

4. T Miyazawa, T Otomatsu, Y Fukui, T Yamada, S Kuwata. Racemization-free and efficient peptide synthesis by the carbodiimide method using 1-hydroxybenzotriazole and copper(II) chloride simultaneously as additives. *J Chem Soc Chem Commun* 419, 1988.
5. T Miyazawa, T Otomatsu, Y Fukui, T Yamada, S Kuwata. Simultaneous use of 1-hydroxybenzotriazole and copper(II) chloride as additives for racemization-free and efficient synthesis by the carbodiimide method. *Int J Pept Prot Res* 39, 308, 1992.
6. T Miyazawa, T Donkai, T Yamada, S Kuwata. Effect of copper(II) chloride on suppression of racemization in peptide synthesis by the mixed-anhydride and related methods. *Int J Pept Prot Res* 40, 49, 1992.
7. JC Califano, C Devin, J Shao, JK Blodgett, RA Maki, KW Funk, JC Tolle. Copper(II)-containing racemization suppressants and their use in segment coupling reactions, in J Martinez, J-A Fehrentz, eds. *Peptides 2000, Proceedings of the 26th European Peptide Symposium,* Editions EDK, Paris, 2001, pp. 99-100.

7.3 MIXED ANHYDRIDES: PROPERTIES AND THEIR USE

Coupling by the mixed-anhydride reaction (see Section 2.6; Figure 7.4) involves addition of chloroformate **2** to a solution of acid **1** and tertiary amine **3** to form anhydride **4** followed by the addition of amine **5** to give peptide **6**, path A. Urethane **7** may be produced by aminolysis at the wrong carbonyl of the anhydride, path B, or by reaction of **5** with unconsumed chloroformate. The latter is avoided by use of a slight excess of starting acid **1**. Optimum performance is achieved by use of a nonhindered tertiary amine such as *N*-methylmorpholine or *N*-methylpiperidine. The electrophile that reacts with deprotonated acid **1** to generate anhydride **4** is probably the acyloxymorpholinium or acyloxypiperidinium ion (see Section 2.6); a hindered

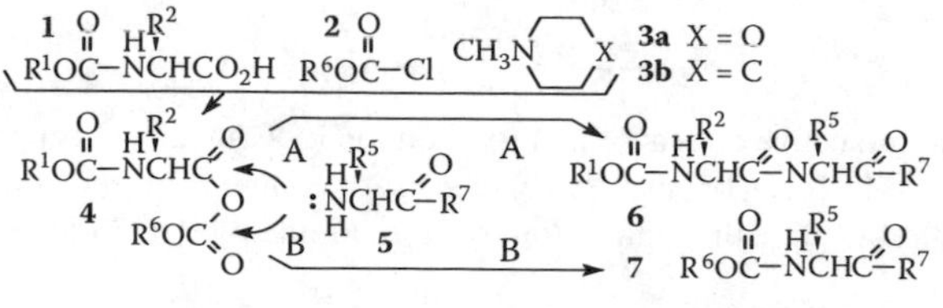

1 $R^1OC(=O)-NHCH(R^2)CO_2H$ **2** $R^6OC(=O)-Cl$ $CH_3N(CH_2CH_2)_2X$ **3a** X = O **3b** X = C

4 $R^1OC(=O)-NHCH(R^2)C(=O)-O-C(=O)OR^6$ + **5** $:NH-CH(R^5)C(=O)-R^7$

A → **6** $R^1OC(=O)-NHCH(R^2)C(=O)-NHCH(R^5)C(=O)-R^7$

B → **7** $R^6OC(=O)-NHCH(R^5)C(=O)-R^7$

FIGURE 7.4 The mixed-anhydride reaction (see Section 2.6). Preparation of mixed anhydride **4** followed by its aminolysis (A) producing peptide **6** and the major side reaction of aminolysis at the wrong carbonyl (B) generating urethane **7**.

base will form the acyloxy quaternary ion less readily. Good solvents for the reaction are tetrahydrofuran, dichloromethane, dimethylformamide, and 2-propanol. Reactions are traditionally carried out below 0°C under anhydrous conditions, but neither of these is essential. Aminolysis occurs efficiently in dimethylformamide-water (4:1). The anhydrides can be prepared in dichloromethane at ambient temperature and purified by washing with aqueous solutions (see Section 2.8). Several are crystalline compounds; some are stable at 5°C for at least 5 days. They are inert to weak nucleophiles such as methanol or 4-nitrophenol. They can also be obtained in high yield from the acid and corresponding pyrocarbonate **9** (see Figure 7.7), including di-*tert*-butylpyrocarbonate [$(Boc)_2O$] in the presence of *N*-methylpiperidine at ambient temperature, or by the use of EEDQ (1-ethoxycarbonyl-2-ethoxy-1,2-dihydroquinoline; see Section 2.15). Demonstration of the latter established that the intermediate in EDDQ-mediated reactions indeed is the mixed anhydride as postulated. There is no obvious advantage to isolating a mixed anhydride before employing it for a reaction; however, there may be cases where it is beneficial. The mixed-anhydride reaction is a quick and inexpensive method for peptide-bond formation, as well as for making esters (see Section 7.8). With one exception (see Section 7.4), the side products are easy to dispose of. They can be employed for preparing amides and activated esters (see Section 7.8), for reacting with the activated ester of an amino acid to produce a protected activated dipeptide (see Section 7.10), and for reacting with an amino acid whose carboxyl group is not protected (see Section 7.21). A variant known as REMA (repetitive excess mixed anhydride) involves the use of a 50% excess of anhydride, with the excess then being destroyed by potassium hydrogen carbonate. The mixed-anhydride reaction is applicable to the coupling of segments, with isopropyl chloroformate being the reagent of choice for preservation of chirality (see Section 7.5).[8,9]

8. HC Beyerman, EWB De Leer, J Floor. On the repetitive excess mixed anhydride method for the synthesis of peptides. Synthesis of the sequence 1-10 of human growth hormone. *Rec Trav Chim Pays-Bas* 92, 481, 1973.
9. FMF Chen, NL Benoiton. The preparation and reactions of mixed anhydrides of *N*-alkoxycarbonylamino acids. *Can J Chem* 65, 619, 1987.

7.4 SECONDARY REACTIONS OF MIXED ANHYDRIDES: URETHANE FORMATION

The major side reaction associated with the use of mixed anhydrides is aminolysis at the carbonyl of the carbonate moiety (Figure 7.4, path B). The product is a urethane that resembles the desired protected peptide in properties, except that the amino-terminal substituent is not cleaved by the usual deprotecting reagents. Hence, its removal from the target product is not straightforward. The problem is serious when the residues activated are hindered (Val, Ile, MeXaa), where the amounts can be as high as 10%. Other residues generate much less, but the reaction cannot be avoided completely, with the possible exception of activated proline (see Section 7.22). This is one reason why mixed anhydrides are not employed for solid-phase synthesis.

More urethane is generated at ambient temperature than at 5°C; toluene and dichloromethane are the best solvents for minimizing urethane formation. Isobutyl chloroformate leads to slightly less urethane than ethyl or isopropyl chloroformate. Isopropyl chloroformate leads to less mixed carbonate than isobutyl chloroformate when the nucleophile is a phenolic anion. Anhydrides made from pivaloyl chloride (see Section 8.15) are less selective for the incoming nucleophile than are those made from chloroformates. The worst-case scenario is the use of triethylamine in a halogen-containing solvent; this combination drastically reduces the rate of formation of the anhydride. Unconsumed chloroformate generates the same urethane; thus, it is imperative that the reagent be consumed before the aminolyzing component is introduced. For this reason, it is prudent to employ a small excess of the acid, as well as sufficient time to allow the reaction generating the anhydride to go to completion, which may be up to 10 minutes. Activated Boc-amino acids lead to less urethane than Fmoc-amino acids, which lead to less than Cbz-amino acids. With few exceptions, the nature of the side chain of the aminolyzing residue has little influence on urethane formation. Proline does not distinguish between the two carbonyls of a mixed anhydride. This can be attributed to the higher basicity of the nucleophile, which is a secondary amine. The obstacle can be circumvented by the addition of 1-hydroxybenzotriazole after the mixed anhydride has been formed. A large amount of urethane is also produced when the aminolyzing residue is *N*-methylaminoacyl. One approach that allows simple purification of a product from a mixed-anhydride reaction is the use of a chloroformate which generates a urethane that is selectively cleavable. As an example, a coupling of Boc-isoleucine employing methanesulfonylethyl chloroformate gave an 80% yield of pure product after destruction of the urethane by β-elimination (Figure 7.5; see Section 3.10).

There have been reports that urethane was produced when the mixed-anhydride method was employed for the coupling of segments. However, studies on urethane formation during the aminolysis of mixed anhydrides of peptides have never been carried out. The anhydrides are too unstable to be isolated. The activated moiety of the peptide cyclizes too quickly to the 2,4-dialkyl-5(4*H*)-oxazolone (see Section 2.23), and since the time allowed to generate the anhydride in segment couplings is always limited to avoid epimerization, one cannot exclude the possibility that the urethane that was produced originated by aminolysis of unconsumed chloroformate.

Boc-Ile-OH
Msc-Cl
NMM
CH_2Cl_2–5°
Boc-Ile-OMsc
H-Lys(Z)-OMe
90% yd → Boc-Ile-Lys(Z)-OMe
(30:1)
Msc-Lys(Z)-OMe
NaOMe 5'
H-Lys(Z)-OMe
wash with dil HCl
80% yd → Boc-Ile-Lys(Z)-OMe
in aqueous layer

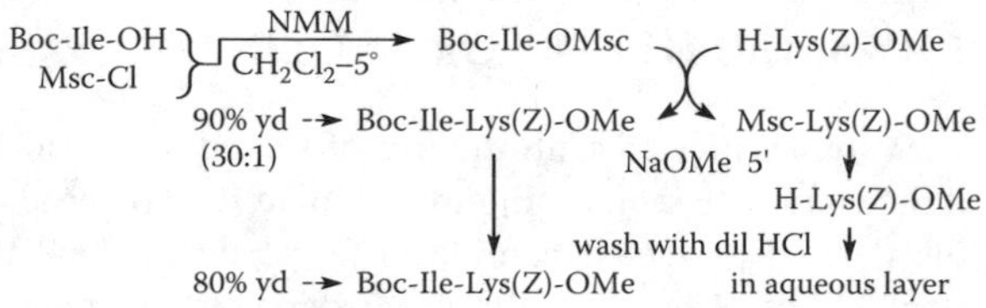

FIGURE 7.5 Preparation of a protected dipeptide by the mixed-anhydride method, employing a chloroformate that generates a cleavable urethane.[13] The urethane impurity is destroyed by a β-elimination reaction. NMM = *N*-methylmorpholine, Msc = methanesulfonylethoxycarbonyl.

In fact, recent work indicates that less urethane is formed by aminolysis of an activated peptide than by aminolysis of an activated *N*-alkoxycarbonylamino acid. Such results are consistent with the postulate that the intermediate undergoing aminolysis during the mixed anhydride reaction of a segment is the 2,4-dialkyl-5(4*H*)-oxazolone, a tenet that is indicated by the high rate at which the activated peptide is converted to the oxazolone.[10–13]

10. M Bodanszky, JC Tolle. Side reactions in peptide synthesis. V. A reexamination of the mixed anhydride method. *Int Pept Prot Res* 10, 380, 1977.
11. FMF Chen, R Steinauer, NL Benoiton. Mixed anhydrides in peptide synthesis. Reduction of urethane formation and racemization using *N*-methylpiperidine as the tertiary amine base. *J Org Chem* 48, 2939, 1983.
12. KU Prasad, MA Iqbal, DW Urry. Utilization of 1-hydroxybenzotriazole in mixed anhydride reactions. *Int J Pept Prot Res* 25, 408, 1985.
13. FMF Chen, Y Lee, R Steinauer, NL Benoiton. Mixed anhydrides in peptide synthesis. A study of urethane formation with a contribution on minimization of racemization. *Can J Chem* 65, 613, 1987.

7.5 DECOMPOSITION OF MIXED ANHYDRIDES: 2-ALKOXY-5(4*H*)-OXAZOLONE FORMATION AND DISPROPORTIONATION

The decomposition of mixed anhydrides is a complicated issue. Mixed anhydrides decompose by formation of the oxazolone (see Section 2.8) or generation of the symmetrical anhydride, and the course of events depends on many factors. The anhydrides are stable in dichloromethane at 0°C. At 20°C, mixed anhydrides of *N*-alkoxycarbonylvaline and isoleucine are stable, whereas those of alanine, leucine, and phenylalanine are not. For the latter, symmetrical anhydrides are detectable in minutes. Less symmetrical anhydride is formed from anhydrides of Boc-amino acids than from anhydrides of Cbz-amino acids; more is formed when the other moiety is ethoxycarbonyl than when it is isobutoxycarbonyl. Decomposition is faster in tetrahydrofuran and dimethylformamide, indicating that these solvents participate in the breakdown. Tertiary amine promotes decomposition to the symmetrical anhydride and the oxazolone, with triethylamine having a greater effect than *N*-methylmorpholine. The alcohol that is liberated from the carbonate moiety esterifies the protected amino acid moiety. *N*-Methylmorpholine hydrochloride, which is an acidic salt, does not promote the decomposition. The extent of symmetrical anhydride formation cannot be correlated with the ease of formation of the oxazolone. Hence, it has been postulated that there are two mechanisms involved in the decomposition of mixed anhydrides. One would involve generation of the oxazolone with release of the carbonic acid in equilibrium with its anion (Figure 7.6, path A). The anion then attacks a second molecule of mixed anhydride (path B) to produce the pyrocarbonate, and a chain reaction (paths C and B) ensues. Promotion of the reaction by the oxygen-containing solvents can be attributed to formation of the acyloxonium ions (Figure 7.6, D), which are more highly activated than the anhydride. A second mechanism must be invoked to account for the resistance of the mixed anhydrides

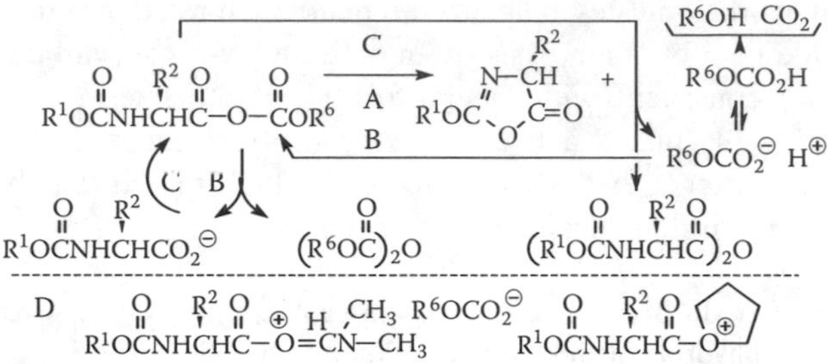

FIGURE 7.6 Decomposition of a mixed anhydride (A) to the 2-alkoxy-5(4*H*)-oxazolone and the alkyl carbonate.[9] The latter is in equilibrium with the anion whose reaction (B) with a second molecule of anhydride produces pyrocarbonate and the acid anion whose reaction (C) with a third molecule produces the symmetrical anhydride. The oxazolone eventually reacts with the alcohol to give the ester. (D) Acyloxonium ions formed by reaction of the anhydride with dimethylformamide and tetrahydrofuran.

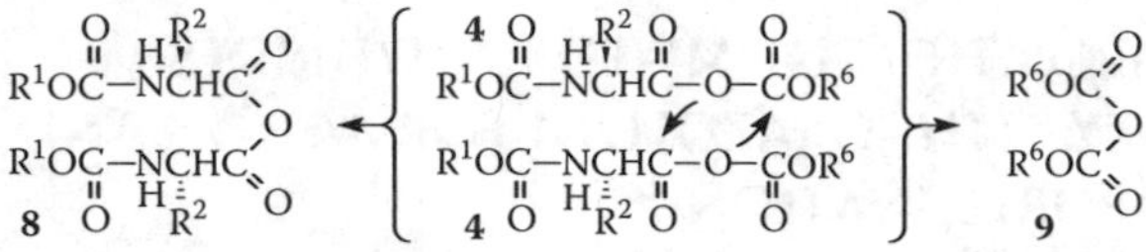

FIGURE 7.7 Disproportionation of a mixed anhydride.[14] Two molecules of anhydride generate symmetrical anhydride **8** and pyrocarbonate **9** probably via a bimolecular reaction.[18]

of valine and isoleucine to generate symmetrical anhydrides under conditions in which the anhydrides of alanine, leucine, and phenylalanine generate symmetrical anhydrides. The bimolecular mechanism (Figure 7.7), where reaction is prevented by the bulkiness of side chains of the residues, has been postulated to explain the apparent anomaly. The transformation of two molecules of mixed anhydride into symmetrical anhydride and pyrocarbonate (Figure 7.7), regardless of the mechanism by which it occurs, has traditionally been referred to as disproportionation.

The ease with which a mixed anhydride generates a 2-alkoxy-5(4*H*)-oxazolone can be correlated with the extent of epimerization in the coupling of segments by the mixed-anhydride method. Mixed anhydrides incorporating a secondary alkyl group in the carbonate moiety generated oxazolone more slowly than those incorporating a primary alkyl group (see Section 2.23). Consonant with this, protected dipeptides coupled employing isopropyl chloroformate as the reagent epimerized three to four times less than those coupled with ethyl and isobutyl chloroformates.[9,14–18]

9. FMF Chen, NL Benoiton. The preparation and reactions of mixed anhydrides of *N*-alkoxycarbonylamino acids. *Can J Chem* 65, 619, 1987.
14. DS Tarbell, EJ Longosz. Thermal decomposition of mixed carboxylic-carbonic anhydrides. Factors affecting ester formation. *J Org Chem* 24, 774, 1959.
15. DS Tarbell. The carboxylic carbonic anhydrides and related compounds. *Acc Chem Res* 2, 296, 1969.

16. NL Benoiton, Y Lee, FMF Chen. Studies on the disproportionation of mixed anhydrides of *N*-alkoxycarbonylamino acids. *Int J Pept Prot Res* 41, 338, 1993.
17. NL Benoiton, Y Lee, FMF Chen. Isopropyl chloroformate as a superior reagent for mixed anhydride generation and couplings in peptide synthesis. *Int J Pept Prot Res* 31, 577, 1988.
18. NL Benoiton, FMF Chen. Preparation of 2-alkoxy-4-alkyl-5(4*H*)-oxazolones from mixed anhydrides of *N*-alkoxycarbonylamino acids. *Int J Pept Prot Res* 42, 455, 1993.

7.6 ACTIVATED ESTERS: REACTIVITY

Carboxyl groups are prevented from participating in reactions by their conversion into esters (see Section 3.17). Carboxyl groups are induced into participating in peptide-bond-forming reactions by their conversion into activated forms (see Section 2.1). One of these activated forms is an ester that reacts by virtue of the electron-withdrawing character of the substituent. There are two types of common activated esters: those with an aryloxy substituent on the carbonyl, and those with a substituted aminoxy substitutent on the carbonyl (see Section 2.9). The esters react reversibly with an amine nucleophile to form a tetrahedral zwitter-ionic intermediate (Figure 7.8). Decomposition of the intermediate by expulsion of the aryloxy or aminoxy anion is the rate-limiting step of the reaction. The less congested the intermediate, the easier it is for reaction, with aminolysis rates being in the order 1-butylamine > piperidine > diethylamine > isopropylamine. Aminolysis of activated esters can be forced to completion by use of an excess of one of the components. The reaction is catalyzed by mild acids such as acetic acid and 1-hydroxybenzotriazole, which probably form a hydrogen bond with the oxygen atom of the carbonyl. Hydroxyl groups in the absence of base are not acylated by activated esters, which undergo hydrolysis very slowly or not at all. The reactivity of activated esters derived from hydroxamic acids depends on the electron-withdrawing character of the substituent, as well as the anchimeric assistance that it provides (see Section 2.10). The reactivity of activated esters derived from phenols varies roughly with the acidity of the phenol. The order of reactivity is pentafluorophenyl > pentachlorophenyl > 2,4,5-trichlorophenyl > 4-nitrophenyl > 2,4,6-trichlorophenyl. Reactivity is diminished by the presence of two *o*-substituents on the aryl rings and by a branch at the β-carbon atom of the activated residue. Hence, esters of valine and isoleucine are aminolyzed more slowly than those of the other amino acids. There is little difference in the inductive effects of pentafluorophenyl and pentachlorophenyl. The greater reactivity of pentafluorophenyl esters can be attributed to the smaller size of the *o*-substituents and an alteration of the electronic distribution of the ring resulting from interaction between the σ- and π-electrons. 4-Oxo-3,4-dihydroxybenzotriazin-3-yl esters are

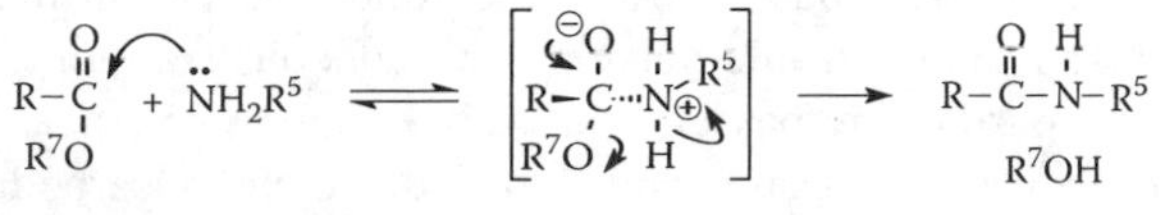

$$\mathrm{R{-}C(=O){-}OR^7} + \mathrm{\ddot{N}H_2R^5} \rightleftharpoons \left[\mathrm{R{-}C(O^{\ominus})(OR^7){\cdots}N^{\oplus}(H)(H)R^5}\right] \longrightarrow \mathrm{R{-}C(=O){-}N(H){-}R^5} + \mathrm{R^7OH}$$

FIGURE 7.8 Aminolysis of an activated ester produces a tetrahedral intermediate, the decomposition of which is the rate-limiting step of the reaction.

five times as reactive as pentafluorophenyl esters, and succinimido esters are about

Chloroform	Dioxane	Dioxane-water (4:1)	Pyridine	Ethyl acetate	Dimethyl-formamide
23	4.9	3.3	3.1	2.3	0.3

FIGURE 7.9 Half-life in minutes of *Z*-Phe-$OC_6H_2Cl_3$ (10^{-4} *M*) in the presence of benzylamine (10^{-2} *M*) in various solvents at 25°C.[20] $C_6H_2Cl_3$ = 2,4,5-trichlorophenyl.

as reactive as pentachlorophenyl esters. Benzotriazolyl esters of *N*-alkoxycarbonyl-*N*-methylamino acids (see Section 8.15) have greatly diminished reactivity relative to those of unmethylated derivatives. However, the succinimido esters react sufficiently to be of practical use. Reaction rates increase by two for a 20°C increase in temperature and, especially for the aryl esters, are greatly affected by the nature of the solvent, increasing by as much as 100 times in going from chloroform to dimethylformamide (Figure 7.9). Ethyl acetate and dioxane are good solvents, with dioxane-water, 4:1, being better than the latter. Activated esters that are left in the presence of tertiary amine undergo a gradual loss of their stereochemistry, except for pyridino esters (–$CO_2NC_5H_5$), which are unique in this regard. If tertiary amine is added to an aminolysis reaction mixture to neutralize the phenol or hydroxamate that is liberated, caution is in order if there are any carboxy-containing compounds present, as the activated ester may react with the carboxy anions.[19–22]

19. G Kupryszewski, M Formela. Amino acid chlorophenyl esters. III N-Protected amino acid pentachlorophenyl esters. *Rocz Chem* 35, 931, 1961.
20. J Pless, RA Boissonnas. On the velocity of aminolysis of a variety of new activated N-protected α-amino-acid phenyl esters, in particular 2,4,5-trichlorophenyl esters. *Helv Chim Acta* 46, 1609, 1963.
21. WA Sheppard. Pentafluorophenyl group. Electronic effect as a substituent. *J Am Chem Soc* 92, 5419, 1970.
22. GW Kline, SB Hanna. The aminolysis of *N*-hydroxysuccinimide esters. A structure-reactivity study. *J Am Chem Soc* 109, 3087, 1987.

7.7 PREPARATION OF ACTIVATED ESTERS USING CARBODIIMIDES AND ASSOCIATED SECONDARY REACTIONS

Activated esters of *N*-alkoxycarbonylamino acids are prepared by two approaches, activation of the acid followed by reaction with the hydroxy compound, and trans-esterification. Most of the products are stable enough to be purified by washing a solution of the ester in an organic solvent with aqueous solutions. A few that are not crystalline are purified by passage through a column of silica. The commonly used method for their preparation is addition of dicyclohexylcarbodiimide to a cold mixture of the reactants in dimethylformamide or ethyl acetate. The first Boc–amino acid nitrophenyl esters were obtained using pyridine as solvent. Pyridine generates the nitrophenoxide ion that is more reactive. For one type of ester, 2-hydroxypyridino

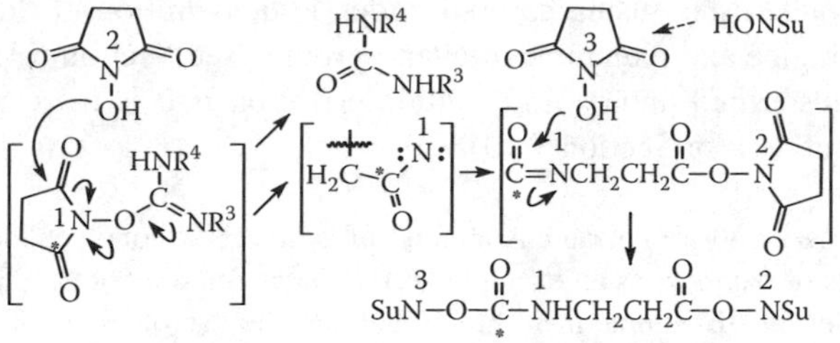

FIGURE 7.10 Formation of the succinimido ester of *N*-succinimidoxycarbonyl-β-alanine by reaction of three molecules of *N*-hydroxysuccinimide (HONSu) with one molecule of dicyclohexylcarbodiimide.[25] The first molecule (N^1) reacts to form the *O*-succinimido-isourea. The second molecule (N^2) ruptures the ring by attack at the carbonyl, generating a nitrene that rearranges to the esterified carboxyalkyl isocyanate. The third molecule (N^3) attacks the carbonyl of the latter. $R^3 = R^4$ = cyclohexyl; SuN– = succinimido.

esters, without the pyridine, the product of the DCC-mediated reaction was the *N*-acylurea. Pyridine also protects activated Boc-amino acids from decomposing (see Section 7.15). An alternative for the pentachloro- and pentafluorophenyl esters is reaction of the acid with a DCC-pentahalophenol (1:3) complex. If the recommended protocols are not observed, the products can be contaminated by secondary products. This pertains especially to succinimido and 4-oxo-3,4-dihydrobenzotriazin-3-yl esters (see Section 2.9). If more than one equivalent of *N*-hydroxysuccinimide is employed for esterification, the carbodiimide can react with the latter to form the equivalent of the *O*-acylisourea, which is attacked by a second molecule of hydroxy compound, generating an activated dimer that reacts with a third molecule to give the activated ester of *N*-protected β-alanine (Figure 7.10). If the contaminant is not removed, the β-alanine is incorporated in the target peptide without interrupting chain assembly.

A similar type of reaction can occur between carbodiimide and two molecules of HOObt, producing an azidobenzoate (Figure 7.11). The azidobenzoate is an acylating agent that causes chain termination if it is not removed from the ester. It is detectable by high-performance liquid chromatography and by the azido absorbance at 2120 cm^{-1} in the infrared spectrum (see Section 7.26). Its formation is avoided if the carbodiimide and acid are left together for 5 minutes before the addition of the hydroxy compound. A third side reaction encountered during prep-

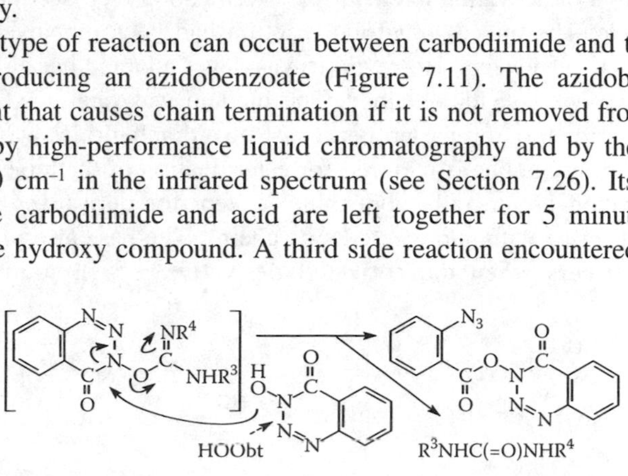

FIGURE 7.11 Formation of 4-oxo-3,4-dihydrobenzotriazine-3-yl 2′-azidobenzoate by reaction of one molecule of dicyclohexylcarbodiimide with two molecules of 3-hydroxy-4-oxo-3,4-dihydrobenzotriazine (HOObt).[26] The intermediate is the equivalent of the *O*-acylisourea. $R^3 = R^4$ = cyclohexyl.

aration of activated esters using carbodiimides is dehydration of the carboxamide groups of asparagine and glutamine to cyano groups (see Section 6.15). Finally, the residue that is esterified undergoes enantiomerization if it is the carboxy-terminal residue of a peptide (see Section 7.10).[23–29]

23. M Bodanszky, V duVigneaud. A method of synthesis of long peptide chains using a synthesis of oxytocin as an example. *J Am Chem Soc* 81, 5688, 1959.
24. JW Anderson, JE Zimmerman, FM Callahan. The use of esters of *N*-hydroxy succinimide in peptide synthesis. *J Am Chem Soc* 86, 1839, 1964.
25. H Gross, L Bilk. On the reaction of *N*-hydroxysuccinimide with dicyclohexylcarbodiimide. *Tetrahedron* 24, 6935, 1968.
26. W König, R Geiger. A new method for the synthesis of peptides: Activation of the carboxyl group with dicyclohexylcarbodiimide and 3-hydroxy-4-oxo-3,4-dihydro-1,2,3-benzotriazine. *Chem Ber* 103, 2034, 1970.
27. L Kisfaludy, M Löw, O Nyeki, T Szirtes, I Schön. Utilization of pentafluorophenyl esters in peptide synthesis (DCC-XPhOH 1:3 complex). *Liebigs Ann Chem* 1421, 1973.
28. A Bodanszky, M Bodanszky, N Chandramouli, JZ Kwei, J Martinez, JC Tolle. Active esters of 9-fluorenylmethyloxycarbonyl amino acids and their application in the stepwise lengthening of a peptide chain. *J Org Chem* 45, 72, 1980.
29. E Atherton, JL Holder, M Meldal, RC Sheppard, RM Valerio. 3,4-Dihydro-4-oxo-1,2,3-benzotriazin-3-yl esters of fluorenylmethoxycarbonyl amino acids as self-indicating reagents for solid phase peptide synthesis. *J Chem Soc Perkin Trans 1* 2887, 1988.

7.8 OTHER METHODS FOR THE PREPARATION OF ACTIVATED ESTERS OF *N*-ALKOXYCARBONYLAMINO ACIDS

A second method of activating the acid for esterification (see Section 7.6) is as the mixed anhydride. The mixed-anhydride reaction had been employed decades ago for preparing activated esters. However, it was never adopted because of its unreliability and the modest yields obtained. The method was fine-tuned (Figure 7.12), after reliable information on the properties of mixed anhydrides was acquired (see Section 2.8). Tertiary amine is required for esterification of the mixed anhydride to occur. The method is generally applicable, except for derivatives of asparagine, glutamine, and serine with unprotected side chains. The base also prevents decomposition that occurs when the activated derivative is a Boc-amino acid (see

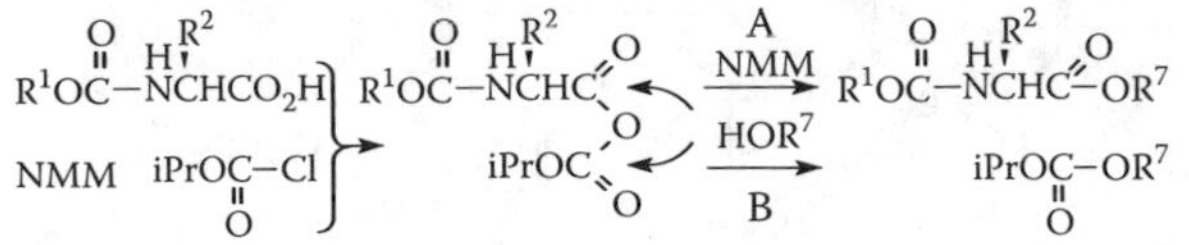

$R^1OC(=O)-NHCHR^2CO_2H$ + NMM + iPrOC(=O)–Cl → $R^1OC(=O)-NHCHR^2C(=O)-O-C(=O)OiPr$ —(A: NMM)→ $R^1OC(=O)-NHCHR^2C(=O)-OR^7$; —($HOR^7$, B)→ iPrOC(=O)–$OR^7$

FIGURE 7.12 Preparation of activated esters of *N*-alkoxycarbonylamino acids by reaction of the hydroxy compound with a mixed anhydride (path A), obtained by leaving the three reagents (NMM = *N*-methylmorpholine) in CH_2Cl_2 at 23°C for 2 minutes.[35] Mixed carbonate that is formed (path B) is readily eliminated by crystallization of the ester.

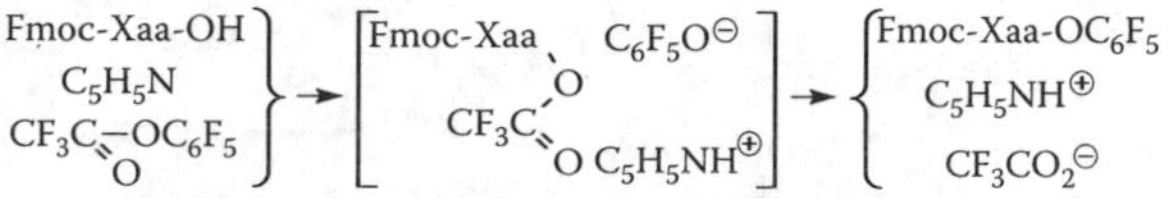

$$\left.\begin{array}{c}\text{Fmoc-Xaa-OH}\\ C_5H_5N\\ CF_3C(=O){-}OC_6F_5\end{array}\right\} \rightarrow \left[\begin{array}{c}\text{Fmoc-Xaa}\quad C_6F_5O^{\ominus}\\ \diagdown O\\ CF_3C(=O)\diagup \quad O\ C_5H_5NH^{\oplus}\end{array}\right] \rightarrow \left\{\begin{array}{c}\text{Fmoc-Xaa-}OC_6F_5\\ C_5H_5NH^{\oplus}\\ CF_3CO_2^{\ominus}\end{array}\right.$$

FIGURE 7.13 Preparation of an Fmoc-amino-acid pentafluorophenyl ester by reaction of the acid with pentafluorophenyl trifluoroacetate in the presence of pyridine.[32]

Section 7.15). Fmoc-amino acids can also be activated for esterification as the acid chlorides (see Section 2.14). Pentafluorophenyl and 4-oxo-3,4-dihydrobenzotriazin-3-yl esters are obtainable in good yield by this approach.

The alternative method for making activated esters is base-catalyzed transesterification. Fmoc-amino acids are esterified in excellent yields by reaction with pentafluorophenyl trifluoroacetate at 40°C in the presence of pyridine (Figure 7.13). A mixed anhydride is formed initially, and the anhydride is then attacked by the pentafluorophenoxy anion that is generated by the pyridine. Succinimido, chlorophenyl, and nitrophenyl esters were made by this method when it was introduced decades ago. A unique variant of this approach is the use of mixed carbonates that contain an isopropenyl group [$CH_3C(=CH_2)O\text{-}CO_2R$]. These react with hydroxy compounds in the presence of triethylamine or 4-dimethylaminopyridine (see Section 4.19) to give the esters and acetone.[30–35]

30. S Sakakibara, N Inukai. The trifluoroacetate method of peptide synthesis. 1. The synthesis and use of trifluoroacetate reagents. *Bull Chem Soc Jpn* 38, 1979, 1965.
31. M Jaouadi, C Selve, JR Dormoy, B Castro, J Martinez. Isopropenyl chloroformate in the chemistry of amino acids and peptides. III. Synthesis of active esters of *N*-protected amino acids. *Tetrahedron Lett* 26, 1721, 1985.
32. M Green, J Berman. Preparation of pentafluorophenyl esters of Fmoc protected amino acids with pentafluorophenyl trifluoroacetate. *Tetrahedron Lett* 31, 5851, 1990.
33. MH Jakobsen, O Buchardt, T Engdahl, A Holm. A new facile one-pot preparation of pentafluorophenyl (Pfp) and 3,4-dihydro-4-oxo-1,2,3-benzotriazine-3-yl (Dhbt) esters of Fmoc amino acids. (acid chlorides). *Tetrahedron Lett* 32, 6199, 1991.
34. K Takeda, A Ayabe, M Suzuki, Y Konda, Y Harigaya. An improved method for the synthesis of active esters of *N*-protected amino acids and subsequent synthesis of dipeptides. (isopropenyl mixed carbonates) *Synthesis* 689, 1991.
35. NL Benoiton, YC Lee, FMF Chen. Preparation of activated esters of *N*-alkoxycarbonylamino and other acids by a modification of the mixed anhydride procedure. *Int J Pept Prot Res* 42, 278, 1993.

7.9 ACTIVATED ESTERS: PROPERTIES AND SPECIFIC USES

Most activated esters are crystalline compounds that can be stored for subsequent use. A variety of properties are exhibited by the various esters. All esters mentioned in this monograph (see Section 2.9) except succinimido esters generate a hydroxy compound that is insoluble in water when aminolyzed. Elimination of this material can be a nuisance in some cases. Nitrophenols are not readily soluble in alkali; a trace is sufficient to produce a yellow color in the solution of the reaction product.

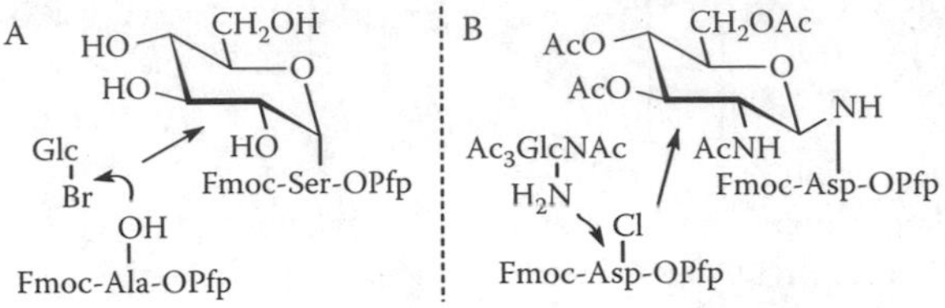

A HO CH$_2$OH O HO HO Glc Br OH Fmoc-Ser-OPfp Fmoc-Ala-OPfp

B AcO CH$_2$OAc O AcO NH Ac$_3$GlcNAc AcNH H$_2$N Cl Fmoc-Asp-OPfp Fmoc-Asp-OPfp

FIGURE 7.14 Activated esters for temporary protection and activation of Fmoc-amino acids for the synthesis of glycopeptides. (A) Reaction of α-D-glucopyranosyl bromide with esterified Fmoc-serine. (B) Reaction of 2-acetamido-2-deoxy-3,4,6-triacetyl-β-D-glucopyranosyl amine with esterified Fmoc-aspartyl chloride.[37,38] Pfp = pentafluorophenyl.

Halophenols are easier to dispose of by extraction. Chlorophenyl esters are resistant to hydrogenation and are hence compatible with benzyloxycarbonyl. Benzotriazolyl esters are unique in that they exist in two different forms (see Section 7.17). They are too reactive for routine use but are often employed without isolation after their preparation, using a carbodiimide or other (see Section 7.20). An exception to this are the benzotriazolyl adducts of *N*-tritylamino acids that are amide oxides (see Section 7.17), which are stable to aqueous sodium hydroxide but undergo aminolysis normally.

Succinimido esters are moderately reactive, relatively stable to water, and release *N*-hydroxysuccinimide, which is soluble in water. This makes them ideally suited for reactions in partially aqueous solution (see Section 7.21), which includes derivatization of amino acids with activated monoalkyl carbonates (see Section 3.15). 4-Nitrophenyl and pentafluorophenyl esters of N^{α}-protected asparagine and glutamine are employed instead of the acids and carbodiimides or other reagents to avoid dehydration to cyanoalanine residues (see Section 6.15). Cyano groups may be formed during preparation of the activated esters, but the required derivatives can be purified before use. Pentafluorophenyl esters of Fmoc-serine and asparagine have been employed for incorporation of the glycosylated residues in a peptide chain. The ester moiety provides protection during derivatization of the side chains of the precursors (Figure 7.14) and an activated carboxyl group for formation of the peptide bond. They are also employed for the incorporation of reversibly alkylated peptide bonds in a chain (see Section 8.5). 4-Oxo-3,4-dihydrobenzotriazol-3-yl esters are employed for solid-phase synthesis in continuous-flow systems (see Section 5.3), as their aminolysis provides a means of monitoring the number of amino groups remaining during a coupling (Figure 7.15). Liberated 3-hydroxy-4-oxo-3,4-dihydrobenzotriazine protonates amino groups, producing the 4-oxo-3,4-dihydrobenzotriazine-3-oxy anion,

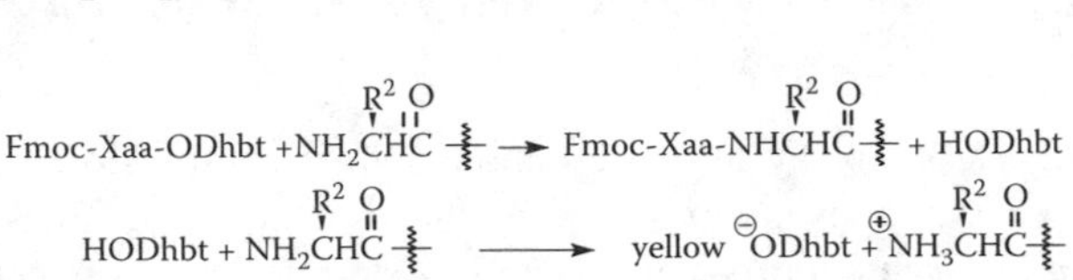

Fmoc-Xaa-ODhbt + $NH_2CH(R^2)C(=O)$– → Fmoc-Xaa-$NHCH(R^2)C(=O)$– + HODhbt

HODhbt + $NH_2CH(R^2)C(=O)$– ⟶ yellow $^{\ominus}$ODhbt + $^{\oplus}NH_3CH(R^2)C(=O)$–

FIGURE 7.15 Aminolysis of 4-oxo-3,4-dihydrobenzotriazin-3-yl esters produces a yellow color, the intensity of which is a measure of the number of amino groups present. The color disappears when the amino groups have been consumed.

which is yellow. The intensity of the yellow color depends on the number of amino groups present. The color decreases as peptide-bond formation proceeds and disappears when there are no amino groups left. There is a report that a 4-oxo-3,4-dihydrobenzotriazol-3-yl ester was more chirally stable than a 7-azabenzotriazoly ester.[23,36–39]

23. M Bodanszky, V duVigneaud. A method of synthesis of long peptide chains using a synthesis of oxytocin as an example. *J Am Chem Soc* 81, 5688, 1959.
36. K Barlos, D Papaioannou, D Theodoropoulus. Preparation and properties of N^{α}-trityl amino acid 1-hydroxybenzotriazole esters. *Int J Pept Prot Res* 23, 300, 1984.
37. M Meldal, B Klaus. Pentafluorophenyl esters for temporary carboxyl group protection in solid phase synthesis of *N*-linked glycopeptides. *Tetrahedron Lett* 48, 6987, 1990.
38. M Meldal, KJ Jense. Pentafluorophenyl esters for temporary protection of the α-carboxy group in solid phase glycopeptide synthesis. *J Chem Soc Chem Commun* 483, 1990.
39. E Atherton, JL Holder, M Meldal, RC Sheppard, RM Valerio. 3,4-Dihydro-4-oxo-1,2,3-benzotriazin-3-yl esters of fluorenylmethoxycarbonyl amino acids as self-indicating reagents for solid phase peptide synthesis. *J Chem Soc Perkin Trans 1* 2887, 1988.

7.10 METHODS FOR THE PREPARATION OF ACTIVATED ESTERS OF PROTECTED PEPTIDES, INCLUDING ALKYL THIOESTERS

Until the 1990s, activated esters of protected peptides were rarely employed for segment condensations because there is no established method by which a segment can be esterified without enantiomerizing the carboxy-terminal residue. Transesterification and the carbodiimide method (see Section 7.7) allow efficient synthesis of activated esters of protected segments terminating with the achiral glycine residue, but these approaches are rarely employed. Recent work has now shown that N^{α}-protected dipeptides can be converted to the succinimido esters by the mixed-anhydride method (see Section 7.8) without any deleterious effect on the chiral integrity of the compounds. The mixed anhydrides were generated by the addition of an equivalent of *N*-methylmorpholine and isopropyl chloroformate to a solution of the N^{α}-protected peptide in ethyl acetate containing a trace of dimethylformamide at 13°C. After 2 minutes, an excess of *N*-hydroxysuccinimide and 0.5 equivalents of *N*-methylmorpholine were added. The low temperature was crucial for preserving the chiral integrity of the products. The mixed anhydride of the peptide cannot be isolated because it is too easily converted to the 5(4*H*)-oxazolone (see Section 2.23).

The traditional method for preparing activated esters of N^{α}-protected dipeptides is combination of the *N*-protected amino acid with the amino acid ester (Figure 7.16). The latter is obtained by *N*-deprotection of the diprotected amino acid in an acidic milieu. Coupling is achievable using the carbodiimide, mixed-anhydride, and acyl-azide methods. Success with this approach indicates that the esterified residues react preferentially with the other derivatives and not among themselves. The chain cannot be extended to the protected tripeptide ester because the dipeptide ester cyclizes too

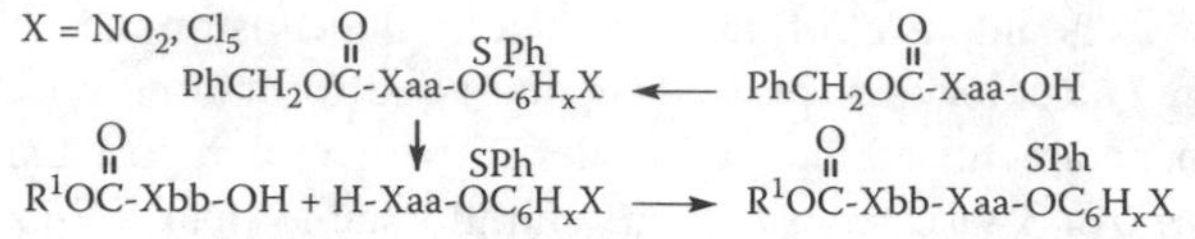

$X = NO_2, Cl_5$

$PhCH_2OC(=O)\text{-Xaa-}OC_6H_xX \xleftarrow{\text{S Ph}} PhCH_2OC(=O)\text{-Xaa-OH}$

$\downarrow$

$R^1OC(=O)\text{-Xbb-OH} + \text{H-Xaa-}OC_6H_xX\ (\text{SPh}) \longrightarrow R^1OC(=O)\text{-Xbb-Xaa-}OC_6H_xX\ (\text{SPh})$

FIGURE 7.16 Activated esters of protected dipeptides obtained by coupling an *N*-alkoxy-carbonylamino acid with an activated amino acid ester. Removal of $PhCH_2CO_2$ with HBr in AcOH or hydrogenation for $X = C_6Cl_5$ provided the activated monomers.[40,41] (Kovacs et al., 1966).

readily to the piperazine-2,5-dione (see Section 6.19). In contrast, a solid-phase variant of this approach that is successful is synthesis on the oxime resin (see Section 5.21). A peptide chain bound to an oxime resin during chain assembly is in fact an activated peptide ester. The activated peptide can be aminolyzed at any stage of chain assembly to produce a larger peptide or, by an activated amino-acid ester such as a phenylthiomethyl ester, to produce another activated peptide.

In the last decade, a novel approach to the coupling of segments through activated esters has emerged. A protected peptide alkyl thioester is assembled on a solid phase, the whole is detached from the support, and the ester is converted into an activated ester in the presence of the aminolyzing peptide segment (Figure 7.17). The synthesis is initiated by securing a spacer such as norleucine or β-alanine to a methylbenzhydrylamine resin, which is followed by coupling γ-mercaptopropanoic acid or, preferably, β-mercapto-β,β-dimethylpropanoic acid to the spacer. The first Boc-amino acid is linked to the support by esterification with the sulfhydryl group. The methyl groups render the thioester bond more stable to trifluoroacetic acid. The spacer serves to stabilize the carboxamido-methyl bond at the linker, which is otherwise sensitized by the thioester. The chain is assembled using Boc-amino acids except for the last derivative, which is an Fmoc-amino acid. The Fmoc-peptide alkylthio ester with spacer attached is released as the amide by hydrogen fluoride-anisole (9:1). The

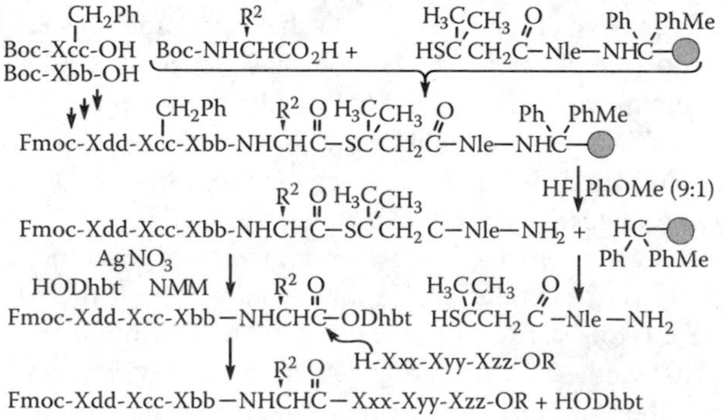

Boc-Xcc(CH_2Ph)-OH, Boc-Xbb-OH, Boc-NHCH(R^2)CO_2H + $HSC(CH_3)_2CH_2C(=O)$–Nle–NHC(Ph)(PhMe)–Ⓡ

$\downarrow$

Fmoc-Xdd-Xcc(CH_2Ph)-Xbb-NHCH(R^2)C(=O)–$SC(CH_3)_2CH_2C(=O)$–Nle–NHC(Ph)(PhMe)–Ⓡ

$\downarrow$ HF/PhOMe (9:1)

Fmoc-Xdd-Xcc-Xbb-NHCH(R^2)C(=O)–$SC(CH_3)_2CH_2$ C–Nle–NH_2 + HC(Ph)(PhMe)–Ⓡ

$\downarrow$ $AgNO_3$, HODhbt, NMM

Fmoc-Xdd-Xcc-Xbb–NHCH(R^2)C(=O)–ODhbt + $HSC(CH_3)_2CH_2C(=O)$–Nle–NH_2

$\downarrow$ H-Xxx-Xyy-Xzz-OR

Fmoc-Xdd-Xcc-Xbb–NHCH(R^2)C(=O)–Xxx-Xyy-Xzz-OR + HODhbt

FIGURE 7.17 Synthesis of a peptide by construction of a peptide alkyl thioester on a solid support, detachment of the assembled molecule that includes the spacer and transesterfication of the ester into an activated ester which is aminolyzed as it is formed. [Aimoto et al., 1991]. Dhbt = 4-oxo-3,4-dihydrobenzotriazin-3-yl.

thioester is converted into the activated ester by addition of silver nitrate in the presence of a hydroxy compound and *N*-methylmorpholine. The silver ion coordinates with the sulfur atom, inducing attack by the nucleophile. A succinimido ester is formed within 10 minutes. Yields are higher with HOBt or HOObt as nucleophiles, and the latter gives rise to the cleanest products. There apparently is no isomerization during the transesterification. Silver nitrate also attacks *S*-acetamidomethyl groups. Use of silver chloride, which is less soluble in dimethylsulfoxide, gives slower coupling but less *S*-Acm cleavage. The method is not compatible with Fmoc-chemistry because the alkyl thioester is sensitive to piperidine, but a modified cleavage mixture that does not affect the alkyl thioester has been developed. It consists of 25% *N*-methylpyrrolidine as base, 2% hexamethyleneimine [c(-CH_2)$_6$NH–] as nucleophile, and 2% HOBt, which reduces the basicity of the nucleophile in *N*-methylpyrrolidone:dimethylsulfoxide (1:1). A variety of large peptides have been obtained by this method. Other methods of preparing the alkyl thioesters are being developed because of their importance for synthesis by chemical ligation (see Section 7.25).[40–46]

40. T Wieland, B Heinke. Peptide synthesis XX. Further experiments with aminoacylphenyl- and aminoacylthiophenyl compounds. *Liebigs Ann Chem* 615, 184, 1958.
41. Goodman, KC Stueben. Peptide synthesis via amino acid active esters. *J Am Chem Soc* 81, 3980, 1959.
42. LM Siemens, FW Rottnek, LS Trzupek. Selective catalysis of ester aminolysis. An approach to peptide active esters. (phenylthiomethyl esters). *J Org Chem* 55, 3507, 1990.
43. H Hojo, S Aimoto. Polypeptide synthesis by use of an *S*-alkyl thio ester of a partially protected segment. Synthesis of the DNA-binding domaine of c-Myb protein (142-1193)-NH_2. *Bull Chem Soc Jpn* 64, 111, 1991.
44. NL Benoiton, YC Lee, FMF Chen. A new coupling method allowing epimerization-free aminolysis of segments. Use of succinimidyl esters obtained through mixed anhydrides, in HLS Maia, ed. *Peptides 1994. Proceedings of the 23rd European Peptide Symposium*, Escom, Leiden, 1995, pp 203-204.
45. L Xiangqun, T Kawakami, S Aimoto. Direct preparation of peptide thioesters using an Fmoc solid-phase method. *Tetrahedron Lett* 39, 8669, 1998.
46. S Aimoto. Polypeptide synthesis by the thioester method. *Biopolymers (Pept Sci)* 51, 247, 1999.

7.11 SYNTHESIS USING *N*-9-FLUORENYLMETHOXYCARBONYLAMINO-ACID CHLORIDES

Acid chlorides were employed for coupling benzyloxycarbonylamino acids in earlier times. However, the technique lost favor with the development of other methods of coupling (see Section 2.14). The use of acid chlorides for peptide-bond formation was resuscitated when the acid-stable Fmoc group was introduced for protection of the α-amino group. Fmoc-amino-acid chlorides are available from the parent acid, using thionyl chloride, or from the mixed anhydride (see Section 2.8), using hydrogen chloride (Figure 7.18). Most are crystalline compounds that are stable in a dry

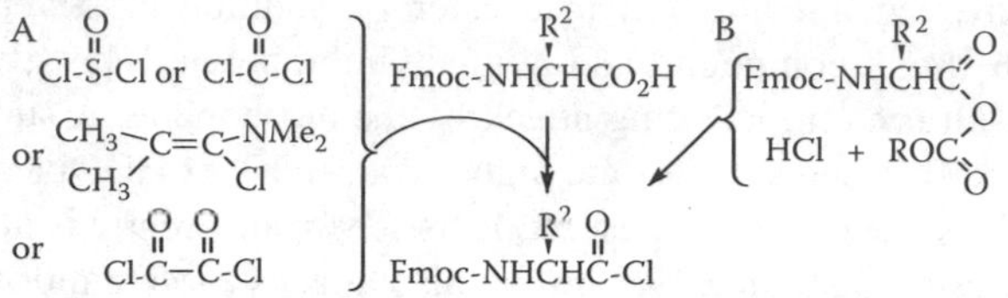

A $Cl\text{-}S(=O)\text{-}Cl$ or $Cl\text{-}C(=O)\text{-}Cl$ or $(CH_3)_2C{=}C(Cl)NMe_2$ or $Cl\text{-}C(=O)\text{-}C(=O)\text{-}Cl$ + Fmoc-NHCH(R^2)CO_2H → Fmoc-NHCH(R^2)C(=O)-Cl

B Fmoc-NHCH(R^2)C(=O)-O-C(=O)OR + HCl → Fmoc-NHCH(R^2)C(=O)-Cl

FIGURE 7.18 Preparation of Fmoc-amino-acid chlorides by reaction (A) of thionyl chloride,[47] phosgene from triphosgene,[54] 1-chloro-2,*N*,*N*-trimethyl-1-propene- 1-amine, [Schmidt et al., 1988] or oxalyl chloride, [Rodriguez, 1997] with the parent acid and (B) of hydrogen chloride with the mixed anhydride.[51]

atmosphere. Their preparation was restricted to derivatives bearing acid-stable protectors on side chains, but this limitation has been eliminated by the use of triphosgene (see Section 7.13) for their production (Figure 7.18). Another reagent suitable for the preparation of Fmoc- and Cbz-amino-acid chlorides is 1-chloro-2,*N*,*N*-trimethyl-1-propene-amine (Figure 7.18). Fmoc-amino-acid chlorides are highly activated molecules, the aminolysis of which generates a strong acid. These two characteristics create a situation that is a major obstacle to their use. A strong base is required to neutralize the hydrogen chloride that is liberated, and the required base causes the acid chloride to cyclize to the 5(4*H*)-oxazolone (see Section 4.17), which is less reactive than the acid chloride and is chirally sensitive to the base (see Section 8.1). In contrast, a favorable feature of Fmoc-amino-acid chlorides is that they are soluble in dichloromethane, which eliminates the need for a polar solvent such as dimethylformamide or *N*-methylpyrrolidone.

For synthesis in solution, the obstacle of 5(4*H*)-oxazolone formation is minimized by operating in a two-phase system of chloroform and aqueous sodium carbonate. The two-phase system keeps the base and the acid chloride separated from each other. In solid-phase synthesis, if triethylamine or diisopropylethylamine is employed as base, the aminolysis does not go to completion, though it does if the coupling mixture contains 1-hydroxybenzotriazole. This means that the acid chloride is converted into the activated ester before it has time to cyclize. Regardless, the best course of action is the use of a scavenger for the acid that is not basic. The molecule of choice is the potassium salt of 1-hydroxybenzotriazole, as it neutralizes the liberated acid without creating a basic environment. A second major obstacle arises during synthesis in solution: disposition of the products generated by deprotection of the α-amino groups. Dibenzofulvene and the adduct produced by cleavage of Fmoc-NH with piperidine (see Section 3.11) stay in the organic layer with the peptide product. This situation is averted by employing an amine with a handle that solubilizes the adduct in aqueous acid (see Section 7.26). 4-Methylaminopiperidine [$CH_3NHC(CH_2CH_2)_2N$] or tris(2-aminoethyl)amine [$(NH_2CH_2CH_2)_3N$] form adducts of unestablished structures with the liberated tricyclic moiety of the protector, which can be extracted out of chloroform by a phosphate buffer of pH 5.5; the peptide ester remains in the organic layer. The amine also serves the purpose of destroying any unreacted acid chloride. This protocol cannot be employed at the dipeptide ester stage, however, unless the ester is *tert*-alkoxy (OtBu, OTrt), because other dipeptide esters cyclize to the piperazine-2,5-diones (see Section 6.19) in alkaline media.

Because of the various obstacles, Fmoc-amino-acid chlorides are not employed for routine use. However, they are particularly suited for couplings of hindered residues. This is exemplified by the 69% yield obtained for the synthesis in solution of Fmoc-Aib-Aib-Aib-Aib-OCH_2Ph (Aib = aminoisobutyric acid), using Fmoc-amino-acid chlorides with KOBt followed by 4-methylaminopyridine for deprotection. Fmoc-*N*-methylamino acids have been coupled efficiently as the chlorides prepared *in situ* using triphosgene (see Section 7.14), with 2,4,6-trimethylpyridine (collidine) as a base to neutralize the hydrogen chloride generated by aminolysis by a resin-bound function. The latter can be effected in tetrahydrofuran, dioxane, or 1,3-dichloropropane, but not in *N*-methylpyrrolidone, which participates in a reaction that enantiomerizes the residue. Trityl-protected side chains are excluded because they are unstable to triphosgene. The issue of 5(4*H*)-oxazolone formation disappears when *p*-toluenesulfonylamino acids are coupled as the acid chlorides; however, removal of the protector is not simple. Esterification of the first residue to a hydroxymethyl-linker-resin can be achieved using an Fmoc-amino-acid chloride dissolved in pyridine-dichloromethane (2:3).[47–54]

47. LA Carpino, BJ Cohen, KE Stephens, SY Sadat-Aalaee, J-J Tien, DC Langridge. ((9-Fluorenylmethyl)oxy)carbonyl (Fmoc) amino acid chlorides. Synthesis, characterization, and application to the rapid synthesis of short peptide segments. *J Org Chem* 51, 3732, 1986.
48. M Beyermann, M Bienert, H Niedrich, LA Carpino, D Sadat-Aalaee. Rapid continuous peptide synthesis via FMOC amino acid chloride coupling and (4-aminomethyl)piperidine deblocking. *J Org Chem* 55, 721, 1990.
49. K Akaji, H Tanaka, H Itoh, J Imai, Y Fujiwara, T Kimura, Y Kiso. Fluoren-9-ylmethyloxycarbonyl (Fmoc) amino acid chloride as an efficient reagent for anchoring Fmoc amino acid to 4-alkoxybenzyl resin. *Chem Pharm Bull (Jpn)* 38, 3471, 1990.
50. LA Carpino, H Chao, M Beyermann, M Bienert. ((9-Fluorenylmethyl)oxy)carbonyl amino acid chlorides in solid-phase synthesis. *J Org Chem* 56, 2635, 1991.
51. FMF Chen, YC Lee, NL Benoiton. Preparation of *N*-9-fluorenylmethoxycarbonyl amino acid chlorides from mixed anhydrides by the action of hydrogen chloride. *Int J Pept Prot Res* 38, 97, 1991.
52. KM Sivanandaiah, VV Suresh Babu, C Renukeshwar. Fmoc-amino acid chlorides in solid phase synthesis of opioid peptides. *Int J Pept Prot Res* 39, 201, 1992.
53. VV Suresh Babu, HN Gopi. Rapid and efficient synthesis of peptide fragments containing α-aminoisobutyric acid using Fmoc-amino acid chlorides/potassium salt of 1-hydroxybenzotriazole. *Tetrahedron Lett* 39, 1049, 1998.
54. E Falb, T Yechezkel, Y Salitra, C Gilon. *In situ* generation of Fmoc-amino acid chlorides using *bis*-(trichloromethyl)carbonate and its utilization for difficult couplings in solid-phase peptide synthesis. *J Pept Res* 53, 507, 1998.

7.12 SYNTHESIS USING *N*-ALKOXYCARBONYLAMINO ACID FLUORIDES

The greater stability of alkoxycarbonyl fluorides relative to alkoxycarbonyl chlorides prompted researchers to examine *N*-alkoxycarbonylamino-acid fluorides as activated forms for peptide-bond formation. The fluorides possess several distinct, favorable features in comparison with acid chlorides. First, the particular nature of the acid–fluoride bond, in part because of the smaller size of the leaving group, imparts equal or greater reactivity toward anionic and amine nucleophiles, yet greater stability toward oxygen nucleophiles such as water or methanol. Second, the acid fluorides have a much lesser tendency to cyclize to the 2-alkoxy-5(4*H*)-oxazolones. Third, they can be aminolyzed in the absence of base. Finally, the usual methods for their preparation allow access to 9-fluorenylmethyl-, benzyl-, and *tert*-butyl based derivatives. These characteristics make the acid fluorides unique activated forms of the protected amino acids. Their comportment, in fact, resembles that of activated esters and not of acid halides. The first acid fluorides were prepared employing the corrosive cyanuric fluoride (Figure 7.19) in a solvent containing pyridine. An alternative synthesis employs diethylaminosulfur trifluoride (Note: this can cause burns on skin), which generates water-soluble secondary products. Both are well-established fluorinating reagents. The acid-sensitive derivatives are prepared at low temperatures (–20° or –30°C). Products are generally purified by washing with water, and most are shelf-stable compounds. Aminolysis can be effected in the absence of base — a reaction that is only slightly less efficient than in the presence of diisopropylethylamine. The resistance to cyclization of the acid fluorides permits their use in solution in a one-phase system. The 30% excess of Fmoc-amino-acid fluoride that is recommended is destroyed by *tris*-(2-aminoethyl)amine (see Section 7.11) at the same time as the protector is removed. It is interesting to note that aminolysis of an acid chloride stops at 50% in the absence of base and that triethylamine binds three atoms of fluoride so that only a fraction of tertiary amine is required to neutralize the fluoride anion. A third reagent, TFFH ($Me_4N_2CF^{+}\cdot PF_6^{-}$; Figure 7.19), is employed for generating the fluorides in the presence of the incoming nucleophile, though separate activation is sometimes preferable. The activation is effected in dimethylformamide in the presence of two equivalents of diisopropylethylamine. The extra base has no deleterious effect on the chiral integrity of ordinary residues that are activated; however, it can induce enantiomerization of the residue if the aminolysis

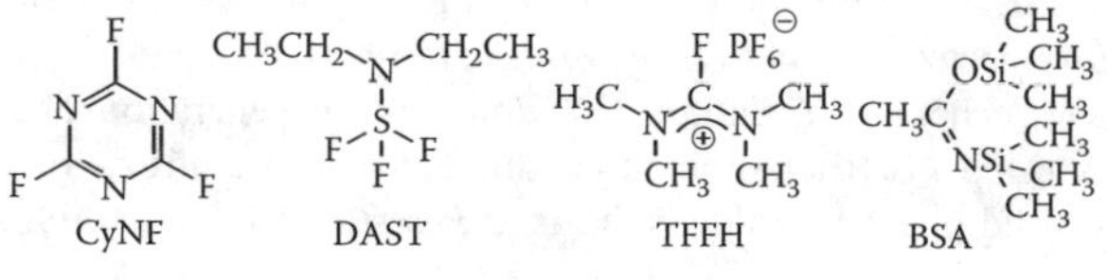

CyNF DAST TFFH BSA

FIGURE 7.19 Reagents for preparing *N*-protected amino-acid fluorides,[55,56,61] Boc_2-amino-acid fluorides (CyNF),[58,59] and a nonbasic acid scavenger (BSA).[62] CyNF = cyanuric fluoride; DAST = diethylaminosulfur trifluoride; TFFH = tetramethylfluoroformamidinium hexafluorophosphate; BSA = *bis*(trimethylsilyl)acetimide.

is impeded as a result of the nature of the reacting residues. Acid fluorides are particularly effective for couplings between hindered residues, but the danger of isomerization in these cases must be kept in mind. Slow aminolysis of Fmoc-amino-acid fluorides in the presence of base can also lead to partial deblocking of the amino function. *bis*(Trimethylsilyl)acetamide (Figure 7.19) is a nonbasic acid scavenger that can be employed instead of tertiary amine to avoid premature deprotection. The neutralizing effect is provided by the ether function. An alternative to avoid tertiary amine is the use of powdered zinc, which destroys the hydrogen fluoride generated by reducing the proton to hydrogen (see Section 3.15).

A different approach to coupling that avoids 5(4*H*)-oxazolone formation is the use of disubstituted derivatives such as *bis*Boc-amino acids. These derivatives react more sluggishly than monosubstituted derivatives when couplings are attempted by the usual methods. The efficacy of acid fluorides for amide-bond formation is dramatically illustrated by the success achieved in reactions between *bis*Boc-amino-acid fluorides and the anion of pyrrole-2-carboxylic acid. Previous attempts to couple amino acids with pyrrole-2-carboxylic acid by other methods had failed. The use of TFFH allows couplings of arginine and histidine derivatives as the fluorides; the fluorides of these derivatives cannot be obtained using other reagents. Recent studies of the use of fluorides for couplings of hindered residues indicate that in some cases, dichloromethane with pyridine may be a better solvent–base combination than dimethylformamide with diisopropylethylamine. The conditions that provide optimum reactivity between two reactants depend on the natures of the latter.[55–63]

55. LA Carpino, D Sadat-Aalaee, HG Chao, RH DeSelms. ((9-Fluorenylmethyl)oxy)carbonyl (FMOC) amino acid fluorides. Convenient new peptide coupling reagents applicable to the FMOC/*tert*-butyl strategy for solution and solid-phase syntheses. *J Am Chem Soc* 112, 9651, 1990.
56. J-N Bertho, A Loffet, C Pinel, F Reuther, G Sennyey. Amino acid fluorides: their preparation and use in peptide synthesis. *Tetrahedron Lett* 32, 1303, 1991.
57. LA Carpino, EME Mansour, D Sadat-Aalaee. *tert*-Butoxycarbonyl and benzyloxycarbonyl amino acid fluorides. New, stable rapid-acting acylating agents for peptide synthesis. *J Org Chem* 56, 2611, 1991.
58. J Savrda, M Wakselman. *N*-Alkoxycarbonylamino acid *N*-carboxyanhydrides and *N,N*-dialkoxycarbonyl amino acid fluorides from *N,N*-diprotected amino acids. *J Chem Soc Chem Commun* 812, 1992.
59. LA Carpino, EME Mansour, A El-Faham. Bis(BOC) amino acid fluorides as reactive peptide coupling reagents. *J Org Chem* 58, 4162, 1993.
60. C Kaduk, H Wenschuh, M Beyermann, K Forner, LA Carpino, M Bienert. Synthesis of Fmoc-amino acid fluorides via DAST, an alternative fluorinating agent. *Lett Pept Sci* 2, 285, 1995.
61. LA Carpino, A El-Faham. Tetramethylfluoroformidinium hexafluorophosphate: a rapid-acting peptide coupling reagent for solution and solid phase peptide synthesis. *J Am Chem Soc* 117, 5401, 1995.
62. SA Triolo, D Ionescu, H Wenschuh, NA Solé, A El-Faham, LA Carpino, SA Kates. Recent aspects of the use of tetramethylfluoroformamidinium hexafluorophosphate (TFFH) as a convenient peptide coupling reagent, in R Ramage, R Epton, eds. *Peptides 1966. Proceedings of the 24th European Peptide Symposium*, Mayflower, Kingswoodford, 1998, pp 839-840.

63. H Wenschuh, D Ionescu, M Beyermann, M Bienert, LA Carpino. Peptide bond formation *via* Fmoc amino acid fluorides in the presence of silylating agents, in R Ramage, R Epton, eds. *Peptides 1966. Proceedings of the 24th European Peptide Symposium*, Mayflower, Kingswoodford, 1998, pp 907-908.

7.13 AMINO-ACID *N*-CARBOXYANHYDRIDES: PREPARATION AND AMINOLYSIS

A unique activated form of amino acids that was developed by H. Leuchs a century ago is the anhydride that is formed between the carboxyl group of the amino acid and a carboxyl group that is bound to the amino group. These anhydro *N*-carboxyamino acids are simultaneously activated and protected at the amino group. They are now known as amino acid *N*-carboxyanhydrides. They undergo aminolysis readily without isomerization. However, their routine use is limited because they have a tendency to oligomerize, as the protecting substituent on the amino group becomes labile once aminolysis has occurred. There are two approaches for the preparation of *N*-carboxyanhydrides: reaction of the amino acid or its cupric salt (see Section 6.24) with phosgene (Figure 7.20, paths AB), or its equivalent (paths EDB and CDB), and cyclization of an *N*-alkoxycarbonylamino-acid halide (Figure 7.20, path F). The original anhydrides were obtained from *N*-methoxycarbonylamino-acid chlorides. Cbz-amino acids are the common starting materials, with the chlorides being obtained using thionyl chloride (see Section 6.14) or phosphorus pentachloride. Cyclization with expulsion of the benzyl cation occurs spontaneously at ambient temperature. The bromides obtained using the corresponding reagent cyclize more readily. Conditions for reaction of the unsubstituted amino acid with phosgene have been refined. Excellent yields are obtained by adding an excess of phosgene dissolved in benzene or tetrahydrofuran to a suspension of the amino acid in tetrahydrofuran, with the mixture being left at 65°C for 1.5 hours.

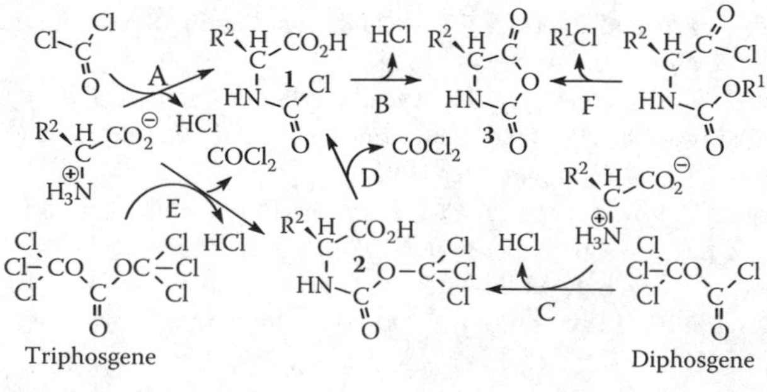

FIGURE 7.20 Reactions giving rise to amino-acid *N*-carboxyanhydrides. [Leuchs, 1906]. (A) Amino acid plus phosgene produces *N*-chlorocarbonyl intermediate **1,** which cyclizes (B) to anhydride **3**. (C) Amino acid plus diphosgene produces *N*-trichloromethoxycarbonyl intermediate **2,** which expels phosgene (D), giving intermediate **1**. (E) Amino acid plus triphosgene produces the same intermediate **2**.[70] (F) *N*-Alkoxycarbonylamino-acid chloride cyclizes to the *N*-carboxyanhydride at ambient temperature. $R^1 = PhCH_2$, tBu.

The chlorocarbonyl derivative **1** (Figure 7.20) is first formed, after which cyclization occurs. Hydrogen chloride is released at each step. Side-chain-blocked amino acids that might be damaged by the acid are first converted to the *N,O*-bis(trimethylsilyl) derivatives, which release trimethylsilyl chloride instead of the acid during the phosgenation. There are two phosgene substitutes that can be employed instead of the extremely hazardous phosgene. Trichloromethyl chloroformate (diphosgene) is a liquid that reacts with an amino acid (Figure 7.20, paths CDB) — not efficiently, but generating 2 moles of phosgene quickly in the presence of activated charcoal. *bis*(trichloromethyl) carbonate (triphosgene) mp 80°C, easily prepared by bubbling chlorine through an irradiated solution of dimethyl carbonate in carbon tetrachloride, is a better substitute that undergoes nucleophilic attack at the carbonyl, producing the same intermediate **2** (Figure 7.20), with the trichloromethoxy leaving group generating chloride and phosgene, which also immediately reacts with the amino acid (path A). The liberated acid tends to stop the reaction prematurely by protonating the amino group; this is averted by occasional sparging of the solution with nitrogen. Triphosgene also efficiently converts Boc-amino acids to the anhydrides, similar to path E, at ambient temperature. Two unique *N*-carboxyanhydrides are those of aspartic and glutamic acids (see Section 6.12), which are obtained directly from the unprotected amino acids. The former, however, is not easy to handle and is advantageously replaced by the thio analogue that is obtained from aspartic acid anhydride (see Section 6.12) and carbon disulfide.

The aminolysis of *N*-carboxyanhydrides can be effected as usual with carboxy-substituted amino acids in organic solvents (Figure 7.21, path A) or in aqueous solution at pH 10, with substituted or unsubstituted (path B) amino acids. The *N*-carboxyamino acid that is formed is unstable to acid, so the next step in a synthesis involves lowering the pH to allow decomposition and then reelevating it for reaction of the amino group with the next activated residue. Premature loss of carbon dioxide, and hence oligomerization, is not uncommon. Only if the conditions of temperature and pH are rigorously controlled is efficient synthesis achieved, and the optimum conditions depend on the nature of the reacting residues. The bulkier the side chain of the *N*-carboxyanhydride, the less likely oligomerization will occur. The benefits of defining optimum conditions for reactions are well illustrated by the fact that the large amounts of L-alanyl-L-proline and N^{ε}-trifluoroacetyl-L-lysyl-L-proline that are

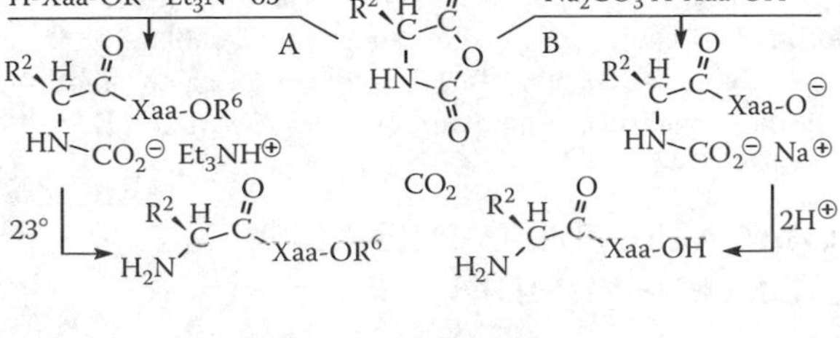

H-Xaa-OR^6 Et$_3$N –65° A; Na$_2$CO$_3$ H-Xaa-OH B; HN–CO$_2^{\ominus}$ Et$_3$NH$^{\oplus}$; Xaa-OR^6; HN–CO$_2^{\ominus}$ Na$^{\oplus}$; Xaa-O$^{\ominus}$; 23°; CO$_2$; 2H$^{\oplus}$; H$_2$N ... Xaa-OR^6; H$_2$N ... Xaa-OH

FIGURE 7.21 Synthesis of a dipeptide by reaction of an amino-acid *N*-carboxyanhydride (A) with an amino-acid ester in tetrahydrofuran[65] and (B) with an amino acid in aqueous solution.[67]

required as starting materials for the preparation of some commercial products are obtained by reaction of proline with the *N*-carboxyanhydrides. A critical feature of the syntheses is the use of carbonates and potassium instead of sodium ion for the aqueous buffer; the potassium salts of the products are more soluble than the sodium salts. A two-phase system of aqueous buffer and acetonitrile provides an efficient coupling environment, with the carbonate stabilizing the carbamate in the aqueous phase and the acetonitrile keeping the anhydride in the organic phase, thus minimizing oligomerization and hydrolysis of the anhydride.

Amino-acid *N*-carboxyanhydrides have been of value for preparing polyamino acids and polylysine in particular. A strong base such as diethylamine initiates polymerization of the anhydride by deprotonating the nitrogen atom of a molecule. They react with alcohols to give esters if hydrogen chloride is present to prevent aminolysis. They serve as reagents for determining enantiomers of amino acids by their conversion into diastereomeric dipeptides (see Section 4.23).[64–73]

64. AC Farthing. Synthetic polypeptides. I. Synthesis of 2,5-oxazolidenediones and a new reaction of glycine. *J Chem Soc* 3213, 1950.
65. JL Bailey. Synthesis of simple peptides from anhydro-*N*-carboxyamino acids. *J Chem Soc* 3461, 1950.
66. R Hischmann, RF Nutt, DF Veber, RA Vitali, SL Varga, TA Jacob, FW Holly, R Denkewalter. Studies on the total synthesis of an enzyme. V. The preparation of enzymatically active material. *J Am Chem Soc* 91, 507, 1969.
67. R Hirschmann, H Schwam, RG Strachan, EF Schoenewaldt, H Barkemeyer, SM Miller, JB Conn, V Garsky, DF Veber, RG Denkewalter. The controlled synthesis of peptides in aqueous medium. The preparation and use of novel α-amino acid *N*-carboxyanhydrides. *J Am Chem Soc* 93, 2746, 1971.
68. WD Fuller, MS Verlander, M Goodman. A procedure for the facile synthesis of amino-acid *N*-carboxyanhydrides. *Biopolymers* 15, 1869, 1976.
69. R Katakai, Y Iizuka. An improved rapid method for the synthesis of *N*-carboxy α-amino acid anhydrides using trichloromethyl chloroformate. (diphosgene). *J Org Chem* 50, 715, 1985.
70. H Eckert, B Forster. Triphosgene, a crystalline phosgene substitute. *Angew Chem Int Ed Engl* 894, 1987
71. TJ Blacklock, RF Shuman, JW Butcher, WE Shearin, J Budavari, VJ Grenda. Synthesis of semisynthetic dipeptides using *N*-carboxyanhydrides and chiral induction on Raney nickel. A method practical for large scale. *J Org Chem* 53, 836, 1988.
72. WH Daly, D Poche. The preparation of *N*-carboxyanhydrides of α-amino acids using bis(trichloromethyl)carbonate. *Tetrahedron Lett* 29, 5894, 1988.
73. RWilder, S Mobashery. The use of triphosgene in preparation of *N*-carboxy-α-amino acid anhydrides. (from Boc-amino acids) *J Org Chem* 57, 2755, 1992.

7.14 *N*-ALKOXYCARBONYLAMINO-ACID *N*-CARBOXYANHYDRIDES

N-Carboxyanhydrides of amino acids (see Section 7.13) can undergo oligomerization if they are aminolyzed under conditions that are not strictly controlled. A variant of the anhydride that is employed because it does not undergo oligomerization during

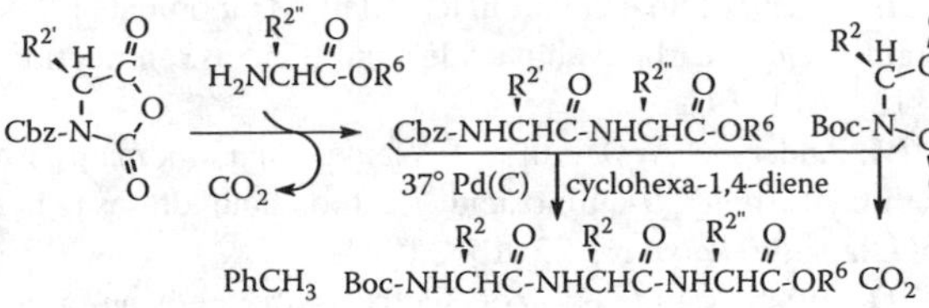

FIGURE 7.22 A one-pot synthesis of a protected tripeptide in tetrahydrofuran using *N*-alkoxycarbonylamino-acid *N*-carboxyanhydrides.[77] The amino group of the second residue is liberated by catalytic transfer hydrogenolysis (see Section 6.21).

aminolysis is the *N*-alkoxycarbonyl-protected derivative (Figure 7.22), referred to as the urethane-protected amino-acid *N*-carboxyanhydride. These anhydrides are usually crystalline compounds, are highly activated, and react with amino groups without isomerization, liberating only carbon dioxide as a secondary product. Their aminolysis occurs efficiently in most solvents, including glacial acetic acid. The Cbz- and Fmoc-derivatives are obtained by acylation of the *N*-carboxyanhydride with the acid chlorides in the presence of *N*-methylmorpholine — a base that does not initiate oligomerization of the anhydride. The Boc-derivatives are obtained using di-*tert*-butyl pyrocarbonate (Boc_2O, see Section 3.16) in the presence of pyridine. Excess tertiary amine is neutralized with a solution of hydrogen chloride in dioxane. Because of the efforts required for their preparation, couplings are not routinely carried out using urethane-protected amino-acid *N*-carboxyanhydrides. However, they can be used to advantage in certain situations (see Section 8.9) including synthesis on a large scale. An interesting example is the synthesis of protected tripeptides involving aminolysis of one anhydride followed by deprotection in the presence of a second anhydride (Figure 7.22), without removal of solvent or transfer of material to a second reaction vessel. They are suitable for anchoring the first residue in a synthesis to a solid support and can be converted into amino alcohols and aldehydes by reduction with appropriate reagents. They are unstable in dichloromethane or dimethylformamide but not tetrahydrofuran in the presence of DBU (see Section 8.12), triethylamine, or diisopropylethylamine. Under these conditions, the Boc-derivatives dimerize with the loss of two moles of carbon dioxide to give 1-Boc-3-BocNH-3,5-dialkylpyrrolidine-2,4-diones (Figure 7.23).[74–80]

74. HR Kricheldorf, M Fehrle. *N*-(2-Nitrophenylsulfenyl)-α-amino acid *N*-carboxyanhydrides. *Chem Ber* 107, 3533, 1974.

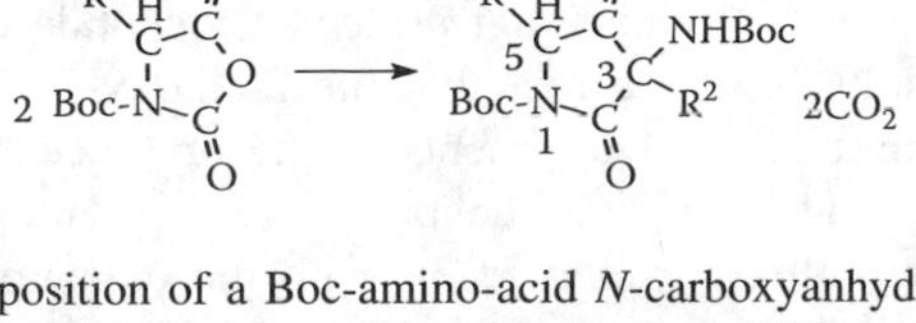

FIGURE 7.23 Decomposition of a Boc-amino-acid *N*-carboxyanhydride by tertiary amine. Two molecules combine with the release of two molecules of CO_2 to form a pyrrolidine-2,4-dione.

75. WD Fuller, MP Cohen, M Shabankareh, RK Blair, M Goodman, FR Naider. Urethane-protected amino acid *N*-carboxyanhydrides and their use in peptide synthesis. *J Am Chem Soc* 112, 7414, 1990.
76. BA Swain, BL Anderson, WD Fuller, F Naider, M Goodman. Esterification of 9-fluorenylmethoxylcarbonyl α-amino acid *N*-carboxyanhydrides to hydroxyl-functionalized resins. *Reactive Polymers* 22, 155, 1994.
77. Y-F Zhu, WD Fuller. Rapid, one-pot synthesis of urethane-protected tripeptides. *Tetrahedron Lett* 36, 807, 1995.
78. WD Fuller, M Goodman, FR Naider, Y-F Zhu. Urethane-protected α-amino acid *N*-carboxyanhydrides and peptide synthesis. *Biopolymers* 40, 183, 1996.
79. JJ Leban, KL Colson. Base induced dimerization of urethane-protected amino acid *N*-carboxyanhydrides. *J Org Chem* 61, 228, 1996.
80. C Pothion, J-A Fehrentz, A Aumelas, A Loffet, J Martinez. Synthesis of pyrrolidine-2,4-diones from urethane *N*-carboxyanhydridess (UNCAs). *Tetrahedron Lett* 37, 1027, 1996.

7.15 DECOMPOSITION DURING THE ACTIVATION OF BOC-AMINO ACIDS AND CONSEQUENT DIMERIZATION

It was noticed in the 1970s that activated Boc-amino acids generate ninhydrin-positive products indicative of free amino groups (see Section 5.4) under conditions in which activated Cbz-amino acids do not. Rationalization of the observations emerged later from studies on the properties of mixed anhydrides. Attempts to react the mixed anhydride of Boc-valine with methanol in dichloromethane produced primarily the Boc-dipeptide ester and not the expected Boc-valine ester. Further studies revealed that dimerization also occurred in varying amounts (4–20%) during the carbodiimide-mediated reactions of Boc-valine with phenol, *p*-nitrophenol, and 1-hydroxybenzotriazole. No dipeptide ester was formed in reactions with the stronger nucleophiles *N*-hydroxysuccinimide and 3-hydroxy-4-oxo-3,4-dihydrobenzotriazine (see Section 2.9). When 2-*tert*-butoxy-4-isopropyl-5(4*H*)-oxazolone, the oxazolone from Boc-valine, was left in dichloromethane in the presence of a deficiency of *p*-nitrophenol (HONp), Boc-Val-ONp, Boc-Val-Val-ONp, valine-*N*-carboxyanhydride (see Section 7.13), and *tert*-butyl-*p*-nitrophenyl ether were produced. The corresponding oxazolone from Cbz-valine produced only Cbz-Val-ONp under the same conditions. It transpires that 2-*tert*-butoxy-5(4*H*)-oxazolones (see Section 1.10) do not undergo simple alcoholysis to the ester or hydrolysis to the acid in the presence of protic solvents or water. They decompose partially or completely to the *N*-carboxyanhydride and the *tert*-butyl cation (Figure 7.24, path B). Thus, the generation of ninhydrin-positive products and protected dipeptide esters from activated Boc-amino acids results from the fact that the 2-alkoxy-5(4*H*)-oxazolone that is formed (path A) is unstable and fragments in the presence of protic compounds instead of undergoing alcoholysis. The protonated oxazolone gives rise to the *N*-carboxyanhydride (path C), which reacts with the oxyanion to give the amino acid ester (path D). Aminolysis of activated derivative by the latter (path E) is the source of the dipeptide ester.

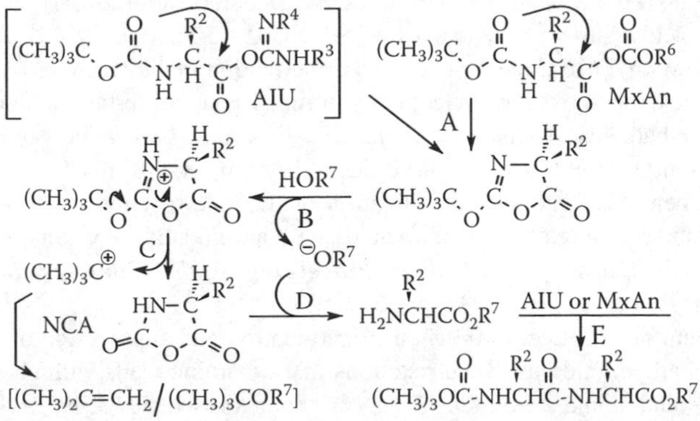

FIGURE 7.24 Dipeptide ester formation during reaction of Boc-amino acids with weak oxygen nucleophiles (HOR^7: $R^7 = CH_3$ or p-$NO_2C_6H_4$).[82] Some activated derivative (AIU = *O*-acylisourea; MxAn = mixed anhydride) cyclizes (A) to the 2-*tert*-butoxy-5(4*H*)-oxazolone, which when protonated by the nucleophile (B) expels the *tert*-butyl cation (C), thus generating *N*-carboxyanhydride (NCA). The NCA reacts with the oxyanion (D), producing the amino acid ester, which reacts with activated derivative (E), yielding the protected dipeptide.

Evidence for protonation of the oxazolone is the fact that a solution of 2-*tert*-butoxy-4-isopropyl-5(4*H*)-oxazolone and a deficiency of *p*-nitrophenol is yellow until the oxazolone has been consumed. The color is a result of the presence of the nitrophenoxide anion. The danger of decomposition during the activation of Boc-amino acids is general, being greater for derivatives of hindered residues that form the oxazolone more readily. Decomposition is suppressed by the presence of a tertiary amine, which prevents protonation of the oxazolone (see Section 2.25). Slight decomposition, up to 1.5%, has also been encountered in peptide-bond forming reactions mediated by carbodiimides. The decomposition was caused by the acidity of the salts, *N*-methylmorpholine hydrochloride and *p*-toluenesulphonate, that resulted from neutralization of the salts of the amino acid esters. Use of excess tertiary amine to prevent decomposition cannot be recommended, as it reduces efficiency by promoting *N*-acylurea formation (see Section 2.12). The addition of pyridine seems to be the best compromise. No decomposition of activated Boc-amino acids is to be expected in solid-phase synthesis, where the acidity is eliminated before coupling is effected. Activation by reagents such as bromotripyrrolidinophosphonium hexaflurophosphate ($Pyr_3PBr^+ \cdot PF_6^-$) that require a tertiary amine and liberate halide also leads to generation of *N*-carboxyanhydride. Decomposition of activated Boc-amino acids is particularly relevant when they are not consumed quickly by aminolysis and when the derivative is that of an *N*-methylamino acid. Cyclization of the latter gives the positively charged oxazolonium ion, which fragments without the need for protonation. The high sensitivity to decomposition of activated Boc-*N*-methylamino acids is illustrated by the finding that a preparation of the mixed anhydride (see Section 2.8) of Boc-*N*-methylvaline contained 20% of *N*-methylvaline *N*-carboxyanhydride. The tendency of activated Boc-*N*-methylamino acids to decompose explains why their aminolysis is not often achieved in high yields (see Section 8.15).[81–85]

81. M Bodanszky, YS Klausner, A Bodanszky. Decomposition of *tert*-butyloxycarbonylamino acids during activation. *J Org Chem* 40, 1507, 1975.
82. NL Benoiton, FMF Chen. Unexpected dimerization in the reactions of activated Boc-amino acids with oxygen nucleophiles in the absence of tertiary amine, in JE Rivier, GR Marshall, eds. *Peptides, Structure and Function. Proceedings of the 11th American Peptide Symposium*, Escom, Leiden, 1990, pp 889-891.
83. FMF Chen, NL Benoiton. Identification of the side-reaction of Boc-decomposition during the coupling of Boc-amino acids with amino acid ester salts, in JA Smith, JR Rivier, eds. *Chemistry and Biology. Proceedings of the 12th American Peptide Symposium*, Escom, Leiden, 1992, pp. 542-543.
84. NL Benoiton, YC Lee, FMF Chen. Identification and suppression of decomposition during carbodiimide-mediated reactions of Boc-amino acids with phenols, hydroxylamines and amino acid esters. *Int J Pept Prot Res* 41, 587, 1993.
85. J Coste, E Frérot, P Jouin, B Castro. NCA: A troublesome by-product in difficult amino acid coupling reactions, in CH Schneider, AN Eberle, eds. *Peptides 1992. Proceedings of the 22nd European Peptide Symposium*, Escom, Leiden 1993, pp 245-246.

7.16. ACYL AZIDES AND THE USE OF PROTECTED HYDRAZIDES

The acyl-azide method (see Section 2.13), which dates back to the beginnings of peptide synthesis, is not employed routinely for chain assembly because its execution is not simple. The traditional method involves several steps. However, there are variants and refinements of the procedure that render the technique valuable for specific situations, and the unique characteristic that the coupling of segments by the acyl-azide method is just about certain not to be accompanied by epimerization (see Section 2.23) if performed properly has ensured the method's survival. Any loss in chirality is a result of the use of excess base and not formation of the 5(4*H*)-oxazolone (see Section 2.23). The precursor of the azide is the hydrazide, which traditionally has been obtained by hydrazinolysis of esters of protected amino acids and peptides (Figure 7.25, A). This has been extended to hydrazinolysis of peptide benzyl esters attached to resins, as well as to SASRIN-supported peptides (see Section 5.21). Various solvents and temperatures are employed, as the rates of hydrazinolysis vary considerably with the nature of the substrates. A *tert*-butyl ester is stable to hydrazinolysis, but trifluoroacetamido [–Lys(Tfa)–], *o*-nitrophenysulfanyl-N [–His(Nps)–], and nitroguanidino [–Arg(NO_2)–] are not. Hydrazine also promotes transpeptidation at –Gly-Asx– sequences (see Section 6.13). Hydrazinolysis occurs with preservation of chirality (see Section 7.26). There are several ways of circumventing the destructive effects of hydrazine. Hydrazides can be prepared by combining the acid and hydrazine using a carbodiimide assisted by 1-hydroxybenzotriazole (Figure 7.25, B) again with preservation of chirality. A completely different approach involves starting a synthesis with an amino acid derivative, the carboxyl group of which has been replaced by a carbazide ($O{=}CN_2H_3$) function that is protected (Figure 7.25, C); that is, an N^{α}-protected amino acid *N*-protected hydrazide. The starting material is obtained by routine methods (Figure 7.25, C), and the peptide chain is assembled by routine methods including solid-phase

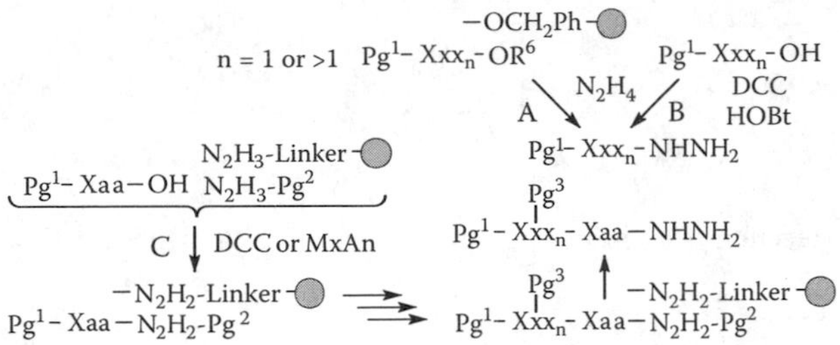

$$\text{n = 1 or >1}\quad Pg^1\text{-}Xxx_n\text{-}OR^6 \ (-OCH_2Ph\text{-resin}) \xrightarrow[A]{N_2H_4} Pg^1\text{-}Xxx_n\text{-}NHNH_2 \xleftarrow[B]{DCC,\ HOBt} Pg^1\text{-}Xxx_n\text{-}OH$$

$$Pg^1\text{-}Xaa\text{-}OH + N_2H_3\text{-}Pg^2 \ (N_2H_3\text{-Linker-resin}) \xrightarrow[C]{DCC\ or\ MxAn} Pg^1\text{-}Xaa\text{-}N_2H_2\text{-}Pg^2 \ (-N_2H_2\text{-Linker-resin}) \Longrightarrow Pg^1\text{-}Xxx_n(Pg^3)\text{-}Xaa\text{-}N_2H_2\text{-}Pg^2 \ (-N_2H_2\text{-Linker-resin}) \rightarrow Pg^1\text{-}Xxx_n(Pg^3)\text{-}Xaa\text{-}NHNH_2$$

FIGURE 7.25 Preparation of N^{α}-protected amino-acid and peptide hydrazides. Pg = protecting group. (A) Hydrazinolysis of esters. (B) Carbodiimide-mediated coupling of the acid with hydrazine assisted by 1-hydroxybenzotriazole.[92] (C) Chain assembly starting with an amino acid derivative combined with protected hydrazine followed by removal of the protector.[86] MxAn = mixed anhydride. Typical protecting combinations: Pg^1 = Z, Pg^2 = Boc; Pg^1 = Boc, Pg^2 = Z, Pg^3 = CF_3CO, tBu; Pg^1 = Bpoc, Pg^2 = Trt, Pg^3 = tBu. This approach avoids side reactions caused by hydrazinolysis.

synthesis. For the latter, the carbazide function can be attached via the terminal nitrogen atom to the tertiary carbon of a 2-chlorotrityl resin [–CONHNHC(Ph,C_6H_4Cl)-Ph-resin; see Section 5.23] or to a carbonyl linked to the oxymethyl of a Wang resin (–CONHNH-$CO_2CH_2C_6H_4OCH_2$Ph-resin; see Section 5.21) or other. Peptide hydrazides containing unprotected serine, threonine, histidine, asparagine, and nitroarginine can be assembled by this approach.

For coupling, the hydrazides of N^{α}-protected amino acids are converted to the azides by sodium nitrite in aqueous acetic or hydrochloric acids. The azide derivative is extracted into an organic solvent and subjected to aminolysis in dimethylformamide or dimethyl sulfoxide at temperatures of below 5°C. Low temperature is required to avoid the side reaction of isocyanate (–CHR-N=C=O; see Section 2.13) formation. An in-depth study of the side reactions led to a simplification of the procedure in which the acyl azide is not isolated and that is applicable to segments. The azide is generated from the hydrazide in a concentrated organic solution using *tert*-butyl nitrite and hydrogen chloride in dioxane (1 *M*) at 25°C for 15 minutes, the temperature is lowered to –65°C, the solution is brought to neutral pH with *N*-methylmorpholine, and the amino-containing component is added. Peptide is produced efficiently after 2–3 days. The side reaction of amide (–CHR-$CONH_2$) formation is suppressed because the system is homogenous. In another variant, the azide is transformed into an activated ester as quickly as it is produced. This is achieved by employing *tert*-butyl nitrite and 1-hydroxy-7-azabenzotriazole or 4-ethoxycarbonyl-1-hydroxy-1(*H*)-1,2,3-benzotriazole instead of hydrogen chloride. Activated ester is generated at room temperature within 30 minutes; peptide is produced by aminolysis within 1–4 hours. This approach eliminates the major unattractive features of the acyl-azide method — side reactions and slow couplings. Yet another variant exists where an acyl azide is prepared without the hydrazide being its precursor. Diphenyl phosphorazidate (Figure 7.26), a reagent that predates the onium salt-based reagents, reacts with a carboxylate anion in a similar manner to generate a penta-substituted phosphorus intermediate that rearranges with expulsion

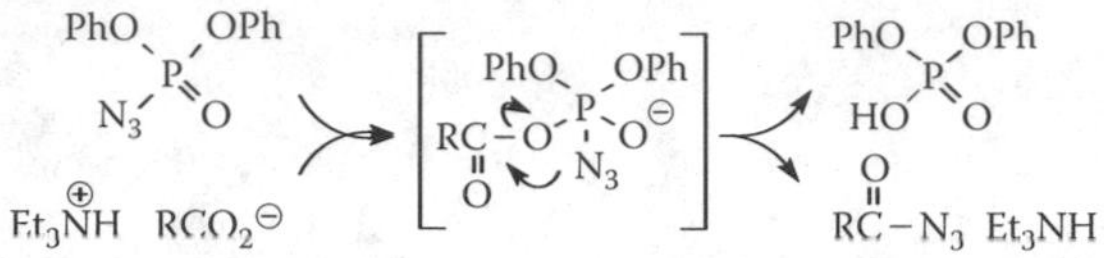

FIGURE 7.26 Reaction of diphenyl phosphorazidate (DPPA) with a carboxylate anion to give the acyl azide.[90]

of the phosphate moiety to produce the acyl azide. The activation is carried out in the presence of the amino group–containing moiety. This reagent has been used advantageously for the cylization of peptides, though the newer 7-azabenzotriazole-containing reagents (see Section 7.19) appear to be more efficient for the purpose.[86–93]

86. K Hofmann, A Lindenmann, MZ Magee, NH Khan. Studies on polypeptides III. Novel routes to α-amino acid and peptide hydrazides. (protected hydrazides). *J Am Chem Soc* 74, 470, 1952.
87. J Honzl, J Rudinger. Amino acids and peptides. XXXIII. Nitrosyl chloride and butyl nitrite as reagents in peptide synthesis by the azide method: suppression of amide formation. *Coll Czech Chem Comm* 26, 2333, 1961.
88. M Ohno, CB Anfinsen. Removal of protected peptides by hydrazinolysis after synthesis by solid-phase. *J Am Chem Soc* 89, 5994, 1967.
89. L Kisfaludy, O Nyeki. Racemization during peptide azide coupling. *Acta Chim Acad Sci Hung* 72, 75, 1972.
90. T Shiori, T Ninomia, S Yamada. Diphenylphosphoryl azide. A new convenient reagent for a modified Curtius reaction and for peptide synthesis. *J Am Chem Soc* 94, 6203, 1972.
91. JK Chang, M Shimuzu, S-S Wang. Fully automated synthesis of fully protected peptide hydrazides on recycling hydroxymethyl resin. *J Org Chem* 41, 3255, 1976.
92. S-S Wang, ID Kulesha, DP Winter, R Makofske, R Kutny, J Meienhofer. Preparation of protected peptide hydrazides from the acids and hydrazine by dicyclohexylcarbodiimide-hydroxybenzotriazole coupling. *Int J Pept Prot Res* 11, 297, 1978.
93. P Wang, R Layfield, RJ Mayer, R Ramage. Transfer active ester condensation: a novel technique for peptide segment coupling. *Tetrahedron Lett* 39, 8711, 1998.

7.17 *O*-ACYL AND *N*-ACYL *N′*-OXIDE FORMS OF 1-HYDROXYBENZOTRIAZOLE ADDUCTS AND THE URONIUM AND GUANIDINIUM FORMS OF COUPLING REAGENTS

1-Hydroxybenzotriazole (HOBt) reacts with activated acyl groups to form activated molecules that are referred to as activated esters (see Section 2.9). However, the original work on the subject established that solutions containing the products actually contained two compounds — one with the acyl group on the oxygen atom and the other with the acyl group on a nitrogen atom (Figure 7.27). The former is an ester, whereas the latter is an acyl amide with the oxygen atom transformed into the *N*-oxide. The two forms are readily distinguishable by their absorbances in the infrared spectrum (Figure 7.27). Preparation of derivatives such as acetyl,

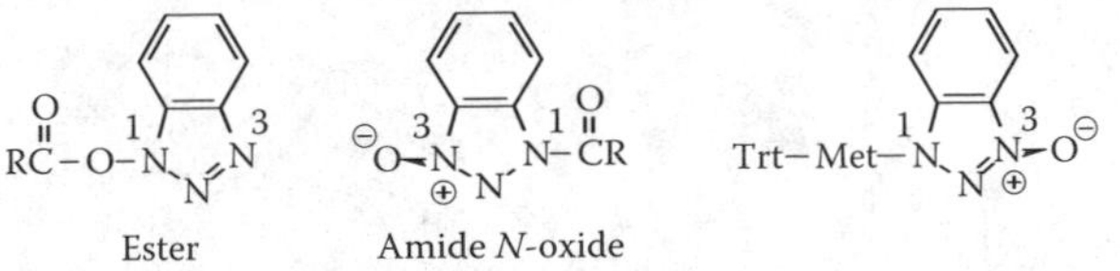

Ester

Amide *N*-oxide

FIGURE 7.27 Ester (IR: 1825-1815 cm^{-1}) and amide *N*-oxide (amide 3-oxide) (IR: 1750-1730 cm^{-1}) forms of benzotriazole adducts.[94] $R = R^1O$ and $R^1OC(=O)$-Xaa. Compounds with R = tBuO and $PhCH_2$ are amide forms.[97] The product from reaction of Trt-methionine and HOBt is the amide 3-oxide[96] (Trt = trityl = triphenylmethyl). Note that the atoms bearing the oxygen atoms are numbered differently in the two compounds.

phenacetyl, *N*-alkoxycarbonylaminoacyl, and *N*-tritylaminoacyl by the usual methods gives solutions containing the two forms. Either form produces an equilibrium mixture when dissolved in a solvent, with the ester converting to the amide more quickly than the reverse. Polar solvent and tertiary amine favor generation of the amide form; the rate of conversion also depends on the nature of the substituent on the amino group of the aminoacyl moiety. Crystalline compounds result from a shift in the equilibrium that occurs once a crystal appears. The first definitive proof of the structure of an acylamide *N*-oxide was obtained by x-ray analysis of the derivative of *N*-trityl-L-methonine, which showed the carbonyl of the methionine residue linked to N-1 of the ring (Figure 7.27). The Boc and Cbz derivatives are also *N*-acyl 3-oxides (Figure 7.27). In each case, two different methods of synthesis produced different forms of the products. Only the ester forms of Boc- and Cbz-valine are produced when 2-*tert*-butoxy- and 2-benzyloxy-4-isopropyl-5(4*H*)-oxazolone are added to HOBt in dichloromethane. The different forms can be monitored by thin-layer chromatography and high-performance liquid chromatography (see Section 7.26). The ester forms of *N*-tritylamino-acid derivatives undergo aminolysis more quickly than the acylamide 3-oxide forms. It is logical to infer that the same holds for the two forms of *N*-alkoxycarbonylamino-acid derivatives.

There are many coupling reagents such as BOP [$BtOP^+(NMe_2)_3 \cdot PF_6^-$] and HBTU [$BtOC^+(NMe_2)_2 \cdot PF_6^-$] that are composed of an epimerization-suppressing additive and a multialkylamino-substituted charged atom (see Section 2.16). Since their development, it has been assumed that the charged atoms of these reagents were linked to the oxygen atoms of the additives. The resulting molecules are substituted ureas that are charged, and hence are designated as uronium salts (Figure 7.28). Discussion of the mechanism of action of these reagents has been based on them having the uronium structure (see Sections 2.17–2.20). It was shown in 1994 by x-ray crystallographic analysis, however, that the reagent known as HBTU does not have the uronium structure but, instead, has the benzotriazole moiety linked directly with the carbon atom of tetramethylurea. The molecule is a substituted guanidine that is charged and hence has the guanidinium salt structure (Figure 7.28). The oxygen atom is in the form of the oxide. The chemical names of the two molecules are as dictated by *Chemical Abstracts*. Analogous reagents HATU [$aBtOC^+(NMe_2)_2 \cdot PF_6^-$], HBPyU [$BtOC^+(Pyr_2)_2 \cdot PF_6^-$], and HAPyU [$aBtOC^+(Pyr_2)_2 \cdot PF_6^-$]; aBt = 7-azabenzotriazole; Pyr = pyrollidino, see Section 2.27) have also been shown to possess the guanidinium structures. Just recently, a method has been found that allows preparation of the uronium forms of the reagents. The two

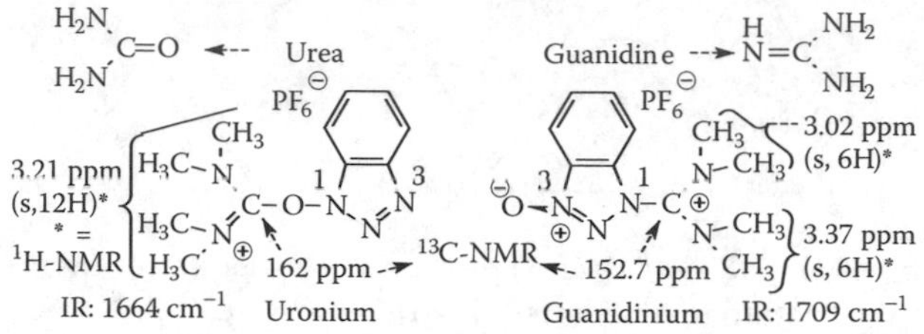

FIGURE 7.28 Structures of the two forms of reagent HBTU with spectroscopic data.[99] Uronium form: *N*-[1*H*-benzotriazol-l-yl-oxy)(dimethylamino)methylene]-*N*-methylmethanaminium hexafluorophosphate (the compound is named as if the charge is on the nitrogen atom). Guanidinium form: 1-[*bis*(dimethylamino)methylene]-1*H*-benzotriazolium hexafluorophosphate 3-oxide. Note that the atoms bearing the oxygen atoms are numbered differently in the two compounds.

forms of a reagent are readily distinguishable by the differences in their spectral data, as exemplified by the ^{13}C nuclear magnetic resonance data for HBTU (Figure 7.28), which are representative of the reagents. The most characteristic feature is a singlet for the 12 methyl protons in the nuclear magnetic resonance spectra of the uronium compounds. The same protons present as two singlets in the spectra of the guanidinium compounds, which is consistent with the effects of hindered rotation in related systems. The guanidinium form of HBTU is obtained by reacting a tetramethylchloroformamidinium salt [$(Me_2N)_2C^+Cl{\cdot}Cl^-$ or $\cdot PF_6^-$] with HOBt in the presence of tertiary amine. Reaction in the presence of potassium carbonate instead of tertiary amine (i.e., with KOBt) with a quick work-up gives the uronium form of HBTU. The uronium form is easily isomerized to the guanidinium form in the presence of triethylamine. In model experiments, the uronium forms of HBTU and HATU reacted more quickly with carboxylic acids in the presence of tertiary amine than the guanidinium forms and led to less epimerization in the preparation of a peptide. It would thus seem that the uronium forms of the reagents will give the best results in couplings.[36,94–99]

36. K Barlos, D Papaioannou, D Theodoropoulus. Preparation and properties of N^{α}-trityl amino acid 1-hydroxybenzotriazole esters. *Int J Pept Prot Res* 23, 300, 1984.
94. W König, R Geiger. A new method for the synthesis of peptides: activation of the carboxyl group with dicyclohexylcarbodiimide and 1-hydroxybenzotriazole. *Chem Ber* 103, 788, 1970.
95. K Horiki. Behavior of acylated 1-hydroxybenzotriazole. *Tetrahedron Lett* 1897, 1977.
96. K Barlos, D Papaioannou, S Voliotis. Crystal structure of 3-(N^{α}-tritylmethionyl)benzotriazole 1-oxide, a synthon for peptide synthesis. *J Org Chem* 50, 696, 1985.
97. J Singh, R Fox, M Wong, TP Kissick, JL Moniot. The structure of alkoxycarbonyl, acyl, and sulfonate derivatives of 1-hydroxybenzotriazole: N- vs O-substitution. *J Org Chem* 53, 205, 1988.
98. I Abdelmoty, F Albericio, LA Carpino, BM Foxman, SA Kates. Structural studies of reagents for peptide bond formation: crystal and molecular structures of HBTU and HATU. *Lett Pept Sci* 1, 57, 1994.

99. LA Carpino, H Imazumi, A El-Faham, FJ Ferrer, C Zhang, Y Lee, BM Foxman, P Henklein, C Hanay, C Mügge, H Wenschuh, J Klose, M Beyermann, M Bienert. The uronium/guanidinium peptide coupling reagents: finally the true uronium salts. *Angew Chem Intl Edn* 41, 442, 2002.

7.18 PHOSPHONIUM AND URONIUM/AMINIUM/GUANIDINIUM SALT-BASED REAGENTS: PROPERTIES AND THEIR USE

Dozens of phosphonium and carbenium salt-based reagents have been investigated for their peptide-bond forming capabilities. More commonly known are the original oxybenzotriazole (BtO)-containing reagents and their 7-aza equivalents (aBt = A): BOP = $BtOP^+(NMe_2)_3 \cdot PF_6^-$ (see Section 2.17), AOP = $aBtOP^+(NMe_2)_3 \cdot PF_6^-$, PyBOP = $BtOP^+(Pyr)_3 \cdot PF_6^-$ (see Section 2.19), PyAOP = $aBtOP^+(Pyr)_3 \cdot PF_6^-$, HBTU = $BtOC^+(NMe_2)_2 \cdot PF_6^-$ (see Section 2.18), HATU = $aBtOC^+(NMe_2)_2 \cdot PF_6^-$, HBPyU = $BtOC^+(Pyr)_2 \cdot PF_6^-$, and HAPyU = $aBtOC^+(Pyr)_2 \cdot PF_6^-$ (see Figure 7.29) (Pyr = pyrollidino, see Section 2.27). Physical contact with these reagents, HBTU, TBTU, and HATU in particular, can cause severe reactions (rhinitis, swelling, breathing difficulties), so they should be handled with extreme care. The phosphonium salts are more reactive than the uronium salts, and, hence, the latter are more suitable for solid-phase synthesis, where the storage of solutions of reagents is pertinent. The 7-aza analogues are less stable than the parent compounds, especially in the presence of the tertiary amine that is required for initiating reactions, but they are generally more efficient than the parent compound. HBTU is the most stable of the above. TBTU = $BtOC^+(NMe_2)_2 \cdot BF_4^-$ (see Section 2.18) is cheaper than HBTU; HBPyU is more expensive than HBTU. Couplings with these reagents can be achieved in most solvents, but there is one side reaction associated with the highly activated PyBroP = $BrP^+(Pyr)_3 \cdot PF_6^-$. In dimethylformamide, this reagent engenders reaction between the solvent and amino groups to give Me_2N^+=CH-substituted nitrogen. The efficacy of reagents is not affected by the presence of water of hydration of derivatives. Uronium salts containing 4-oxo-3-oxy-3,4-dihydrobenzotriazine (see Section 7.29) are not reliable because the side reaction of aminolysis at the carbonyl of the ring (see Section 7.7) is significant. Phosphonium salts do not react with amino groups or with trifluoroacetic acid. Uronium salts react with this acid as well as with amino groups, producing the Schiff's base $(Me_2N)_2C$=N–. For this reason, phosphonium salts such as PyBOP and PyAOP are preferable to uronium salts for cyclization of a peptide chain (see Section 5.24). Phosphonium salts are also ideally suited for solid-phase synthesis, using Boc/Bzl chemistry, because final traces of trifluoroacetate do not have to be removed before proceeding to the next coupling, and the alkaline milieu of the reaction prevents decomposition of the activated derivative (see Section 7.15).

The phosphonium and carbenium salts are efficient reagents for activating and coupling *N*-alkoxycarbonylamino acids as well as peptide acids. However, the requirement for tertiary amine to effect the reaction has several implications. The base renders hydroxyl groups subject to acylation. Hence, the side chains of serine and threonine and any hydroxymethyl groups of a resin that have not been derivatized should be blocked before acylation of a moiety containing these is effected. However, derivatives of serine and threonine with unprotected side chains can be coupled successfully. The basic milieu also promotes imide formation at asparagine and glutamine residues (see Section 6.13), so their side chains must be protected when these reagents are employed. The reagents also lead to dehydration of the carboxamide groups when employed to activate derivatives of asparagine and glutamine with unprotected side chains (see Section 6.15). The greatest implication, however, is on the chiral integrity of the activated residue. Tertiary amine is required to effect activation, but tertiary amine promotes enantiomerization of the residue of a peptide that is activated. Hence, a delicate balance must be found between the basicity of the amine required to achieve efficient coupling and the nature of the amine required to minimize epimerization. In this regard, the more basic (p*K* 10.1) and more hindered diisopropylethylamine is superior to the less basic (p*K* 7.38) and less hindered *N*-methylmorpholine (see Section 2.22). 2,4,6-Trimethylpyridine (p*K* 7.43), with the shielded nitrogen, is effective particularly with 7-azabenzotriazole-derived reagents. Excess of any base promotes isomerization. Inclusion of an additive in a coupling mixture may be beneficial, but it has increased epimerization in some cases. The tendency for isomerization is particularly pertinent, as always, to the coupling of peptides (see Sections 1.9 and 2.23). Activation and coupling of *N*-alkoxycarbonylamino acids mediated by onium salt-based reagents proceeds as expected without isomerization (see Section 1.10), provided the normal protocol (see Section 7.20) is executed. However, excess tertiary amine that is unnecessary can result in the production of epimeric products (see Sections 4.17 and 8.1) in some syntheses.[100–105]

100. R Steinauer, FMF Chen, NL Benoiton. Studies on racemization associated with the use of benzotriazol-1-yl-*tris*(dimethylamino)phosphonium (BOP). *Int J Pept Prot Res* 34, 295, 1989.
101. H Gausepohl, U Pieles, RW Frank. Schiff base analog formation during *in situ* activation by HBTU and TBTU, in JA Smith, JR Rivier, eds. *Peptides Chemistry and Biology. Proceedings of the 12th American Peptide Symposium*, Escom, Leiden, 1992, pp 523-524.
102. LA Carpino, A El-Faham, F Albericio. Racemization studies during solid-phase peptide synthesis using azabenzotriazole-based coupling reagents. *Tetrahedron Lett* 35, 2279, 1994.
103. LA Carpino, A El-Faham, F Albericio. Efficiency in peptide coupling: 1-hydroxy-7-azabenzotriazole *vs* 3,4-dihydro-3-hydroxy-4-oxo-1,2,3-benzotriazine. *J Org Chem* 60, 3561, 1995.
104. A Stierandová, R Safár. Unexpected reactivity of PyBrop towards *N,N*-disubstituted formamides and its application, in HLS Maia, ed. *Peptides 1994. Proceedings of the 23rd Peptide Symposium*, Escom, Leiden, 1995, pp 183-184.
105. LA Carpino, D Ionescu, A El-Faham. Peptide segment coupling in the presence of highly hindered tertiary amines. *J Org Chem* 61, 2460, 1996.

7.19 NEWER COUPLING REAGENTS

There are dozens of new reagents that have been developed and examined for their coupling efficiency. Most are composed of a good leaving group derived from a

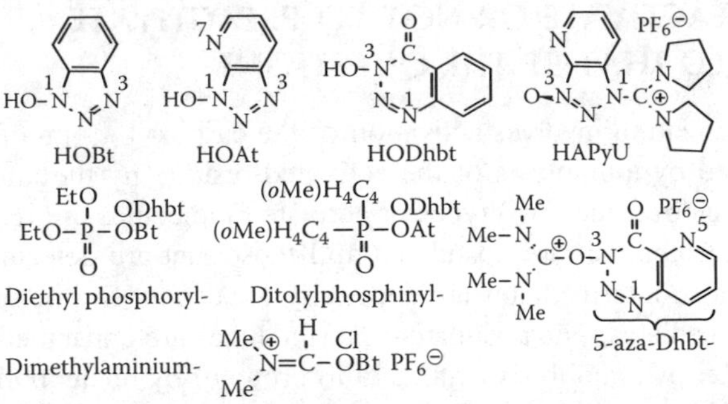

FIGURE 7.29 Abbreviated structures of some new coupling reagents with complete names omitted. Diethylphosphoryl-,[106,108] ODhbt, HAPyU,[107] dimethylaminium-,[109] ditolylphosphinyl-, and 5-aza-Dhbt-.[110]

halide, pentafluorophenol, or one of the common epimerization-suppressing additives *N*-hydroxysuccinimide, 1-hydroxybenzotriazole and its aza-equivalent, and 3-hydroxy-4-oxo-3,4-dihydrobenzotriazine (Figure 7.29). In many reagents, these are linked to a carbon, nitrogen, or phosphorus atom that is substituted by dialkylamino or pyrollidino groups that are insufficient in number, such that the atom is positively charged. Other reagents are activated esters of disubstituted phosphoric and phosphinic acids or monosubstituted sulfonic acids. Reagents that will survive the test of time are those that are crystalline and stable and that effect coupling efficiently with minimum or no enantiomerization at the activated residue. Comparison of the merits of different reagents is a difficult task that is beyond the scope of this work. 7-Azabenzotriazole-derived reagents have been shown to be superior to benzotriazole-derived reagents in many cases. Fewer data are available on the relative performance of the former and 4-oxo-3,4-dihydrobenzotriazine-derived reagents. More attention is now being devoted to the latter.[106–111]

106. S Kim, A Chang, YK Ko. Benzotriazol-1-yl diethyl phosphate. A new convenient coupling reagent for the synthesis of amides and peptides. *Tetrahedron Lett* 26, 1341, 1985.
107. A Ehrlich, M Brudel, M Beyermann, R Winter, LA Carpino, M Bienert. Cyclization of all L pentapeptides by means of HAPyU. *Peptides 1994. Proceedings of the 23rd Peptide Symposium*, Escom, Leiden, 1995, pp 167-168
108. Y-H Ye, C-X Fan, D-Y Zhang, H-B Xie, X-L Hao, G-L Tian. Application of novel organophosphorus compounds as coupling reagents for the synthesis of bioactive peptides, in JP Tam, PTP Kaumaya, eds. *Peptides Frontiers of Peptide Science. Proceedings of the 15th American Peptide Symposium*, Klewer, Dordrecht, 1999, pp 337-338.
109. P Li, J-C Xu. New and highly efficient immonium type peptide coupling reagents: synthesis, mechanism and application. *Tetrahedron* 56, 4437, 2000.
110. LA Carpino, J Xia, A El-Faham. 3-Hydroxy-4-oxo-3,4-dihydro-5-aza-benzo-1,2,3-triazine. *J Org Chem* 69, 54, 2004.
111. LA Carpino, J Xia, C Zhang, A El-Faham. Organophosphorus and nitro-substituted sulfonate esters of 1-hydroxy-7-azabenzotriazole as highly efficient fast-acting peptide coupling reagents. *J Org Chem* 69, 62, 2004.

7.20 TO PREACTIVATE OR NOT TO PREACTIVATE: SHOULD THAT BE THE QUESTION?

Peptide bond formation involves activation of the carboxyl group of an amino acid residue, followed by aminolysis of the activated residue by the amino group of a second amino acid residue. Two types of activated molecules are recognized: those that are not detectable but are postulated and those that are detectable and can be isolated. Postulated intermediates are necessary to account for the formation of the detectable intermediates. The postulated intermediates are consumed as fast as they are formed, either by aminolysis by an amino group or by nucleophilic attack by an oxygen nucleophile, which produces activated molecules that are also immediate precursors of the peptide. More than one activated compound may be generated by a postulated intermediate. Activated esters, acyl halides and azides, and mixed and symmetrical anhydrides are isolatable activated compounds that are generated from postulated intermediates. Peptides are produced by one of three ways:

Method 1: addition of a coupling reagent (carbodiimide, EEDQ, phosphonium and carbenium salts, trisubstituted phosphates, etc.) and tertiary amine, if necessary, to a mixture of the acid and the amine nucleophile that are to be combined

Method 2: addition of the amine nucleophile to one of the activated forms of the acid (activated ester, acyl azide, anhydrides, etc.) to which it is to be combined.

Method 3: addition of the amine nucleophile to a solution of a coupling or other reagent and the acid after having allowed the two to react to generate an activated compound.

The option of adding the acid to a mixture of the reagent and the amine nucleophile is inapplicable because the reagent usually reacts with the nucleophile, giving an adduct that does not react with the acid. The typical example of method 3 is the mixed-anhydride procedure, in which the chloroformate and acid are allowed to react together before the amine nucleophile is added. The term "preactivation," which originates from descriptions of this procedure, is employed to express this generation of activated intermediates before addition of the nucleophile. In this case, preactivation is necessary — the reaction cannot be carried out otherwise. Carbodiimides, EEDQ, and the initial phosphonium and carbenium salt-based reagents (BOP, HBTU, PyBOP, etc.; see Section 7.18) were developed for use according to method 1. If a carbodiimide is employed with preactivation (method 3), the immediate precursor of the peptide is the symmetrical anhydride (Figure 7.30, path C; see Section 2.5). There are good reasons why the symmetrical anhydride may be allowed to form before the primary amine is introduced when employing carbodiimides. The anhydride is less reactive, and hence more selective, than the *O*-acylisourea, and its generation removes acid (R^1OCO-Xaa-OH) that causes terminal glutaminyl to cyclize to pyroglumate (see Section 6.16). Phosphonium and carbenium salts mediate reactions (method 1) that produce peptide bonds efficiently and very quickly. If used with preactivation (method 3), the precursor of the peptide (path C) is the

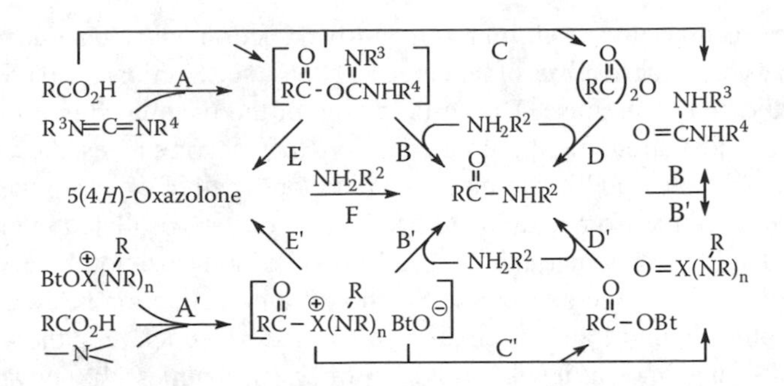

FIGURE 7.30 Reactions of a carbodiimide (A) and a phosphonium or carbenium salt (A′) with an acid (R = R^1OCO-Xaa-OH or R^1OCO-Xaa-Xxx_n-OH) to give postulated intermediates that are aminolyzed (B,B′) to give the peptide. In the absence of amine NH_2R^2, the postulated intermediates generate the symmetrical anhydride (C) and benzotriazolyl ester (C′), which undergo aminolysis (D, D′) to give the peptide. 5(4*H*)-Oxazolone may be formed (E,E') from the postulated intermediates; it would also generate peptide (F).

benzotriazolyl ester that is formed by reaction of the postulated intermediate (path B) with the benzotriazolyloxy anion that has been ejected during the formation of the intermediate. The ester, and not the anhydride, is generated because the benzotriazolyloxy anion is a better nucleophile than the carboxylate anion. Preactivation is optional.

Preactivation may be deemed preferable, or it may be imposed by the type of technology employed, such as a continuous-flow system. For carbodiimide-mediated reactions, the *O*-acylisourea that is postulated as the first intermediate is recognized as the activated form that undergoes aminolysis to give peptide. Symmetrical anhydride may be generated and aminolyzed if the *O*-acylisourea is not consumed quickly (see Section 2.2). However, for phosphonium and carbenium salt-mediated reactions, for reasons that are difficult to understand, production of peptide is attributed to aminolysis of the benzotriazolyl ester and not the acyl-phosphonium/carbenium adduct that is postulated as the first intermediate. It must be inferred from this that the benzotriazolyoxy anion is such a good nucleophile, compared to the amino group, that it attacks the postulated intermediate before it can be aminolyzed. This writer is not aware that this has been established.

A fundamental question emerges: Why do several reagents derived from the same additive, for example BOP, PyBOP and HBTU, perform differently if the peptide originates in each case by aminolysis of the benzotriazolyl ester? Why is HAPyU a more efficient coupling reagent than HATU if the 7-azabenzotriazolyl ester is the molecule that is aminolyzed in both cases? The different performances cannot be attributed to the presence of different amides, hexamethylphosphoric triamide, tripyrrolidino phosphate, and tetramethylurea in the coupling mixtures. The different performances must be attributed to the different natures of the activated forms of the acyl moieties that are generated (Figure 7.30, path C). Failed attempts to detect the intermediate are not proof that it is not a precursor of the peptide. At least some, if not much or all, of the peptide produced in these salt-mediated reactions must originate by aminolysis (path B) of the postulated intermediates. The high speed of the reactions, the results of a competitive reaction, and the reduced reactivity of activated esters of derivatives of hindered residues (see Section 2.20) are consistent with and validate this premise. The participation of the postulated intermediate also provides an explanation for the epimerization that accompanies the coupling of peptides. Just as for carbodiimide-mediated reactions (path E), it is simpler and more reasonable to attribute isomerization to cyclization of the postulated intermediate to the chirally labile oxazolone (path E) than to invoke isomerization by conversion of the activated ester into the oxazolone. So should a preactivation be executed when employing phosphonium and carbenium salt-based reagents? For the coupling of segments, absolutely not, as it will promote epimerization unless the activated residue is proline or glycine. It can also cause enantiomerization of amino acid derivatives (see Section 8.1). If it is unavoidable, yes. If it has been demonstrated to produce better results, yes. But take note that preactivation dispossess the reagent of the unique and favorable properties that reside in the intermediate that is formed when it combines with the acid. The activated ester that is generated is less reactive than the postulated intermediate. Take advantage of the reagent's unique properties and do not preactivate unless there is a good reason to do so.[103,112]

103. LA Carpino, A El-Faham, F Albericio. Efficiency in peptide coupling: 1-hydroxy-7-azabenzotriazole *vs* 3,4-dihydro-3-hydroxy-4-oxo-1,2,3-benzotriazine. *J Org Chem* 60, 3561, 1995.
112. 1-53. GE Reid, RJ Simpson. Automated solid-phase peptide synthesis: use of 2-(1*H*-benzotriazol-1-yl)-1,1,3,3-tetramethyluronium tetrafluoroborate for coupling of *tert*-butyloxycarbonyl amino acids. *Anal Biochem* 200, 301, 1992.

7.21 AMINOLYSIS OF SUCCINIMIDO ESTERS BY UNPROTECTED AMINO ACIDS OR PEPTIDES

The objective of peptide synthesis is to produce pure peptides in the simplest manner possible. This is often achieved by reaction of an amino acid residue that is activated with the amino group of an amino acid whose carboxyl group is not protected. The amino group of the zwitter-ion is deprotonated by the addition of a base. Reaction of the *N*-carboxyanhydride of an amino acid with a second amino acid (see Section 7.13) is the prime example. The more common instance, however, of this strategy of minimum protection (see Section 6.1) involves reaction of the deprotonated amino acid with the activated ester of an amino acid derivative. For practical reasons, succinimido esters are the derivatives of choice (Figure 7.31). They are relatively resistant to hydrolysis by the partially aqueous solvent that is required to solubilize the amino acid, and the *N*-hydroxysuccinimide that is liberated is easy to dispose of because it is soluble in water. Maximum consumption of the activated derivative by aminolysis is desirable because any that is hydrolyzed gives the parent acid whose physical properties resemble those of the N^{α}-protected dipeptide. Ideal consumption is achieved by employing an excess of the aminolyzing component. The excess also serves to prevent enantiomerization of the activated residue by ensuring its quick consumption (see Section 4.20). Sodium carbonate is preferable to sodium bicarbonate or tertiary amine as the base. Optimum conditions effectively suppress

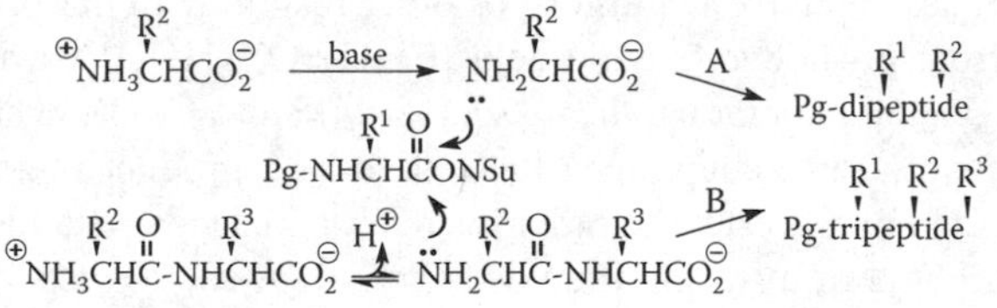

$$\overset{\oplus}{N}H_3\overset{R^2}{C}HCO_2^{\ominus} \xrightarrow{base} NH_2\overset{R^2}{C}HCO_2^{\ominus} \xrightarrow{A} \text{Pg-dipeptide } (R^1\ R^2)$$

$$\text{Pg-NH}\overset{R^1}{C}\text{H}\overset{O}{C}\text{ONSu}$$

$$\overset{\oplus}{N}H_3\overset{R^2}{C}H\overset{O}{C}\text{-NH}\overset{R^3}{C}HCO_2^{\ominus} \rightleftharpoons (H^{\oplus})\ NH_2\overset{R^2}{C}H\overset{O}{C}\text{-NH}\overset{R^3}{C}HCO_2^{\ominus} \xrightarrow{B} \text{Pg-tripeptide } (R^1\ R^2\ R^3)$$

FIGURE 7.31 Aminolysis of a succinimido ester by (A) an amino-acid anion generated by base [Anderson et al., 1974] and (B) a peptide anion that is in equilibrium with the peptide.[114] Pg = protecting group. Appropriate solvents are tetrahydrofuran, acetone, or dimethylformamide with water.

isomerization that might be caused by the base (see Section 4.20). When the aminolyzing moiety is a peptide, however, no base is required to induce aminolysis. Unprotonated amino groups are available without the addition of base because the p*K* of the amino group of a peptide is significantly lower than that of an amino acid (see Section 1.3). The p*K* of the amino group of a peptide is close to 7, with the result that an equilibrium exists between the zwitter-ionic form of the peptide and the anionic form of the peptide (Figure 7.31). See Section 7.22 for a side reaction that might occur during the aminolysis of activated proline.[113–115]

113. JW Anderson, JE Zimmerman, FM Callahan. The use of esters of *N*-hydroxy succinimide in peptide synthesis. *J Am Chem Soc* 86, 1839, 1964.
114. L Moroder, W Göhring, P Lucietto, J Musiol, R Schaarf, P Thamm, G Bovermann, G Wünsch, J Lundberg, G Tatemoto. Synthesis of porcine intestinal peptide PHI and its 24-glutamine analog. (aminolysis without base). *Hoppe Seyler's Z Physiol Chem* 364, 1563, 1983.
115. NL Benoiton, YC Lee, FMF Chen. Racemization during aminolysis of activated esters of *N*-alkoxycarbonylamino acids by amino acid anions in partially aqueous solvents and a tactic to minimize it. *Int J Pept Prot Res* 41, 512, 1993.

7.22 UNUSUAL PHENOMENA RELATING TO COUPLINGS OF PROLINE

Proline is unique in structure, in that it contains a secondary instead of a primary amino group, and the second substituent on the nitrogen atom is the carbon atom at the end of the side chain of the amino acid. The result is an imino acid with a five-membered ring (see Section 1.4), endowed with unique chemical properties. The ring prevents the residue that is activated from undergoing the side reaction of cyclization to the 5(4*H*)-oxazolone at a significant rate. The result is that the coupling of a segment with an activated proline residue is not accompanied by epimerization, as is the coupling of other segments (see Section 2.23), and a derivative such as Fmoc-Pro-Cl is stable to aqueous washes when other Fmoc-amino-acid chlorides are not (see Section 6.16). For a reason that is not obvious, aminolysis of a mixed anhydride of proline (–Pro-O-CO_2R) does not lead to the side reaction of aminolysis at the carbonyl of the carbonate moiety, as it does for aminolysis of anhydrides of other residues (see Section 7.4). Activated esters of proline formed from hydroxamic acids seem to be less reactive than those of other residues. In the presence of base, esters formed from pivalohydroxamic acid [tBuC(=O)NHOH] partially undergo a Lossen rearrangement, liberating the isocyanate (tBuN=C=O), which attacks the incoming nucleophile of a coupling. Much more side product is generated by the ester of proline than by the esters of other residues, and a side reaction of aminolysis at the wrong carbonyl of a succinimido ester occurs under conditions when it does not for esters of other residues (see below). Analogous unusual behavior is observed when proline is the aminolyzing residue in a coupling.

A terminal proline residue does not distinguish between the two carbonyls of a mixed anhydride (see Section 7.4); thus, mixed anhydrides cannot be employed for coupling to the nitrogen atom of proline. Aminolysis of an activated peptide by a proline ester or a peptide-containing proline at the amino terminus leads to much more epimerization than does aminolysis by other amino acid esters or peptides. Enantiomerization of the aminolyzing residue occurs if 1-hydroxybenzotriazole is added to the EDC (soluble carbodiimide)-mediated coupling of a Boc-amino acid with proline phenacyl ester [H-Pro-OCH$_2$C(=O)Ph]. Isomerization is explained on the basis of the acid-catalyzed intramolecular formation of a Schiff's base. A proline residue in a piperazine-2,5-dione is chirally sensitive to alkali (see Section 8.14). Finally, the side reaction of aminolysis at the carbonyl of the succinimido function of Boc-proline succinimido ester can occur during its reaction with proline (Figure 7.32) or *N*-methylglycine in dimethylformamide in the presence of triethylamine. The side reaction is avoided by use of benzyltrimethylammonium hydroxide as the base.[7,116–119]

7. JC Califano, C Devin, J Shao, JK Blodgett, RA Maki, KW Funk, JC Tolle. Copper(II)-containing racemization suppressants and their use in segment coupling reactions, in J Martinez, J-A Fehrentz, eds. *Peptides 2000, Proceedings of the 26th European Peptide Symposium,* Editions EDK, Paris, 2001, pp. 99-100.
116. TR Govindachari, S Rajappa, AS Akerkar, VS Iyer. Hydroxamic acids and their derivatives: IV. Further studies on the use of esters of pivalohydroxamic acid for peptide synthesis. *Tetrahedron* 23, 4811, 1967.
117. J Savrda. An unusual side reaction of 1-succinimidyl esters during peptide synthesis. *J Org Chem* 42, 3199, 1977.
118. H Kuroda, S Kubo, N Chino, T Kimura, S Sakakibara. Unexpected racemization of proline and hydroxy-proline phenacyl ester during coupling reactions of Boc-amino acids. *Int J Pept Prot Res* 40, 114, 1992.
119. J Ottl, HJ Musiol, L Moroder. Heterotrimeric collagen peptides containing functional epitopes. Synthesis of single-stranded type 1 peptides related to the collagenase cleavage site. (elimination of side reaction) *J Pept Sci* 5, 103, 1999.

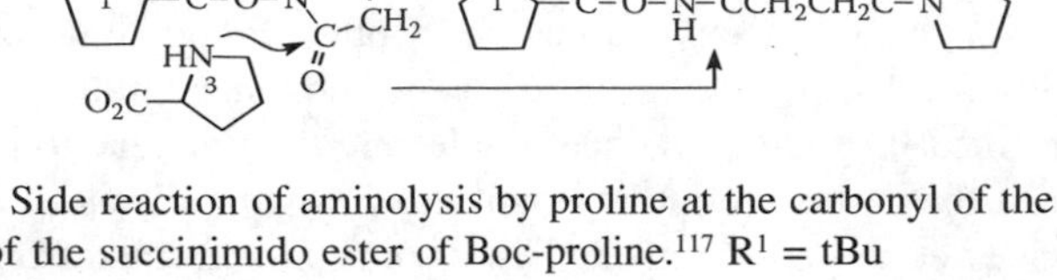

FIGURE 7.32 Side reaction of aminolysis by proline at the carbonyl of the pyrrolidine-2,5-dione moiety of the succinimido ester of Boc-proline.[117] R^1 = tBu

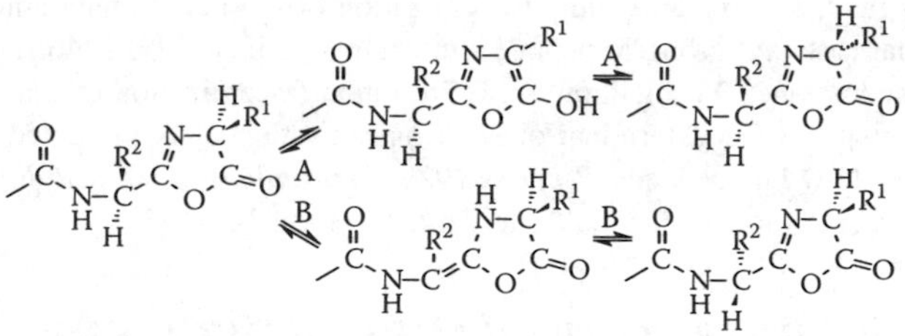

FIGURE 7.33 Enantiomerization (A) of the terminal residue and (B) of the penultimate residue of an activated peptide by tautomerization of the 5(4*H*)-oxazolone of the terminal residue. [Bergmann and Zervas, 1928].

7.23 ENANTIOMERIZATION OF THE PENULTIMATE RESIDUE DURING COUPLING OF AN N^{α}-PROTECTED PEPTIDE

2-Alkyl-5(4*H*)-oxazolones were first recognized by the developers of the benzyloxycarbonyl group (see Section 3.3), who found that heating an amino acid in acetic acid in the presence of acetic anhydride produced the racemic *N*-acetylamino acid. Isomerization was attributed to formation of the oxazolone. Treatment of a *Z*-dipeptide under the same conditions produced the protected dipeptide that had undergone isomerization at both residues. It transpires that the oxazolone of a peptide can tautomerize in two ways — by transfer of the α-proton of the terminal residue (Figure 7.33, path A; see Section 4.4), and by transfer of the α-proton of the adjacent residue (path B). Each event results in enantiomerization of the residue. The residue adjacent to the activated residue of a peptide is referred to as the penultimate residue. The phenomenon of path B was investigated four decades later, using dipeptides with L-isoleucine as the penultimate residue. This allowed determination of isomerization by analysis for D-alloisoleucine, the diastereoisomer of isoleucine (see Section 1.4), after hydrolysis. A carbodiimide-mediated coupling of *Z*-Ile-Phe-OH with H-Val-OtBu gave a peptide containing 3% alloisoleucine. Others found amounts of 1–2% of D-enantiomer at the penultimate residue after activation of tripeptides by the methods available in the 1970s. Noteworthy was the demonstration that 8–25% of D-enantiomer was produced from mixed anhydrides kept for 2 hours in dimethylformamide at 15°C in the presence of triethylamine. Much less enantiomerization occurs at the penultimate residue when the activated residue is glycine. Activated glycine has a lesser tendency to cyclize to the oxazolone because of the absence of a side chain. However, when the activated residue is aminoisobutyroyl, there is an unusually high tendency for cyclization to occur, with a possible deleterious effect on the stereochemistry of the adjacent residue. Thus, the penultimate residue of a peptide that is activated can also undergo enantiomerization during activation under conditions that are not controlled. The phenomenon is not a major problem, but the possibility that it might occur should not be disregarded.[120,121]

120. F Weygand, A Prox, W König. Racemization of the penultimate amino acid with a terminal carboxyl group in peptide synthesis. *Chem Ber* 99, 1446, 1966.
121. M Dzieduszycka, M Smulkowski, E Taschner. Racemization of amino acid residue penultimate to the C-terminal one during activation of N-protected peptides, in H Hanson, H-D Jakubke, eds. *Peptides 1972. Proceedings of the 12th European Peptide Symposium*, North-Holland, Amsterdam, 1973, pp 103-107.

7.24 DOUBLE INSERTION IN REACTIONS OF GLYCINE DERIVATIVES: REARRANGEMENT OF SYMMETRICAL ANHYDRIDES TO PEPTIDE-BOND-SUBSTITUTED DIPEPTIDES

There are two possible side reactions associated with coupling at activated glycine. There is the realistic but slight chance that the penultimate residue of the segment might undergo enantiomerization (see Section 7.23), and there is the potential that two residues instead of one might be inserted into a chain when an activated glycine derivative is aminolyzed. It was observed five decades ago that a mixed-anhydride coupling of *Z*-glycine with glycine ethyl ester produced as contaminant a product containing two *Z*-glycine moieties for each glycine ester moiety. The reaction was studied in detail later, when it was found that the synthesis of Leu-Ala-Gly-Val on a solid support using *N*-biphenylisopropoxycarbonyl(Bpoc)-amino acids and the mixed-anhydride procedure produced in addition to the target peptide a product, in about 4% yield, containing two residues of glycine instead of one. The authors eliminated the possibility that the product resulted from acylation of the terminal urethane moiety or the amide bond of the growing chain. Similar results were obtained using the mixed anhydride of *Z*-glycine, which was allowed to stand before use. The conclusion was made that two adjacent residues of glycine had been inserted into the peptide because the mixed anhydrides of the glycine derivatives had disproportionated (see Section 7.5) to the symmetrical anhydrides during the unnecessary preactivation periods (see Section 7.20), and the latter had rearranged to the *N,N'*-disubstituted dipeptides (Figure 7.34, path B) R^2 = H, which were activated by reaction with the mixed or symmetrical anhydrides. Only one residue of glycine was incorporated when the coupling was effected under the recommended conditions of operation. The reaction is general, as some insertion of dimer could be observed in reactions of Boc-alanine and after a carbodiimide-mediated reaction, and recently in the base-catalyzed esterification reactions of Fmoc-amino acids (see Sections 4.19 and 5.22). The rearrangement of a symmetrical anhydride had been demonstrated decades ago, when *Z*-glycylglycine was prepared by the action of tertiary amine on the anhydride of *Z*-glycine. The mechanism of dimerization was confirmed by a study of the comportment of the symmetrical anhydride of *N*-methoxycarbonyl-L-valine in the presence of a base. The anhydride underwent an acyl transfer from the oxygen atom to the urethane, producing a compound that was shown by x-ray crystal analysis to be the *N,N'*-disubstituted dipeptide (path B). The event is sometimes referred to as urethane acylation.

Rearrangement of symmetrical anhydrides occurs even in the absence of base, but at a much slower rate. Insertion of two residues during synthesis occurs by

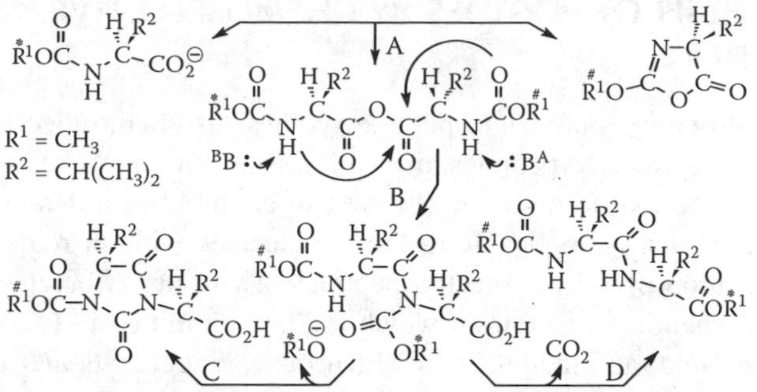

FIGURE 7.34 Decomposition of the symmetrical anhydride of *N*-methoxycarbonyl-valine ($R^1 = CH_3$) in basic media.[2] (A) The anhydride is in equilibrium with the acid anion and the 2-alkoxy-5(4*H*)-oxazolone. (B) The anhydride undergoes intramolecular acyl transfer to the urethane nitrogen, producing the *N,N′-bis*methoxycarbonyldipeptide. (A) and (B) are initiated by proton abstraction. Double insertion of glycine can be explained by aminolysis of the *N,N′*-diprotected peptide that is activated by conversion to anhydride Moc-Gly-(Moc)Gly-O-Gly-Moc by reaction with the oxazolone. (C) The *N,N′*-diacylated peptide eventually cyclizes to the *N,N′*-disubstituted hydantoin as it ejects methoxy anion or (D) releases methoxycarbonyl from the peptide bond leading to formation of the N^{α}-substituted dipeptide ester.

aminolysis of the substituted dipeptide, which is activated (Figure 7.34) by reaction most likely with the 5(4*H*)-oxazolone, which is in equilibrium with the anhydride (path A; see Section 4.17). The substituent on the amide nitrogen of the intermediate is labile in basic solution, decomposing in two ways depending on the solvent, expelling methoxy anion in aqueous dimethylformamide as it cyclizes to the hydantoin (path C), and losing carbon dioxide in anhydrous solvent, producing the corresponding ester, path D. Hydantoin formation is a known side reaction that occurs during the saponification (see Section 3.9) of N^{α}-protected dipeptide esters if the solution is not cold or there is an excess of base.[122–126]

122. K Kopple, RJ Renick. Formation of an *N*-acylamide in peptide synthesis. *J Org Chem* 23, 1565, 1958.
123. H Kotake, T Saito. The rearrangement of acid anhydrides by a *t*-amine. The preparation of glycylglycine from *N*-benzyloxycarbonylglycine anhydride. *Bull Chem Soc Jpn* 39, 853, 1966.
124. RB Merrifield, AR Mitchell, JE Clarke. Detection and prevention of urethane acylation during solid-phase peptide synthesis by anhydride methods. *J Org Chem* 39, 660, 1974.
125. FR Ahmed, FMF Chen, NL Benoiton. The crystal structure of *N,N′*-bismethoxycarbonyl-L-valyl-L-valine, a product of the rearrangement of the symmetrical anhydride of *N*-methoxycarbonyl-L-valine. *Can J Chem* 64, 1396, 1986.
126. NL Benoiton, FMF Chen. Symmetrical anhydride rearrangement leads to three different dipeptide products, in D Theodoropoulus, ed. *Peptides 1986. Proceedings of the 19th European Peptide Symposium.* Walter de Gruyter, Berlin, 1987, pp 127-130.

7.25 SYNTHESIS OF PEPTIDES BY CHEMOSELECTIVE LIGATION

An altogether different approach to peptide synthesis is chemoselective ligation, in which two fully unprotected peptides, each possessing a uniquely reactive functional group at one of the two termini, are allowed to combine in a denaturing aqueous solvent. The approach is applicable to large molecules without requiring handling and characterization of fully protected peptides. There are two types of chemical ligation: native chemical ligation, in which the bond joining the two segments is a normal peptide bond, and the other, in which the two segments are linked through a nonnatural bond. Bond formation by native chemical ligation requires a peptide with the carboxy terminus in the form of a thioester [–C(=O)SR] and a peptide bearing a cysteine residue at the amino terminus (Figure 7.35). Thioesters are known to be relatively inert to amino groups but sensitive to sulfhydryls (–SH). The two peptides are mixed in the presence of thiophenol and benzyl mercaptan at a pH slightly above neutrality. The thiophenol converts the thioester into a more activated phenyl thioester (path A). The sulfhydryl of the terminal residue reacts with the activated ester (path B), producing the *S*-acyl-peptide that spontaneously undergoes an *S*-to-*N*-acyl shift (path C), analogous to the *O*-to-*N*-acyl shift of substituted serine (see Section 6.6), to produce the *N*-acyl-peptide, which is the two peptides combined. Some sulfhydryls other than the one at the terminus will have reacted with the activated ester or thiophenol. Benzyl mercaptan serves as a reducing agent (path D) that keeps all sulfhydryls in the reduced state through disulfide interchange (see Section 6.18). Very little hydrolysis of the activated ester occurs at the operating pH. Ligation is much less efficient when the thioester is that of proline, valine, or isoleucine. Several linkers have been developed to allow access to alkyl thioesters

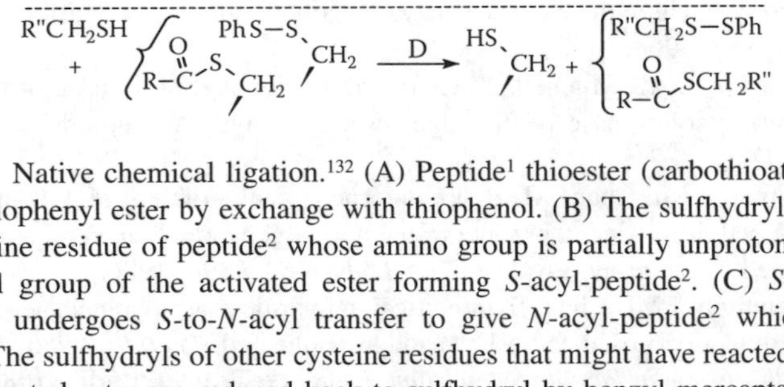

FIGURE 7.35 Native chemical ligation.[132] (A) Peptide[1] thioester (carbothioate) is activated further to a thiophenyl ester by exchange with thiophenol. (B) The sulfhydryl of the amino-terminal cysteine residue of peptide[2] whose amino group is partially unprotonated displaces the thiophenyl group of the activated ester forming *S*-acyl-peptide[2]. (C) *S*-Acyl-peptide[2] spontaneously undergoes *S*-to-*N*-acyl transfer to give *N*-acyl-peptide[2] which is peptide[1] peptide[2]. (D) The sulfhydryls of other cysteine residues that might have reacted with thiophenol or the activated ester are reduced back to sulfhydryl by benzyl mercaptan. *S*-to-*N*-acyl transfer was recognized previously. [Wieland et al., 1953].

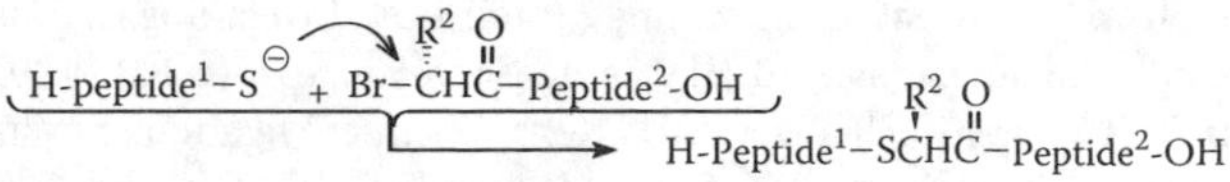

$$\text{H-peptide}^1\text{–S}^{\ominus} + \text{Br–CH(R}^2\text{)C(=O)–Peptide}^2\text{-OH} \longrightarrow \text{H-Peptide}^1\text{–SCH(R}^2\text{)C(=O)–Peptide}^2\text{-OH}$$

FIGURE 7.36 Thioester-forming ligation.[129] The carbothioate anion of peptide[1] reacts with bromoacyl-peptide[2] to give a product made up of the two peptides joined through a stable thioester moiety. When $R^2 \neq H$, the S_N2 reaction produces a thioester moiety with configuration opposite to that of the bromoacyl moiety.

by solid-phase synthesis, with 3-sulfanylpropionic acid ($–SCH_2CH_2CO_2H$, see Section 7.10) being a popular thioester moiety. An interesting variant of the conjoining reaction is ligation at a homocysteine residue instead of at a cysteine residue. Methylation with methyl *p*-nitrobenzenesulfonate of the sulfhydryl after ligation produces a peptide with a methionine residue at the juncture.

There are a variety of chemistries that can be employed for effecting ligation through an unnatural (not peptide) bond. The original and simplest is thioester-forming ligation involving reaction between a peptide thioacid and an α-bromoacetyl-peptide (Figure 7.36), $R^2 = H$. The reaction is carried out in a 6 *M* guanidine chloride-acetate buffer of pH 4. At this pH, the only nucleophile on the peptides is the carbothioate anion, which attacks the α-carbon of bromoacetyl, displacing bromide to give the two peptides conjoined through thioacetyl. The reaction can be effected with an α-bromoacyl-peptide (Figure 7.36); the result is a juncture composed of a thioacyl moiety ($SCHR^2C=O$) with a configuration opposite to that of the α-bromoacyl moiety because of the S_N2 (see Section 3.5) nature of the coupling reaction. The thioacids are prepared by solid-phase synthesis, starting with Boc-Xaa-SC(Ph)-$C_6H_4OCH_2CO_2H$ derivatives that are attached to NH_2CH_2Ph-resin, which is analogous to synthesis on a phenylacetamidomethyl resin (see Section 5.17). Fmoc/tBu chemistry is not compatible with thioesters because they are sensitive to piperidine. Bromoacyl peptides are obtained by carbodiimide-mediated reaction of the bromoalkanoic acid with the amino group of a peptide. Solid-supported peptides can also be employed for chemical ligation, but these must be assembled on a support that is resistant to the hydrogen fluoride that is employed for deprotecting the side chains. Other chemistries producing unnatural bonds by ligation involve reaction of a carboxy-terminal aldehyde ($–CO_2CH_2CH=O$) with the amino groups of a variety of amino-terminal residues. The moieties that join the two peptides are heterocyclic rings of various natures. As is nearly always the case, developments in technology make use of information that was previously available. In addition to *O/S*- to *N*-acyl shifts and disulfide interchange, the synthesis of peptide thioacids and studies on thiol capture ligation between two functionalized reacting groups preceded the work on chemical ligation.[127–135]

127. J Blake. Peptide segment coupling in aqueous medium: silver ion activation of the thiolcarboxylic group. *Int J Pept Prot Res* 17, 273, 1981.
128. DS Kemp, S-L Leung, DJ Kerkman. Models that demonstrate peptide bond formation by prior thiol capture. 1. Capture by disulfide formation. *Tetrahedron Lett* 22, 181, 1981.

129. M Schnölzer, SBH Kent. Constructing proteins by dovetailing unprotected synthetic peptides: backbone-engineered HIV protease. *Science* 256, 221, 1992.
130. M Baca, TW Muir, M Schnölzer, SBH Kent. Chemical ligation of cysteine-containing peptides: synthesis of a 22 kDa tethered dimer of HIV-1 protease. *J Am Chem Soc* 117, 1881, 1995.
131. JP Tam, Y-A Lu, C-F Liu, JJ Shao. Peptide synthesis using unprotected peptides through orthogonal coupling methods. *Proc Natl Acad Sci USA* 92, 12485, 1995.
132. PE Dawson, MJ Churchill, MR Ghadiri, SBH Kent. Modulation of reactivity in native chemical ligation through the use of thiol additives. *J Am Chem Soc* 119, 4325, 1997.
133. JA Camerero, GJ Cotton, A Adeva, TW Muir. Chemical ligation of unprotected peptides directly from a solid support. *J Pept Res* 51, 303, 1998.
134. JP Tam, Q Yu. Methionine ligation strategy in the biomimetic synthesis of parathyroid hormones. *Biopolymers* 46, 319, 1998.
135. JP Tam, A Yu, A Miao. Orthogonal ligation strategies for peptide and protein. *Biopolymers (Pept Sci)* 51, 311, 1999.

7.26 DETECTION AND QUANTITATION OF ACTIVATED FORMS

Peptide-bond formation involves generation of an activated form of the carboxylic acid, followed by its reaction with the amino group of another residue (see Section 1.5). The activated form may be partly or wholly transformed into a second or third activated form (see Section 2.10) before it is aminolyzed. The more known about the course of a reaction, the more likely it is that a protocol can be developed for effecting an efficient reaction. Determination of the appearance and disappearance of activated forms is central to the acquisition of this knowledge. The amino-acid residue that is activated contains a carbonyl whose stretching band absorbs light at a certain wavelength of the infrared spectrum. The intensity of absorbance varies with the concentration of the compound in solution. Hence the change in intensity of absorbance of a solution at that wave length is a reflection of the appearance or disappearance of the compound. The wave lengths of maximum absorbance depend on the level of activation and are roughly as follows (cm^1): *N*-carboxyanhydride, 1864–1852; acyl fluoride, 1859–1833; 2-alkoxy/alkyl-5(4*H*)-oxazolone, 1845–1830 (the oxazolone has an additional C=N stretching vibration band at 1700–1670); mixed/symmetrical anhydride, 1830–1825; benzotriazolyl ester, 1825–1815; acyl chloride, 1800–1790; *p*-nitrophenyl ester, 1780–1760; mixed carbonate, 1760–1740; amide *N*-oxide (see Section 7.17), 1750–1730; urethane, 1690; and acylamido, 1650 — the latter two being unactivated carbonyls. Whether a mixture of two or more components can be analyzed by infrared spectroscopy depends on the extent of the differences between the wavelengths of maximum absorbance. A 2-alkoxy-5(4*H*)-oxazolone cannot be determined in the presence of a mixed or symmetrical anhydride by infrared spectroscopy, but the oxazolone can be determined in the presence of the corresponding *N*-alkoxycarbonyl substituent by nuclear magnetic resonance spectroscopy because the peak for its alkoxy protons is sufficiently separated from that of uncyclized derivatives (see Section 1.18). A more direct determination of activated forms can be made by high-performance liquid chromatography. Activated forms except transient intermediates such as *O*-acylisoureas (see Section 2.3) and

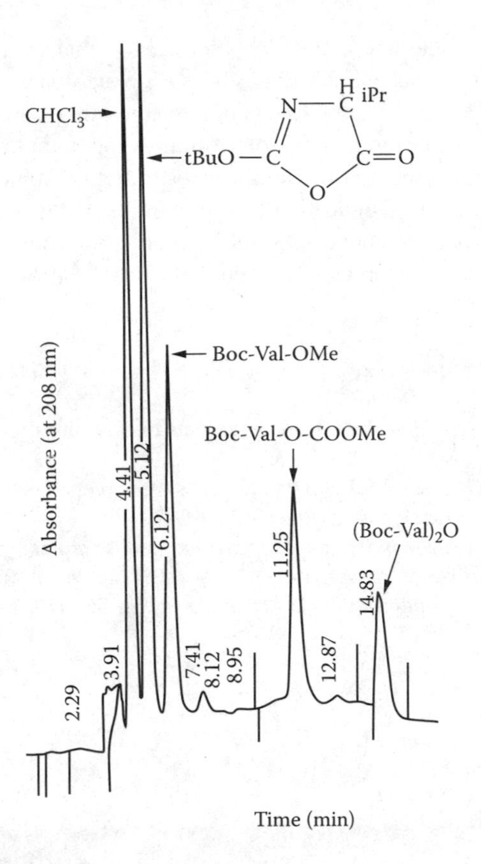

FIGURE 7.37 High performance liquid chromatography profile of a mixture of activated forms of Boc-valine and an ester chromatographed on a 25 cm × 4.0 mm I.D. cartridge containing 7 μm LiChrosorb-CN packing eluted with *tert*-butyl alcohol-hexane (1.5–98.5).[139] Retention times of related compounds: Boc-Val-O-CO_2Et 7.3, Boc-Val-NHMe 18, Val-*N*-carboxyanhydride 37 minutes. The absorbances of the compounds at 208 nm are not identical; hence the relative absorbances of components must be established to analyze a mixture.

O-acyluronium/*N*-acylguanidinium adducts (Section 7.17) can be determined by normal-phase chromatography using a nonaqueous solvent (Figure 7.37). This includes Fmoc-amino-acid chlorides and the two forms (*O*-/*N*-) of acylbenzotriazole (Section 7.17). In contrast to reversed-phase chromatography, in which retention is based on hydrophobic interaction, with the more hydrophobic compound being retained more strongly, in normal-phase chromatography, compounds that are not too different in size emerge from the column in order of decreasing hydrophobicity.

Aminolysis of activated forms by methylamine provides a convenient derivative for characterizing activated forms. Boc-valine methylamide (Figure 7.37) was the derivative made to confirm the structure of the 2-*tert*-butoxy-4-isopropyl-5(4*H*)-oxazolone that had been prepared for the first time (see Section 1.18). However, for activated forms that might undergo enantiomerization or epimerization during aminolysis, hydrazinolysis is the reagent of choice for their derivatization and for establishing the isomeric nature of a substance. Reaction with hydrazine produces a product with enantiomeric composition identical to that of the material hydrazinolyzed. Even chirally sensitive 2,4-dialkyl-5(4*H*)-oxazolones react with hydrazine without undergoing any enolization that generates the other isomer (see Section 1.9). Activated forms sometimes remain in solution after the completion of a reaction. A simple tactic to remove them is the addition of dimethylaminopropylamine to the mixture. The tertiary amine handle on the propyl amide that is produced is protonated at acidic pH, and hence the adduct is soluble in aqueous acid. This allows removal of the compound from an organic solvent by extraction with aqueous acid (see Section 1.2).[136–139]

136. IZ Simeon, J Morawiec. Reaction of azlactones with hydrazine. *Bull Acad Pol Sc, Ser Sc Chim* 12, 295, 1964.
137. M Goodman, CB Glaser. Mechanistic aspects of oxazolone reactions with α-nucleophiles. *J Org Chem* 35, 1954, 1970.
138. FMF Chen, K Kuroda, NL Benoiton, N.L. A simple preparation of 5-oxo-4,5-dihydro-1,3-oxazoles (Oxazolones). *Synthesis* 230, 1979.
139. FMF Chen, NL Benoiton. High-performance liquid chromatography of 2,4-disubstituted-5(4*H*)-oxazolones, anhydrides and other activated forms of *N*-acyl- and *N*-alkoxycarbonylamino acids. *Int J Pept Prot Res* 36, 476, 1990.

8 Miscellaneous

8.1. ENANTIOMERIZATION OF ACTIVATED *N*-ALKOXYCARBONYLAMINO ACIDS AND ESTERIFIED CYSTEINE RESIDUES IN THE PRESENCE OF BASE

Tertiary amines are employed for abstracting protons from *N*-alkoxycarbonylamino acids to initiate onium salt-mediated reactions (see Sections 2.16–2.19) and for neutralizing protons that are generated during aminolysis of the acid halides (see Sections 7.11 and 7.12). At the same time, the amines may abstract protons that were not intended, such as the α-proton of the residue (see Section 4.2) or the NH proton of the urethane that leads to 2-alkoxy-5(4*H*)-oxazolone formation (see Section 1.10). Under ordinary conditions, the secondary reactions do not occur or are of no consequence. However, there are situations in which unwanted deprotonations do occur, and they may be followed by enolization, which means isomerization of the residues. Situations in which the presence of a base can lead to isomerization include reactions of highly activated compounds that are not consumed quickly, tertiary-amine initiated coupling reactions carried out in the presence of excess base, and repeated exposure to secondary amines of peptides that have an esterified cysteine residue at the carboxy terminus. Examples of enantiomerization sustained by compounds left in the presence of basic amines appear in Figure 8.1. The acid halides that generate oxazolone in the presence of a base (see Section 4.17) isomerize very quickly; no isomerization occurs when they are aminolyzed under normal operating conditions. Even Fmoc-Pro-Cl is not immune to the effect of a base (Figure 8.1), having been isomerized during an attempt to react it with the hydroxymethyl group of a linker-resin. Reducing the temperature and amount of pyridine allowed esterification without isomerization.

Compound	Base (equiv)	Solvent	Time	% D
Z-Phe-ONp[a]	Et_3N (0.2)	CH_2Cl_2	18 h	7.9
Boc-Phe-F[b]	Et_3N (2.0)	CH_2Cl_2	5 min	7.2
Fmoc-Val-Cl[b]	Et_3N (1.0)	$CHCl_3$	1 min	3
Fmoc-Pro-Cl[c]	C_6H_5N-CH_2Cl_2	(2:3)	10 h	5–10
Boc-Cys(R)-OCH_2-resin	Various[d]	DMF	16 h	4–15

FIGURE 8.1 Enantiomerization of compounds by basic amines. [a](Benoiton et al., 1993); [b](Carpino et al., 1988, 1991); [c]Ref. 5; [d]Piperazine, morpholine or DBU (see Section 8.12); R = trityl >> acetamidomethyl > tBu.[3]

An exception to the sensitivity of activated amino acid derivatives to tertiary amine are the benzotriazolyl adducts of tritylamino acids (the *N*-oxide Trt-Xaa-BtO; see Section 7.17), which are stable even in hot solvent. A few examples of enantiomerization of *N*-alkoxycarbonylamino acids have surfaced where slow reaction in the presence of tertiary amine seems to be the reason why products containing both isomers were produced. Reaction of *Z*-leucine with the allyl ester of aminoisobutyric acid mediated by bromo-*tris*(dimethylamino)phosphonium hexafluorophosphate $[(Me_2N)_3P^+Br \cdot PF_6^-]$ in the presence of two equivalents of diisopropylethylamine produced a peptide containing 1.5% of D-leucine. Reaction of *Z*-methylvaline with *N*-methylvaline *tert*-butyl ester mediated by *bis*(2-oxo-3-oxazolidino)phosphinic chloride $\{c[\text{-C(=O)OCH}_2\text{CH}_2\text{N-}]_2\text{PCl}\}$ (see Section 8.15) in the presence of two equivalents of triethylamine produced a product containing 4% of epimer. These examples should serve as notice that contact of coupling reactants with basic amines should be avoided if possible. Even couplings that are not retarded by the hindered nature of the reactants can lead to isomerized products if basic amines are present. Derivatives of cysteine and serine in particular are sensitive to base.

Couplings of Fmoc-Cys(Trt)-OH mediated by TBTU $[BtOC^+(NMe_2)_2 \cdot BF_4^-]$ with diisopropylethylamine and 1-hydroxybenzotriazole as additive led to isomerized products. A detailed study on the coupling of Fmoc-cysteine in solid-phase synthesis established that preactivation when using onium salt-based coupling reagents led to enantiomerization of the residue, which is not surprising (see Section 7.20). 2,4,6-Trimethylpyridine and 2,6-di-*tert*-butylpyridine are the safest tertiary amines for use in coupling chirally sensitive residues using the base-requiring reagents. Esterification of Wang resin ($HOCH_2C_6H_4OCH_2$-polystyrene) with Fmoc-Cys(tBu)-F with the aid of the latter amine (see Section 7.12) occurred without isomerization. The presence of tertiary amines can be avoided if they are replaced by the cupric salts of 1-hydroxybenzotriazole or 1-hydroxy-7-azabenzotriazole (see Section 7.2), with the heterocyclic anions serving to deprotonate the carboxyl group of the substrate to initiate the reaction.

In addition to their effect on activated forms of residues, basic amines enantiomerize cysteine residues that are esterified regardless of the N^{α}-substitutent. This is particularly pertinent when a peptide containing a carboxy-terminal cysteine residue is assembled using Fmoc chemistry. Repeated treatment of the peptide with piperidine or another base employed to remove the Fmoc group at the amino terminus of the lengthening chain causes substantial enantiomerization of the residue. The data in Figure 8.1 correspond to 24 cycles of solid-phase synthesis. The problem is avoided by use of 2-chlorotrityl chloride resin (see Section 5.23), which gives cysteine esterified to a tertiary carbon atom. The sensitivity to base of cysteine esters had been observed a half century ago, during experiments on the saponification of esters of cysteine. Cysteine residues linked at the carboxyl group to a nitrogen atom are unaffected by base.[1–10]

1. JA Maclaren. Amino acids and peptides. V. The alkaline saponification of *N*-benzyloxycarbonyl peptide esters. *Aust J Chem* 11, 360, 1958.
2. K Barlos, D Papaioannou, D Theodoropoulos. Preparation and properties of N^{α}-trityl amino acid 1-hydroxybenzotriazole esters. *Int J Pept Prot Res* 23, 300, 1984.

3. E Atherton, PM Hardy, DE Harris, H Matthews. Racemization of C-terminal cysteine during peptide assembly, in E Giralt, D Andreu, eds. *Peptides 1990. Proceedings of the 21st European Peptide Symposium*. Escom, Leiden, 1991, pp 243–244.
4. Y Fujiwara, K Akaji, Y Kiso. Racemization-free synthesis of C-terminal cysteine-peptide using 2-chlorotrityl resin. *Chem Pharm Bull (Jpn)* 42, 724, 1994.
5. F Dick, M Schwaller. SPPS of peptides containing C-terminal proline: racemization-free anchoring of proline controlled by an easy and reliable method, in HLS Maia, ed. *Peptides 1994. Proceedings of the 23rd European Peptide Symposium*. Escom, Leiden, 1995, pp 240–241.
6. NL Benoiton. 2-Alkoxy-5(4H)-oxazolones and the enantiomerization of *N*-alkoxycarbonylamino acids. *Biopolymers (Pept Sci)* 40, 245, 1996.
7. T Kaiser, GJ Nicholson, H Kohlbau, W Voelter. Racemization studies of Fmoc-Cys(Trt)-OH during stepwise Fmoc-solid phase peptide synthesis. *Tetrahedron Lett* 37, 1187, 1996.
8. Y Han, F Albericio, G Barany. Occurrence and minimization of cysteine racemization during stepwise solid-phase synthesis. *J Org Chem* 62, 4307, 1997.
9. A Di Fenza, M Tacredi, C Galoppini, P Rovero. Racemization studies of Fmoc-Ser(tBu)-OH during stepwise continuous-flow solid-phase peptide synthesis. *Tetrahedron Lett* 39, 8529, 1998.
10. W van den Nest, S Yuval, F Albericio. $Cu(OBt)_2$ and $Cu(OAt)_2$, copper(II)-based racemization suppressors ready for use in fully automated solid-phase peptide synthesis. *J Pept Sci* 7, 115, 2001.

8.2 OPTIONS FOR PREPARING *N*-ALKOXYCARBONYLAMINO ACID AMIDES AND 4-NITROANILIDES

Amides of *N*-alkoxycarbonylamino acids can be obtained by any coupling procedure if ammonia or a linker-resin bearing an amino group or an alkylamine or linker-resin bearing an alkylamino group (see Section 8.3) is employed as nucleophile, followed by release of the support where pertinent. The less complicated methods are obviously the more attractive. The mixed-anhydride procedure provides a simple and economical method for preparing amides, with an early example being the synthesis of isoglutamine from protected glutamic acid (Figure 8.2). Tributylamine was an appropriate base at the time. Several variants of the use of unisolated activated esters generated using dicyclohexylcarbodiimide (Figure 8.3) provide access to amides. Ammonium hydroxide, the crystalline salt of diaminomethane that releases ammonia in solution, and salts formed from ammonia or alkylamines and the activating hydroxy compounds can serve as sources of nucleophile. The salts allow accurate measurement of the amounts of amine nucleophile employed. Aminolysis occurs without significant enantiomerization of the amino-acid residues.

Anilides [–C(=O)NHC_6H_4X] are obtainable by the same general approaches. For glutamic and aspartic acids that contain two carboxyl groups it is not always necessary to protect one of them to amidate the other one. A mixture of monoamides is produced by aminolysis of the *N*-protected amino acid anhydrides (see Section 6.12), with the relative amounts of the positional isomers being highly dependent on the polarity of the solvent, and the generation of the α-isomer being favored by

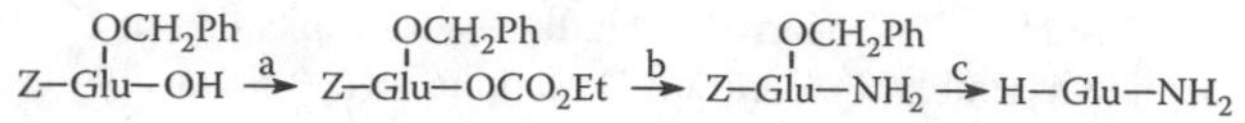

$$Z\text{–}Glu(OCH_2Ph)\text{–}OH \xrightarrow{a} Z\text{–}Glu(OCH_2Ph)\text{–}OCO_2Et \xrightarrow{b} Z\text{–}Glu(OCH_2Ph)\text{–}NH_2 \xrightarrow{c} H\text{–}Glu\text{–}NH_2$$

FIGURE 8.2 Synthesis of isoglutamine by reaction of the mixed anhydride of protected glutamic acid with gaseous ammonia followed by deprotection by hydrogenolysis.[11] a = $EtOCOCl/Bu_3N$ in dioxane; b = anhydrous NH_3; c = $H_2/Pd(CaCO_3)$.

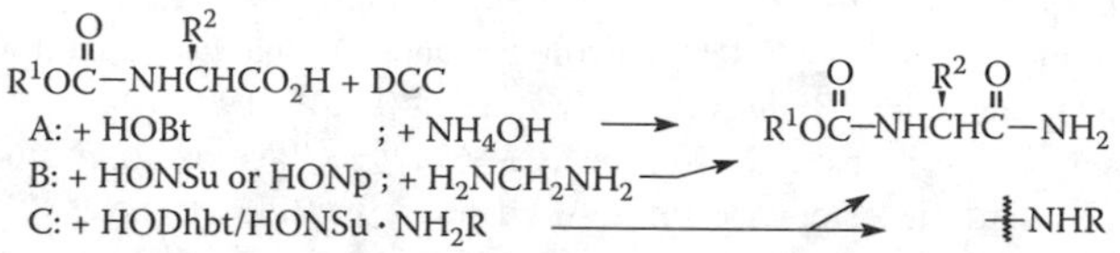

$$R^1OC(=O)\text{–}NHCH(R^2)CO_2H + DCC$$

$$\text{A: + HOBt ; + }NH_4OH \longrightarrow R^1OC(=O)\text{–}NHCH(R^2)C(=O)\text{–}NH_2$$

$$\text{B: + HONSu or HONp ; + }H_2NCH_2NH_2 \longrightarrow R^1OC(=O)\text{–}NHCH(R^2)C(=O)\text{–}NH_2$$

$$\text{C: + HODhbt/HONSu}\cdot NH_2R \longrightarrow R^1OC(=O)\text{–}NHCH(R^2)C(=O)\text{–}NHR$$

FIGURE 8.3 Synthesis of amides by aminolysis of unisolated activated esters obtained using carbodiimide by (A) ammonium hydroxide,[12] (B) diaminomethane dihydrochloride neutralized with Et_3N,[15] and (C) the amine of crystalline salts of "additives," R = H, Me, Et.[13] DCC = dicyclohexylcarbodiimide; HOBt = 1-hydroxybenzotriazole; HONSu = *N*-hydroxysuccinimide; HONp = *p*-nitrophenol; HODhbt = 3-hydroxy-4-oxo-3,4-dihydrobenzotriazine.

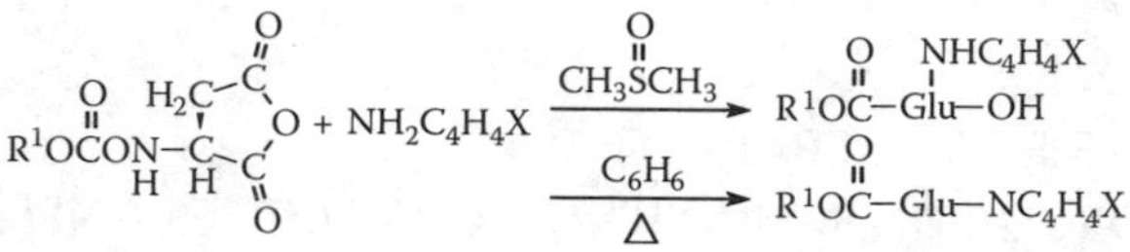

$$R^1OC(=O)N(H)\text{–}CH(\text{anhydride ring: }H_2C\text{–}C(=O)\text{–}O\text{–}C(=O)) + NH_2C_4H_4X \xrightarrow{CH_3S(=O)CH_3} R^1OC(=O)\text{–}Glu(NHC_4H_4X)\text{–}OH$$

$$\xrightarrow[\Delta]{C_6H_6} R^1OC(=O)\text{–}Glu\text{–}NC_4H_4X$$

FIGURE 8.4 Synthesis of anilides of *N*-protected glutamic acid by aminolysis of the anhydride in dimethylsulfoxide, which gives the α-isomers, and in benzene, which gives the α-isomers.[18] X = H, Cl, OMe, NO_2.

the less-polar solvent. Aminolysis of *N*-protected glutamic-acid anhydride by anilines in boiling benzene gives almost exclusively the α-isomers (Figure 8.4). Only *p*-nitroaniline produces significant amounts of the γ-isomer, whose formation can be suppressed by the addition of 2% acetic acid. Aminolysis in dimethylsulfoxide results in a regioselectivity of 1:5 (α:γ), thus allowing obtainment of some of the side-chain substituted anilides in a pure state. The regioselectivity of the reaction is less when the electrophile is *N*-protected aspartic-acid anhydride, but it is still sufficient for preparative purposes in some cases.

The synthesis of *p*-nitroanilides that are used as chromogenic substrates for determining rates of enzymatic reactions is not straightforward. *p*-Nitroaniline is a very weak nucleophile; hence, only very activated forms of amino acid derivatives react efficiently with this aromatic amine. Two types of anhydrides, mixed and the phosphoric anhydride generated using phosphorus oxytrichloride (Figure 8.5), provide access to the *p*-nitroanilides in good yield. A third method involves reaction of the amino acid derivative with *p*-phenylisocyanate, with the former releasing carbon dioxide during the process. When making *p*-nitroanilides of Fmoc-amino acids

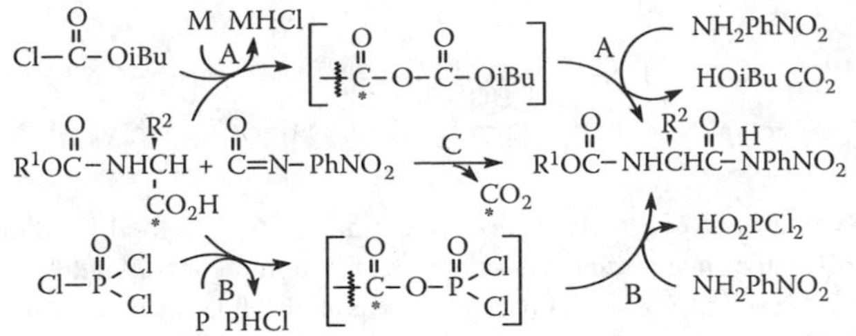

$Cl-C(=O)-OiBu$ —(M, MHCl; A)→ $[-C(=O)^{*}-O-C(=O)-OiBu]$ —(A; NH_2PhNO_2; HOiBu CO_2)→

$R^1OC(=O)-NHCH(R^2)(C^{*}O_2H)$ + $C(=O)=N-PhNO_2$ —(C; $C^{*}O_2$)→ $R^1OC(=O)-NHCH(R^2)C(=O)-N(H)PhNO_2$

$Cl-P(=O)(Cl)Cl$ —(P, PHCl; B)→ $[-C(=O)^{*}-O-P(=O)(Cl)Cl]$ —(B; NH_2PhNO_2; HO_2PCl_2)→

FIGURE 8.5 Synthesis of *p*-nitroanilides by aminolysis (A) of the mixed anhydride obtained with M = *N*-methylmorpholine as base,[14] (B) of the dichlorophosphoric anhydride obtained with P = pyridine as base and solvent, [Rijkers et al., 1991] and (C) by reaction with *p*-nitrophenylisocyanate. (Nishi & Noguchi, 1973).

unreacted aromatic amine is efficiently removed from the organic solution by 2*M* hydrochloric acid.[11–18]

11. M Kraml, LP Bouthillier. Synthesis of L-isoglutamine from γ-benzyl glutamate. Can J Chem 33, 1630, 1955.
12. S-T Chen, S-H Wu, K-T Wang. A simple method for amide formation from protected amino acids and peptides. *Synthesis* 37–38, 1989.
13. C Somlai, G Szókán, L Baláspiri. Efficient, racemization-free amidation of protected amino acids. *Synthesis* 285, 1992.
14. H Nedev, H Naharisoa, T Haertlé. A convenient method for synthesis of Fmoc-amino acid *p*-nitroanilides based on isobutyl chloroformate as condensing agent. *Tetrahedron Lett* 34, 4201, 1993.
15. G Galaverna, R Corradini, A Dossena, R Marchelli. Diaminomethane dihydrochloride, a novel reagent for the synthesis of primary amides of amino acids and peptides from active esters. *Int J Pept Prot Res* 42, 53, 1993.
16. DTS Rijkers, HC Hemker, GHL Nefkens, GI Tesser. A generally applicable synthesis of amino acid *p*-nitroanilides as synthons, in CH Schneider, AN Eberle, eds. *Peptides 1992. Proceedings of the* 22nd *European Peptide Symposium.* Escom, Leiden, 1993, pp 175–176.
17. M Schutkowski, C Mrestani-Klaus, K Neubert. Synthesis of dipeptide 4-nitroanilides containing non-proteinogenic amino acids. (*p*-nitrophenylisocyanate) *Int J Pept Prot Res* 45, 257, 1995.
18. X Huang, X Luo, Y Roupioz, JW Keillor. Controlled regioselective anilide formation from aspartic and glutamic acid anhydrides. *J Org Chem* 62, 8821, 1997.

8.3 OPTIONS FOR PREPARING PEPTIDE AMIDES

The most direct method of synthesizing a peptide amide is the assembly of the chain by stepwise addition of single residues, starting with an amino acid amide. The latter is obtainable from the *N*-protected amino acid (see Section 8.2) by deprotection of the amino group. The amide function may be the desired function, or it may have been created by reacting the *N*-protected amino acid with an appropriate linker that is to be bound or is already bound to a resin. In the latter cases, the target molecule is obtained after chain assembly by detaching the peptide amide from the support.

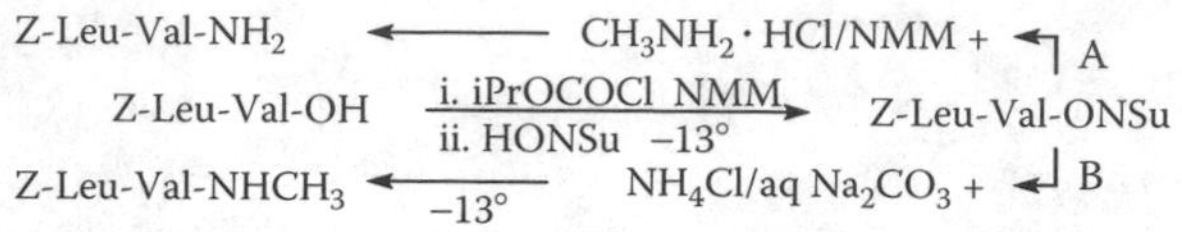

$$\text{Z-Leu-Val-NH}_2 \longleftarrow \text{CH}_3\text{NH}_2\cdot\text{HCl/NMM} + \text{ (A)}$$

$$\text{Z-Leu-Val-OH} \xrightarrow[\text{ii. HONSu } -13^\circ]{\text{i. iPrOCOCl NMM}} \text{Z-Leu-Val-ONSu}$$

$$\text{Z-Leu-Val-NHCH}_3 \xleftarrow[-13^\circ]{} \text{NH}_4\text{Cl/aq Na}_2\text{CO}_3 + \text{ (B)}$$

FIGURE 8.6 Synthesis of *N*-protected dipeptide amides (A) and methylamides (B) by aminolysis of the unisolated succinimido ester obtained through the mixed anhydride in CH_2Cl_2.[22] The products contained <0.5% of epimerized compound. HONSu = *N*-hydroxysuccinimde, NMM = *N*-methylmorpholine.

Benzhydryl- (phenylbenzyl) and 4-methylbenzhydrylamine resins (see Section 5.18) are available for preparing primary amides, using Boc/Bzl chemistry. 4-Methylbenzhydrylamine, Rink amide, Sieber amide and dimethoxytritylamine resins and the linkers PAL and XAL (see Section 5.20) are available for preparing primary amides, using Fmoc/tBu chemistry. Secondary amides can be synthesized by making use of *N*-alkyl-Sieber or *N*-alkyl-PAL linkers.

Some amino acid and peptide amides were previously obtained by ammonolysis of the corresponding methyl or resin-bound benzyl esters in methanol, but the method was often unsatisfactory because the reaction was incomplete or the amidated residue underwent enantiomerization. The recent finding that ammonolysis of esters by gaseous ammonia in the absence of solvent (see Section 5.18) gives single-isomer products in excellent yields will undoubtedly reinvigorate the approach of ammonolysis of esters for the synthesis of amides. The methods described in Section 8.2 for preparing amides of *N*-protected amino acids involving the intermediacy of activated esters obtained using dicyclohexylcarbodiimide have been applied to the synthesis of a few dipeptide amides, but that is the extent of the information on the subject. In contrast, the method describing the synthesis of succinimido esters through the mixed anhydrides (see Section 7.10) has been extended to the preparation of dipeptide amides (Figure 8.6) and a tetrapeptide amide of established chiral purity. An additional approach is available for preparing peptide amides by solid-phase synthesis. The peptide chain is assembled on the substituted-oxime resin (see Section 5.21), and it is displaced from the support by ammonium acetate, an alkylamine, or an amino-acid amide. A few difficulties are associated with some versions of this approach, but it remains a valuable method because of its versatility.[19–24]

19. JT Lobl, L Maggiore. Convenient synthesis of C-terminal peptide analogues by aminolysis of oxime resin-linked protected peptides. *J Org Chem* 53, 1979, 1988.
20. JC Hendrix, JT Jarrett, ST Anisfield, PT Lansbury. Studies related to a convergent fragment-coupling approach to peptide synthesis using the Kaiser oxime resin. *J Org Chem* 57, 3414, 1992.
21. N Voyer, A Lavoie, M Pinette, J Bernier. A convenient solid phase preparation of peptide subtituted amides. (oxime resin) *Tetrahedron Lett* 35, 355, 1994.
22. NL Benoiton, YC Lee, FMF Chen. A new coupling method allowing epimerization-free aminolysis of segments. Use of succinimidyl esters obtained through mixed anhydrides, in HLS Maia, ed. *Peptides 1994. Proceedings of the* 23rd *European Peptide Symposium*. Escom, Leiden, 1995, pp 203–204.

23. L Andersson, L Blomberg, M Flegel, L Lepsa, B Nilsson, M Verlander. Large-scale synthesis of peptides. (gaseous ammonia) *Biopolymers (Pept Sci)* 55, 227, 2000.
24. J Alsina, F Albericio. Solid-phase synthesis of C-terminal modified peptides. *Biopolymers (Pept Sci)* 71, 454, 2003.

8.4 AGGREGATION DURING PEPTIDE-CHAIN ELONGATION AND SOLVENTS FOR ITS MINIMIZATION

Two major obstacles are encountered during the synthesis of larger peptides: decreased solubility of intermediates in organic solvents during synthesis in solution, and reduced reactivity of functional groups during solid-phase synthesis. The latter is caused by association of peptide chains that results from hydrogen bonding between the NH protons of peptide bonds of one chain and the oxygen of the carbonyls of the peptide bonds of a second chain — a phenomenon similar to the formation of β-sheets in proteins. The decrease in solubility of intermediates is attributed to the aggregation of molecules that results from hydrophobic interactions between the protectors of different chains, as well as the formation of β-structures. These changes in properties that accompany the elongation of peptide chains are actualized by a diminution of solvation of the molecules (see Section 5.7). Hydrophobic interactions can be reduced by employing benzyl- and 9-fluorenylmethyl-based protectors that are attached to polyoxyethylene chains (see Section 5.9), as well as by changing the nature of the protectors or employing minimum protection strategies (see Section 6.1). Polar interactions can be reduced by including in solvents compounds known as chaotropic salts, composed of a large anion and a cation that does not complex with the substrates. Prewash of a peptido-resin with 0.4 *N* sodium perchlorate or potassium thiocyanate disrupts secondary structures and encourages faster and complete reactions. A "magic mixture" composed of dichloromethane, *N*-methylpyrrolidone (1-methylpyrrolidine-2-one), and dimethylformamide containing 1% of Triton-X-100 as anionic detergent, and sometimes ethylene carbonate [-$CH_2O)_2C$=O], is particularly effective in normalizing the reactivity of functional groups in difficult sequences. A study of the solubilities of a large number of protected peptides found that nearly all were soluble in *N*-methylpyrrolidone, dimethylformamide, dimethylsulfoxide, or mixtures thereof. The remainder were soluble in a mixture of chloroform and trifluoroethanol; chloroform with phenol has since proven to be the best solvent. These hydroxy-compounds disrupt hydrogen bonds by displacing the NH protons from the carbonyls. The three solvents above are in the order of increasing polarity. An empirical parameter of solvent polarity is as follows: water, 63.1; trifluoroethanol, 59.5; methanol, 55.5; dimethylsulfoxide, 45.0; dimethylformamide, 43.7; *N*-methylpyrrolidone, 42.2; dichloromethane, 41.1; pyridine, 40.2; chloroform, 39.1; tetrahydrofuran, 37.4; dioxane, 36.0; toluene, 33.9.

A completely different tactic is the use of microwave energy; absorption of this energy by a peptide results in decreased aggregation. Lithium salts facilitate the solution of peptides in these polar solvents, increase the rate of swelling of polar resins, and improve coupling yields. Sometimes the order of addition of two solvents to a peptide makes a difference. The inclusion of pyridine can also have a beneficial effect. Most amino acids and side-chain-substituted amino acids are rendered soluble in dimethylformamide by the presence of pyridine trifluoroacetate. The mixture dimethylformamide-pyridine-trifluoroacetic acid (100:19:6, v/v) allows their acylation to give *N*-protected dipeptides in good yields. All Boc- and Fmoc-amino acids are soluble in a toluene-dimethylsulfoxide (3:1) mixture. Reaction times are shortened by a modest increase in temperature, to 40° or 50°C. For solid-phase synthesis, this allows the use a smaller excess of amino acid derivative. The association of peptide chains is eliminated by replacing some of the protons of the peptide bonds by alkyl groups (see Sections 8.5 and 8.6). Finally, it has been demonstrated that the tactic of obtaining an insoluble peptide through the soluble *O*-acyl isopeptide (see Section 6.6) that rearranges to the target peptide at pH above neutral can be employed to prepare a peptide whose synthesis by the conventional approach is inefficient or fails because of the obstacles imposed by aggregation of the peptide chains as they are being extended.[25–33]

25. C Reichardt. Solvent effects in organic chemistry. Verlag Chemie, Weinheim, 1979, p 242.
26. WA Klis, JM Stewart. Chaotropic salts improve SPPS coupling reactions, in JR Rivier, GR Marshall, eds. *Peptides Chemistry, Structure, and Biology. Proceedings of the 11th American Peptide Symposium*. Escom, Leiden, 1990, pp 904–906.
27. A Thaler, D Seebach, FC Cardinaux. Lithium-salt effects in peptide synthesis. Part II. Improvement of degree of resin swelling and efficiency of coupling in solid-phase synthesis. *Helv Chim Acta* 74, 628, 1991.
28. H Kuroda, Y-N Chen, T Kimura, S Sakakibara. Powerful solvent systems useful for synthesis of sparingly-soluble peptides in solution. *Int J Pept Prot Res* 40, 294, 1992.
29. W Rapp, E Bayer. Uniform microspheres in peptide synthesis: ultrashort cycles and synthesis documentation by on-line monitoring as an alternative to multiple peptide synthesis, in CH Schneider, AN Eberle, eds. *Peptides 1992. Proceedings of the 22nd European Peptide Symposium*. Escom, Leiden, 1992, pp 25–26. ("magic mixture").
30. HH Saneii, ML Paterson, H Anderson, ET Healy, RG Bizanek. Advanced automation for peptides and non-peptidic molecules, in PTP Kaumaya, RS Hodges, eds. *Peptides: Chemistry, Structure, and Biology. Proceedings of the 14th American Peptide Symposium*. Mayflower. Kinsgwinford, 1996, pp 136–138. (temperature effects).
31. YV Mitin. An effective organic solvent system for the dissolution of amino acids. *Int J Pept Prot Res* 48, 374, 1996.
32. J Bódi, H Nishio, T Inui, Y Nishiuchi, T Kimura, S Sakakibara. Segment condensation of sparingly soluble protected peptides in chloroform-phenol mixed solvent, in S Bajusz, F Hudecz, eds. *Peptides 1998. Proceedings of the 25th European Peptide Symposium*. Akadémiai Kiadó, Budapest, 1999, pp 198–199.
33. Y Sohma, M Sasaki, Y Hayashi, T Kimura, Y Kiso. Novel and efficient synthesis of difficult sequence-containing peptides through *O-N* intramolecular acyl migration reaction of *O*-acyl isopeptides. *Chem Commun* 124, 2004.

8.5 ALKYLATION OF PEPTIDE BONDS TO DECREASE AGGREGATION: 2-HYDROXYBENZYL PROTECTORS

It was established when solid-phase synthesis was developed that high levels of attachment of the first residue to a polymeric support led to inferior results because of hindrance between reacting species. Superior performance was achieved with lower resin-loading, which allowed freer access of the functional groups to the reagents employed. A second phenomenon of lesser reactivity was often encountered after six or seven residues had been incorporated into a peptide chain. A dramatic reduction of the rates of acylation and deprotection, which was not caused by the nature of the residues implicated, was observed. The classical example of what became known as "difficult sequences" was segment 63–74 of acyl-carrier protein. Reactivity during synthesis of such peptides was normal if *N*-methylglycine or proline was inserted into the chain, but in longer peptides reduced reactivity appeared after several more residues were added. This atypical behavior is attributed to the association of peptide chains that results from hydrogen bonding between the peptide bonds of the chains, a phenomenon similar to the formation of β-sheets in proteins. The association is believed to occur within the resin matrix, causing a shrinkage of the expanded gel that reduces the rate of penetration of the reagents. Dichloromethane favors the onset of association, and polar solvents disfavor the onset of association. Dimethylsulfoxide delays the appearance of the phenomenon for a few residues more than dimethylformamide. The fact that this obstacle to synthesis did not emerge when a secondary amino acid was incorporated into the chain indicated that protection of the peptide bond by a removable substituent would allow chain assembly to occur in the normal manner. A protector designed for the purpose for use with Fmoc/tBu chemistry is 2-hydroxy-4-methoxybenzyl (Hmb; Figure 8.7).

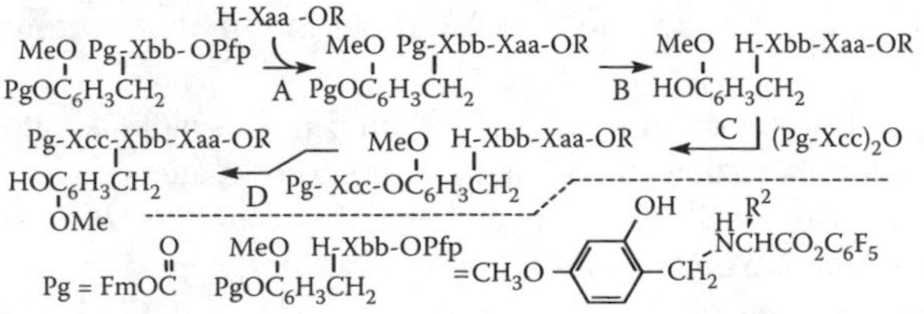

H-Xaa-OR
MeO Pg-Xbb-OPfp / $PgOC_6H_3CH_2$ → (A) MeO Pg-Xbb-Xaa-OR / $PgOC_6H_3CH_2$ → (B) MeO H-Xbb-Xaa-OR / $HOC_6H_3CH_2$
(C) $(Pg\text{-}Xcc)_2O$ → MeO H-Xbb-Xaa-OR / $Pg\text{-}Xcc\text{-}OC_6H_3CH_2$ → (D) Pg-Xcc-Xbb-Xaa-OR / $HOC_6H_3CH_2$ / OMe

Pg = FmOC(=O)   MeO H-Xbb-OPfp / $PgOC_6H_3CH_2$ = CH_3O–C₆H₃(OH)–CH_2–N(H)CH(R^2)$CO_2C_6F_5$

FIGURE 8.7 Synthesis of a protected tripeptide containing a 2-hydroxy-4-methoxybenzyl-protected peptide bond.[38] (A) Acylation of the carboxy-terminal residue, (B) removal of both protecting groups, (C) *O*-acylation of the benzyl-protector by the symmetrical anhydride of the amino-terminal residue, and (D) migration of the protected amino-terminal residue from the oxygen atom to the amino group of the dipeptide ester.

Previous work had established that the presence of protectors such as 2,4,6-trimethoxybenzyl on peptide bonds dramatically increased the solubility of protected peptides that were insoluble in organic solvents as a result of association of the peptide chains (see Section 8.4). However, their use for synthesis was not adopted because coupling to the secondary amine was impeded by the bulkiness of the substituents. The novelty of the new protector resides in the *ortho*-hydroxyl group, which permits easier acylation. The protector is stable to coupling and Fmoc-removal conditions and is sensitized by the methoxy group for cleavage by trifluoroacetic acid at the end of the synthesis. The protector is introduced at the same time as the residue is incorporated into the chain; the *bis*-Fmoc-amino-acid pentafluorophenyl ester with the assistance of 1-hydroxybenzotriazole as catalyst (see Section 7.9) is employed for acylation (Figure 8.7, A). The two Fmoc groups are removed at the same time (Figure 8.7, B), leaving the benzyl-protected amine as nucleophile for the next coupling. A more activated form of the next residue such as the symmetrical anhydride is employed for acylation, but the key is that the *ortho*-hydroxyl group of the protector serves as an acceptor of the protected residue (Figure 8.7, C). The Fmoc-amino-acyl group then spontaneously migrates (Figure 8.7, D) to the substituted amino group by the well-known *O*-to-*N* shift (see Section 6.6). Starting materials Fmoc-(Hmb)Xaa-OH are obtained by reaction of the amino acid with the corresponding benzaldehyde, which gives the Schiff's base [$HOC_6H_4(OMe)C{=}NCHRCO_2H$]. The latter is reduced to H-(Hmb)Xaa-OH with sodium borohydride, which is followed by acylation with Fmoc-chloride. Incorporation of the Hmb group at every fifth or sixth residue in a chain renders so-called "difficult sequences" amenable to synthesis by the solid-phase approach. In addition, the Hmb group helps to solubilize protected or unprotected peptides that are insoluble in organic solvents, and it prevents side reactions that involve a deprotonation such as base-promoted imide formation (see Section 6.13) at the carboxy groups of aspartyl residues [–Asp(X)-(Hmb)Gly- gives no imide]. The equivalent protector for Boc/Bzl chemistry is the same molecule without the methoxy group.[29,34–40]

29. W Rapp, E Bayer. Uniform microspheres in peptide synthesis: ultrashort cycles and synthesis documentation by on-line monitoring as an alternative to multiple peptide synthesis, in CH Schneider, AN Eberle, eds. *Peptides 1992. Proceedings of the 22nd European Peptide Symposium.* Escom, Leiden, 1992, pp 25-26.
34. F Weygand, W Steglich, J Bjarnason, R Akhtar, N Chytil. Easily cleavable protective groups for acid amide groups. II. Comparative studies on the cleavage of substituted benzyl groups from amide nitrogen and possible combination of such radicals with urethane protective groups. *Chem Ber* 101, 3623, 1968.
35. WS Hancock, DJ Prescott, PR Vagelos, GR Marshall. Solvation of the polymer matrix. Source of truncated and deletion sequences in solid phase synthesis. *J Org Chem* 38, 774, 1973.
36. RC de L Milton, SCF Milton, PA Adams. Prediction of difficult sequences in solid-phase peptide synthesis. *J Am Chem Soc* 112, 6039, 1990.
37. H Kuroda, Y-N Chen, T Kimira, S Sakakibara. Powerful solvent systems useful for synthesis of sparingly-soluble peptides in solution. *Int J Pept Prot Res* 40, 294, 1992.

38. C Hyde, T Johnson, D Owen, M Quibell, RC Sheppard. Some "difficult sequences" made easy. A study of interchain association in solid-phase peptide synthesis. *Int J Pept Prot Res* 43, 431, 1994.
39. T Johnson, M Quibell. The *N*-(2-hydroxybenzyl) group for amide bond protection in solid-phase peptide. *Tetrahedron Lett* 35, 463, 1994.
40. T Johnson, M Quibell, RC Sheppard. *N,O*-bisFmoc derivatives of *N*-(2-hydroxy-4-methoxybenzyl)-amino acids: useful intermediates in peptide synthesis. *J Pept Sci* 1, 11, 1995.

8.6 ALKYLATION OF PEPTIDE BONDS TO DECREASE AGGREGATION: OXAZOLIDINES AND THIAZOLIDINES (PSEUDO-PROLINES)

Reversible alkylation of peptide bonds prevents association of peptide chains by eliminating hydrogen bonding (see Section 8.5). A second substitution on the nitrogen atom of peptide bonds that achieves the same purpose is a methyl group that is also linked to the side chain of the residue through the oxygen atom of a serine or threonine residue forming an oxazolidine ring, or through the sulfur atom of a cysteine residue forming a thiazolidine ring (Figure 8.8). The cyclic structures are known as pseudo-prolines. As for an amide bond at prolyl, the amide bond at the nitrogen atom of a pseudo-proline has a preference for the *cis* conformation (see Section 6.19), and, thus, a pronounced effect on the backbone conformation of the chain. This prevents self-association and β-sheet formation and, thus, improves solvation and the kinetics of the coupling reactions. The parent amino acids are regenerated at the end of a synthesis by opening the rings with acid. The thiazolidine ring is 10,000 times more stable than the oxazolidine ring. The stability of the rings is manipulated by insertion of one or two methyl groups at C-2, which increases the sensitivity to acid. The substituents are indicated as superscripts (Figure 8.8). The ease of ring opening is also greatly dependent on the acid–solvent combination. The pseudo-prolines are incorporated into peptide chains as dipeptide units. No isomerization accompanies the couplings because of the cyclic nature of the residues that are activated (see Section 2.23). Compounds suitable for Fmoc chemistry,

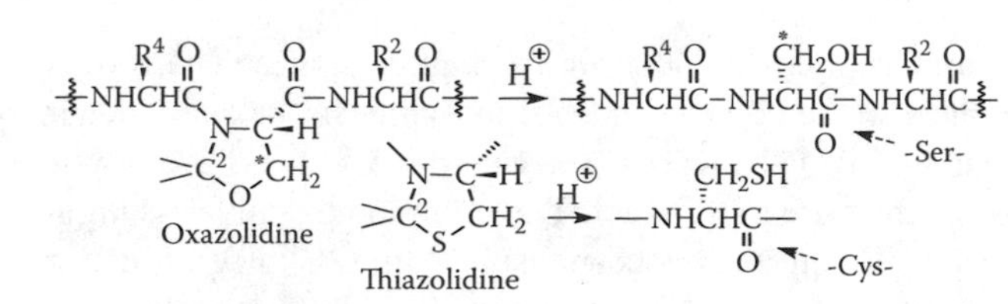

R^4 O O R^2 O $H^{\oplus}$ R^4 O CH_2OH R^2 O
-NHCHC C-NHCHC- → -NHCHC-NHCHC-NHCHC-
N-C-H O -Ser-
C2 *CH_2 N-C-H $H^{\oplus}$ CH_2SH
O C2 CH_2 → -NHCHC-
Oxazolidine S O -Cys-
Thiazolidine

FIGURE 8.8 A peptide sequence containing an oxazolidine ring that gives rise to a serine residue on acidolysis and a thiazolidine ring that gives rise to a cysteine residue on acidolysis. Substituents at C-2 can be H_2, H,Me, or Me_2. A structure with a methyl group at C* produces a threonine residue on acidolysis.

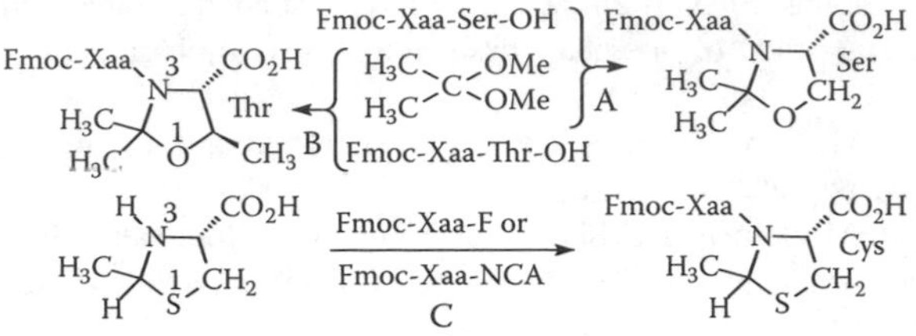

FIGURE 8.9 Synthesis of 2,2-dimethyl-1,3-oxazolidines by reaction of the Fmoc-dipeptide with 2,2-dimethoxypropane (A) giving Fmoc-Xaa-Ser(($\Psi^{Me,Me}$pro)-OH and (B) giving Fmoc-Xaa-Thr($\Psi^{Me,Me}$pro)-OH. Synthesis of a 2-methyl-1,3-thiazolidine by acylation of the cyclic parent (C) giving Fmoc-Xaa-Cys($\Psi^{H,Me}$pro)-OH. NCA = *N*-carboxyanhydride.

meaning that they are sensitive to trifluoroacetic acid, are Fmoc-Xaa-Ser($\Psi^{Me,Me}$pro)-OH, Fmoc-Xaa-Thr($\Psi^{Me,Me}$pro)-OH, and Fmoc-Xaa-Cys($\Psi^{H,Me}$-pro)-OH. Compounds suitable for Boc/Bzl chemistry, meaning that the rings are stable to trifluoroacetic acid, are Boc-Xaa-Ser($\Psi^{H,H}$pro)-OH, Boc-Xaa-Thr($\Psi^{H,H}$pro)-OH and Boc-Xaa-Cys($\Psi^{Me,Me}$pro)-OH. The cyclic structures are accessible by reaction of the amino acid or residue with formaldehyde, acetaldehyde, or a ketone equivalent (Figure 8.9). The dipeptide units are obtained by two routes: acylation of the cyclized amino acid (Figure 8.9, C) or cyclization of the protected dipeptide (Figure 8.9, A, B). The pseudo-prolines are in reality residues with their amino and side-chain functional groups protected by an aldehyde or ketone that is incorporated into a peptide chain. Their value has been demonstrated by their use in the synthesis of various peptides containing sequences that present obstacles to chain assembly.[41,42]

41. M Mutter, A Nefzi, T Sato, X Sun, F Wahl, T Wöhr. Pseudo-prolines (Ψpro) for accessing inaccessible peptides. *Pept Res* 8, 145, 1995.
42. T Wöhr, F Wahl, A Nefzi, B Rodwedder, T Sato, X Sun, M Mutter. Pseudo-prolines as a solubilizing, structure-disrupting protection technique in peptide synthesis. *J Am Chem Soc* 118, 9218, 1996.

8.7 CAPPING AND THE PURIFICATION OF PEPTIDES

The purification of peptides is more an art than a science. The foremost challenge arises for products obtained by solid-phase synthesis because the impurities are peptides containing failed sequences (see Section 5.5) that have physical properties similar to those of the primary product. The common technique is high-performance liquid chromatography on a reversed-phase column. Choice of stationary phase is made among tetramethyl- (C_4H_9), octamethyl- (C_8H_{17}), or octadecamethyl- ($C_{18}H_{37}$) substituted silica, depending on the hydrophobicity of the peptide, with the lesser-substituted support being employed for the more hydrophobic peptides. Elution is with gradients composed of acetonitrile and a buffer containing a volatile salt, such as ammonium acetate or triethylammonium phosphate, and 0.1% trifluoroacetic acid, with inclusion of the latter being common practice for the HPLC of peptides to ensure that the functional groups remain protonated and do not ionize. A tactic to facilitate separation of these impurities that result from incomplete reactions is to

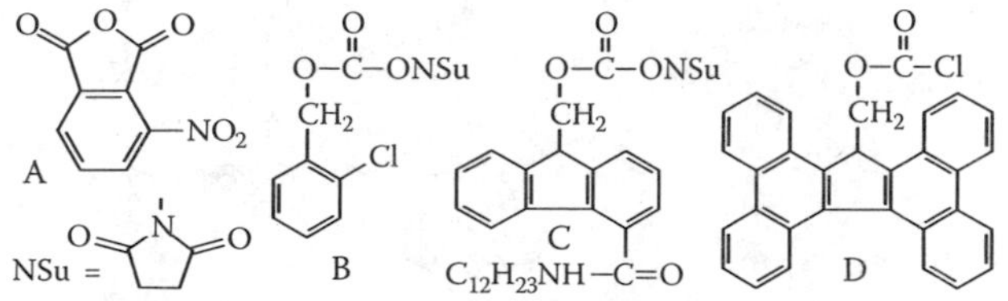

FIGURE 8.10 Reagents for capping: (A) 3-nitrophthalic anhydride[43] and (B) 2-chlorobenzyl succinimido carbonate.[45] Reagents providing hydrophobic base-labile substituents that allow isolation of peptides; (C) 4-dodecylcarbamidofluorenylmethyl succinimido carbonate[48] and (D) Tbfmoc = Fmoc-Cl bearing four benzene rings.[46]

change their physical properties by capping the amino groups after each coupling. A highly activated acylating reagent is employed (Figure 8.10), sometimes in the presence of trifluoroethanol (see Section 8.4). The original reagent was 3-nitrophthalic anhydride, which generates a nitrophenyl carboxylic acid that is strongly acidic, secured to the impurity that is now removable by adsorption by an anion-exchange resin. The common capping reaction is acetylation, using acetic anhydride and triethylamine, which, however, attacks *p*-toluenesulfonyl and nitro substituents on arginine. Acetylimidazole in pyridine is effective without causing any side reactions. *N*-Acetoxysuccinimide is employed when there are unprotected serine or threonine residues because it does not attack hydroxyl groups. 2-Chlorobenzyl succinimido carbonate and benzoic anhydride generate substituents that are hydrophobic and hence facilitate removal of the impurities. Benzoylated peptides are also easier to detect because of their increased absorbance of ultraviolet light. In contrast, an opposite tactic is to attach a very hydrodrophobic substituent to the main chain and then remove it after isolation of the derivatized peptide — in effect a reversible capping. Two reagents designed for this purpose are activated forms of the Fmoc group with rings that have been modified (Figure 8.10). The substituents are labile to base. A different approach involves modification of the main chain with a unique chemistry that allows its separation by affinity chromatography (Figure 8.11). Only the primary product is withheld by the stationary phase. As for the purification of protected peptides, a study of compounds having a free carboxy terminus and being 8–12 residues in length concluded that solvents containing dimethylformamide were the most likely to provide satisfactory results. The use of propionic acid instead of trifluoroacetic acid as modifier for chromatographic solvents ensures that Boc protectors remain intact.[43–50]

43. T Wieland, C Birr, H Wissenbach. 3-Nitrophthalic anhydride as blocking agent in Merrifield polypeptide synthesis. *Angew Chem* 8, 764, 1969.
44. J Rivier, R McClintock, R Galycan, H Anderson. Reversed-phase high-performance liquid chromatography: preparative purification of synthetic peptides. *J Chromatog* 288, 303, 1984.
45. HL Ball, G Bertolini. *N*-(Chlorobenzyloxycarbonyloxy)succinimide as a terminating agent for solid-phase peptide synthesis: application to a one-step purification procedure. *Lett Pept Sci* 2, 49, 1995.

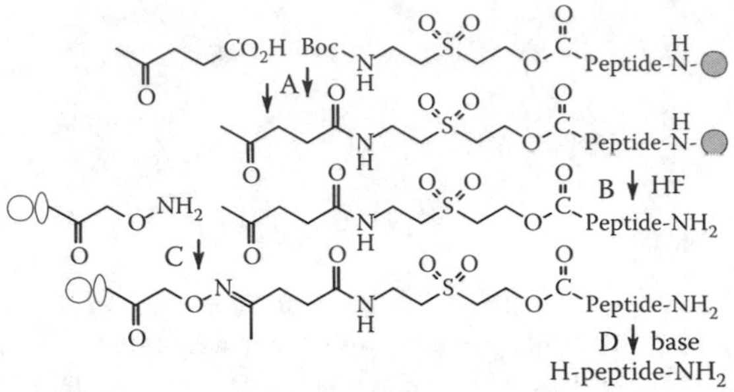

FIGURE 8.11 Purification of a peptide amide by chemoselective affinity chromatography.[50] Peptidamido-resin that has been derivatized by reaction with Boc-$NHC_2H_4SO_2C_2H_4OCO_2Np$ is deprotected, and (A) the amino group is reacted with levulinic acid. (B) The tagged peptide is detached from the resin and (C) chemically bound as it passes through a Sephadex column bearing a hydroxylamine (D). The peptide is released by base and eluted after unbound peptides have been washed through the column. Np = *p*-nitrophenyl.

46. AR Brown, SL Irving, R Ramage, G Raphy. (17-tetrabenzo[a,c,g,i]fluorenyl)methyl chloroformate (Tbfmoc-Cl) a reagent for the rapid and efficient purification of synthetic peptides and proteins. *Tetrahedron Lett* 51, 11815, 1995.
47. M Gairí, P Lloyd-Williams, F Albericio, E Giralt. Convergent solid-phase peptide synthesis. 12. Chromatographic techniques for the purification of protected peptide segments. *Int J Pept Prot Res* 46, 119, 1995.
48. HL Ball, P Mascagni. Chemical synthesis and purification of proteins: a methodology. *Int J Pept Prot Res* 48, 31, 1996.
49. C Miller, J Rivier. Peptide chemistry: development of high performance liquid chromatography and capillary zone electrophoresis. *Biopolymers (Pept Sci)* 40, 265, 1996.
50. RL Winston, J Wilken, SBH Kent. Design and application of a chemoselective affinity column for purifying full-length peptides, in R Ramge, R Epton, eds. *Peptides 1996. Proceedings of the 24th European Peptide Symposium.* Mayflower, Kingswoodford, 1998, pp 915–916.

8.8 SYNTHESIS OF LARGE PEPTIDES IN SOLUTION

The synthesis of peptides containing a large number of residues has been achieved by various approaches (see Sections 2.24, 6.17, and 7.13). The major difficulties encountered were the purification of products assembled by the solid-phase method and the insolubility of protected segments obtained by synthesis in solution. Over two decades ago, the group of Sakakibara in Japan decided to try to develop an approach that would permit the synthesis of peptides of more than 100 residues without having to resort to difficult purification procedures. They reasoned that the purification of segments obtained by assembly in solution was the more promising approach, and they formulated a set of conditions that must be fulfilled to maximize the chances of success. These are that side-chain functional groups must be protected, protectors of main-chain amino and carboxy groups must be removable by different mechanisms (orthogonal), protected segments must be soluble in organic solvents, couplings must be achieved without epimerization, unreacted segments and reagents

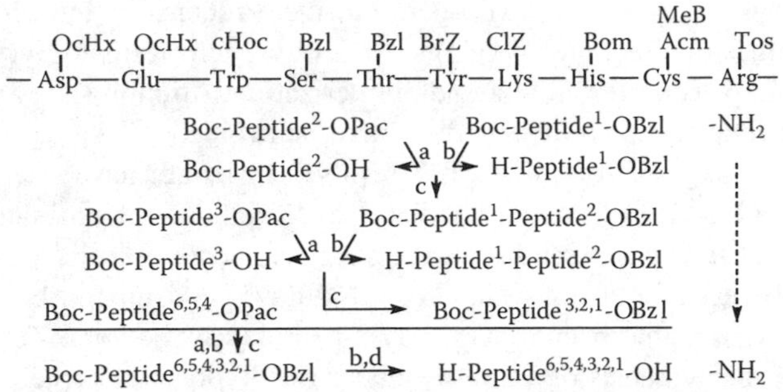

MeB
OcHx OcHx cHoc Bzl Bzl BrZ ClZ Bom Acm Tos
— Asp — Glu — Trp — Ser — Thr — Tyr — Lys — His — Cys — Arg —

Boc-Peptide^2-OPac   Boc-Peptide^1-OBzl   -NH$_2$
Boc-Peptide^2-OH ←a b→ H-Peptide^1-OBzl
c ↓
Boc-Peptide^3-OPac   Boc-Peptide^1-Peptide^2-OBzl
Boc-Peptide^3-OH ←a b→ H-Peptide^1-Peptide^2-OBzl
Boc-Peptide^{6,5,4}-OPac —c→ Boc-Peptide^{3,2,1}-OBzl
a,b ↓ c
Boc-Peptide^{6,5,4,3,2,1}-OBzl —b,d→ H-Peptide^{6,5,4,3,2,1}-OH   -NH$_2$

FIGURE 8.12 Strategy for synthesis of large peptides in solution [Sakakibara et al, 1981, 1996]. Protected peptides[1 & 2], 10–12 residues in length, are assembled by stepwise addition. This is followed by (a) reduction with Zn^0/CH_3CO_2H, (b) acidolysis with CF_3CO_2H, (c) coupling using EDC-HODhbt and subsequent chain extension, and (b, d) Boc removal and final deprotection by acidolysis with HF. An additional step is required to remove the Acm group. Pac = phenacyl (benzoylmethyl), cHx = cyclohexyl, cHoc = cyclohexyloxycarbonyl, Bom = benzyloxymethyl, MeB = 4-methylbenzyl, Acm = acetamidomethyl, Tos = *p*-toluenesulfonyl, EDC = ethyl-(3-dimethylaminopropyl)-carbodiimide (free base), HODhbt = 4-hydroxy-3-oxo-3,4-dihydrobenzotriazine.

must be readily removable, and side-chain protectors must not generate side products that are difficult to eliminate when the functional groups are deprotected. Their experience with hydrogen fluoride and the refinements that had emerged for its use led them to choose this acid for final deprotection and Boc/Bzl chemistry for protecting functional groups (Figure 8.12). The phenacetyl (benzoylmethyl) ester obtained with bromomethylphenyl ketone and base (see Section 3.17) and easily cleaved by reduction with zinc was chosen for α-carboxy protection. Boc groups were removed with trifluoroacetic acid. Peptides 10–12 residues in length were assembled by stepwise addition as usual, starting from the carboxy terminus. Couplings were effected using a soluble carbodiimide to facilitate removal of side products (see Section 1.16). Here, however, they introduced a novelty: instead of employing the hydrochloride of the reagent, the researchers employed the free base (Figure 8.12), with the rationale that this would ensure that any amino groups that might have remained protonated would be deprotonated during the coupling reaction. Deprotection at the appropriate terminal residues of the protected segments provided the reactants for segment coupling (Figure 8.12).

It was found that less epimerization occurred during segment coupling when 3-hydroxy-4-oxo-3,4-dihydrobenzotriazine and not 1-hydroxybenzotriazole was employed as the auxiliary nucleophile. Initial work was carried out with formyl as the protector for tryptophan; cyclohexyloxycarbonyl (see Section 6.9) was developed because of unsatisfactory results with formyl. Benzyloxymethylimidazole of histidine is slightly basic — if it is located in the activated segment, it is neutralized with hydrogen chloride. At the end of the synthesis, the Boc group is removed before final deprotection to avoid *tert*-butylation of sulfhydryls. *S*-Acetamidomethyl is stable to hydrogen fluoride, so another step is required to remove it. However, this provides the advantage that the acetamidomethylated peptide is easier to purify.

S-Methylbenzyl is removed by hydrogen fluoride. Prudence is in order in deciding at which residues to effect the couplings. As always, activation at glycine or proline (see Section 4.16) is preferred to avoid epimerization. Aminolysis by peptides with a hindered residue at the amino terminus is to be avoided. Close to 20 biologically active peptides ranging from 40 to 123 residues in length have been obtained by this approach. Only 1% of the intermediates were not soluble in dimethylsulfoxide. A recent variant includes assembly of the short segments by solid-phase synthesis, employing a base-labile linker, which takes 1–2 days per segment. Bromobenzyloxycarbonyl of tyrosine has been replaced by 3-pentyl (see Section 6.7). The work has culminated in the synthesis of a 238-residue green fluorescent protein that is found in the jellyfish *Aequorea victoria*. The greatest obstacle to the synthesis of large peptides is characterization of the products.[51–54]

51. T Kimura, M Takai, Y Masui, T Morika, S Sakakibara. Strategy for the synthesis of large peptides: an application to the total synthesis of human parathyroid hormone [hPTH (1-84)]. *Biopolymers* 20, 1823, 1981.
52. S Sakakibara. Synthesis of large peptides in solution. *Biopolymers (Pept Sci)* 37, 17, 1995.
53. Y Nishiuchi, H Nishio, T Inui, J Bódi, F Tsuji, T Kimura, S Sakakibara. Total synthesis of green fluorescent protein (GFP), a 238 residue protein from jellyfish *Aequorea victoria*, in S Bajusz, F Hudecz, eds. *Peptides 1998. Proceedings of the 25th European Peptide Symposium*. Akadémiai Kiadó, Budapest, 1999, pp 36–37.
54. Y Nishiuchi, H Nishio, J Bódi, T Kimura. Combined solid-phase and solution approach for the synthesis of peptides and proteins. *J Pept Sci* 6, 84, 2000.

8.9 SYNTHESIS OF PEPTIDES IN MULTIKILOGRAM AMOUNTS

The chemistry for the synthesis of peptides on an industrial scale is not different from the usual, but emphasis is placed on minimizing the expense and time devoted to synthesis. Solid-phase synthesis is the fastest way of acquiring material; however, it is not readily amenable to scale-up as synthesis in solution, which requires longer times to establish optimum conditions for efficient synthesis. Short peptides are more likely to be prepared in solution, employing Boc/Bzl or Cbz/tBu chemistry; peptide-bond formation may include an enzyme-catalyzed reaction. High cost may preclude the use of patented reagents. Synthesis in solution usually involves the strategy of minimum side-chain protection, with its attendant restrictions on the choice of coupling methods and its effects on the solubilities of products (see Section 6.1). Activated derivatives that are suitable for acylating the aminolyzing component with unprotected carboxyl groups (see Sections 7.13 and 7.21) are often employed. Both solid-phase and solution synthesis may be employed for the preparation of a product. Available chemistry allows the assembly of protected segments by solid-phase synthesis (see Section 5.21); these may be combined by reactions on a solid phase or in solution. An example of the synthesis of a commercial product appears in Figure 8.13. Of interest is the fact that five coupling methods are used for the formation of nine peptide bonds, and final deprotection is by reduction with sodium in liquid

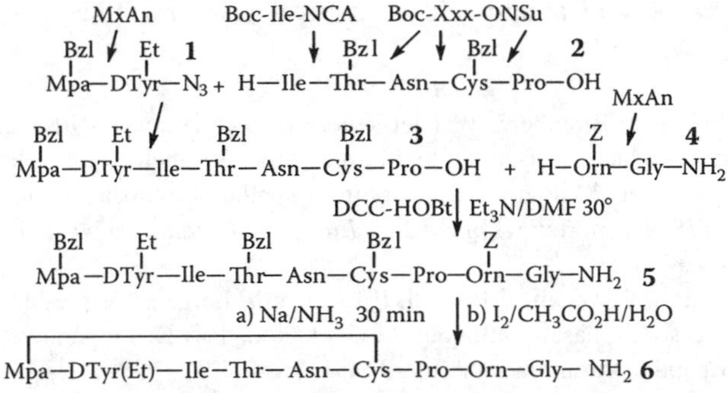

MxAn — Boc-Ile-NCA — Boc-Xxx-ONSu

Mpa(Bzl)—DTyr(Et)—N_3 (1) + H—Ile—Thr(Bzl)—Asn—Cys(Bzl)—Pro—OH (2)

Mpa(Bzl)—DTyr(Et)—Ile—Thr(Bzl)—Asn—Cys(Bzl)—Pro—OH (3) + H—Orn(Z)—Gly—NH_2 (4) [MxAn]

DCC-HOBt | Et_3N/DMF 30°

Mpa(Bzl)—DTyr(Et)—Ile—Thr(Bzl)—Asn—Cys(Bzl)—Pro—Orn(Z)—Gly—NH_2 **5**

a) Na/NH_3 30 min | b) $I_2/CH_3CO_2H/H_2O$

Mpa—DTyr(Et)—Ile—Thr—Asn—Cys—Pro—Orn—Gly—NH_2 **6** (Mpa–Cys disulfide bridge)

FIGURE 8.13 Synthesis of the oxytocin antagonist Atosiban in solution,[55] using Boc-amino acids with the activated form or coupling method indicated for each peptide bond that is formed. Mpa- = mercaptopropionyl (desaminocysteinyl) = $HSCH_2CH_2C(=O)$-; MxAn = mixed anhydride; NCA = *N*-carboxyanhydride; NSu = succinimido. Acyl azide **1** is obtained from the hydrazide that was obtained from the ester; **4** is obtained from the Boc-dipeptide amide. Amino groups are generated for coupling by removal of the Boc groups.

ammonia. Assembly of peptide **2** is achieved by aminolysis of succinimido esters, except for the last residue, which is incorporated by use of the Boc-amino-acid *N*-carboxyanhydride (see Section 7.14). Aminolysis of the succinimido ester of Boc-isoleucine by the tetrapeptide had failed, most likely because two hindered residues, isoleucine and *O*-protected threonine (see Section 1.4), were involved in the reaction. Coupling of acid **3** with dipeptide amide **4** is hastened by the presence of triethylamine and warming; no isomerization occurs because the activated residue is proline (see Section 4.16). Oxidation after deprotection of peptide **5** on a half-kilo scale is achieved with iodine; oxidization of the sulfhydryls by air is too slow. Purification by ion-exchange chromatography followed by reversed-phase HPLC gives a product of >98% purity. No purification of intermediates by column chromatography is required. A second example of synthesis of a commercial product involves reaction between Glp-Glu(OBzl)-Asp(Bzl)-OH (Glp = pyroglutamyl, see Section 6.16) and the amino groups of 2,7-diaminooctanedioic acid *bis*[Lys(Z)-OBzl]. Of interest is the fact that 36 methods were examined to find the most efficient way of coupling. The authors found that TBTU [$BtOC^+(NMe_2)_2 \cdot BF_4^-$] was superior to HBTU [$BtOC^+(NMe_2)_2 \cdot PF_6^-$], and that the 4-oxo-3,4-dihydrobenzotriazine equivalent of TBTU [$DhbtOC^+(NMe_2)_2 \cdot BF_4^-$) gave the best results (98% yield, 1.08% epimer). Thirty-five-gram batches of product were purified by reversed-phase HPLC. Various modifications to the technology of solid-phase synthesis have been developed. An example is several columns of resin in series in a continuous-flow system (see Section 5.3). Another is a vertical rotating cylindrical basket reactor positioned inside a stationary housing. A slurry of resin is fed through a central inlet, forming a 2- to 6-cm cake of uniform depth that is held by centrifugal force on a filter cloth on the perforated surface of the drum. Solvents containing activated derivatives and reagents are sprayed onto the resin from the center of the reactor. Liquids are removed by drainage at the bottom and recycled or discarded. In solid-phase synthesis, an excess of reactants is always employed. A cost analysis performed by other researchers showed that recovery of amino acid derivatives after a synthesis was not worthwhile.[55–59]

55. C Johansson, L Blomberg, W Hlebowicz, H Nicklasson, B Nilsson, L Andersson. Industrial production of an oxytocin antagonist: Synthetic approaches to the development of a multi-kilogram scale solution synthesis. (Atosiban) In: HLS Maia, ed. *Peptides 1994. Proceedings of the 23rd European Peptide Symposium*. Escom, Leiden, 1995, pp 34–35.
56. V Caciagli, F Cardinali, F Bonelli, P Lombardi. Large-scale production of peptides using the solid phase continuous flow method. Part 2: Preparative synthesis of a 26-mer peptide thrombin inhibitor. *J Pept Sci* 4, 327, 1998.
57. J Heibl et al. Large-scale synthesis of hematoregulatory nonapeptide SK&F 1077647 by fragment coupling. *J Pept Res* 54, 54, 1999.
58. N Stepaniuk, K Tomazi, MC Stapleton. Reactor and method for solid phase peptide synthesis. (basket reactor) US Patent 6,028,172. February 22, 2000.
59. L Andersson, L Blomberg, M Flegel, L Lepsa, B Nilsson, M Verlander. Large-scale synthesis of peptides. *Biopolymers (Pept Sci)* 55, 227, 2000.

8.10 DANGERS AND POSSIBLE SIDE REACTIONS ASSOCIATED WITH THE USE OF REAGENTS AND SOLVENTS

The dangers associated with the use of carbodiimides (see Section 7.1) and onium salt-based reagents (see Section 7.18) have been alluded to. All reagents should be handled with caution, and protective clothing should be worn. Hydrohalo and trifluoroacetic acids are corrosive. Contact with hydrogen fluoride is extremely dangerous. A spill of hydrogen fluoride on trousers can be fatal, as can a spill of molten 2,4-dichlorophenol on skin. Skin affected by hydrogen fluoride should be treated immediately with calcium gluconate gel or a cold hyamine (an anionic detergent) solution. Benzyl chloride is lacrimatory. Some solvents including ethyl acetate can be the cause of skin sensitivities, as well as side reactions. Dichloromethane slowly reacts with amines; hence, prolonged contact between the two should be avoided. It is also carcinogenic. An alternative for organic reactions is benzotrifluoride (trifluoromethylbenzene).

Chloroform generates phosgene on storage. It is safer to use ethanol-preserved chloroform than pentene-preserved chloroform, as the alcohol reacts with the phosgene. Chloroform should not be included in the solvent mixture for reductions using zinc and acetic acid if Boc-amino acids are involved. The hydrogen chloride generated decomposes the Boc-group. Halogen-containing solvents should not be used if triethylamine is employed for mixed-anhydride reactions because the anhydride is formed too slowly with this reagent-solvent combination (see Section 7.4). It is also uncertain as to whether halo-solvents are compatible with azides. Dioxane is readily oxidized to peroxides that are explosive. It should be purified by pouring through a column of alumina (Al_2O_3) and stored over alumina. The less flammable and higher-boiling methyl-*tert*-butyl ether forms peroxides much faster than diethyl ether. Dimethylformamide is not a stable solvent. It generates dimethylamine, which is a base and a nucleophile in the presence of moisture. Both are removed by leaving the solvent over potassium hydroxide pellets, decanting it, distilling it with benzene, and then placing it under reduced pressure. Reactions carried out in dimethylacetamide can be accompanied by acetylation. *N*-Methylpyrrolidone causes the residue to enantiomerize if it is the solvent employed to convert an Fmoc-amino acid into the chloride using triphosgene. Hexamethylphosphoric triamide may slow down coupling reactions and is a carcinogen. A recommended replacement with similar physical properties is the cyclic dimethylpropyleneurea DMPU = 1,3-dimethyl-3,4,5,6-tetrahydro-2(1*H*)-pyrimidone (*N,N'*-dimethylurea with nitrogen atoms linked through $-(CH_2CH_2CH_2)-$]. Dimethylformamide is more readily removed by extraction from an organic solvent by water than by aqueous bicarbonate. Traces of these high-boiling solvents can be readily removed by depositing the solution of peptide on a reversed-phase column and washing it with aqueous buffer before eluting the peptide. A solvent that is aprotic but has the same polarity as water is ethylene carbonate [$-(CH_2O)_2C{=}O$].

Hexafluoro-2-propanol has superior dissolving power to that of trifluoroethanol but is too nucleophilic for use as solvent for coupling reactions. Carbodiimide-mediated reactions effected in trifluoroethanol produce benzotriazolyl esters if 1-hydroxybenzotriazole is employed as additive. Less esterification occurs if 3-hydroxy-4-oxo-3,4-dihydrobenzotriazine is the additive. Trifluoroacetic acid may contain acetaldehyde that binds to the side chain of tryptophan. The side reaction is avoided by storing the acid over indole (1 mg/mL). Trifluoroacetic acid forms an azeotrope with toluene and, hence, can be removed by distilling with toluene (see Section 8.11). Bottles of formic acid should have an escape valve, as pressure builds up with time. When used as solvent for palladium-catalyzed hydrogenation, methanol or ethanol can be a source of alkylation if the alcohol is anhydrous and oxygen has not been completely removed from the system (see Section 6.20). Carboxylic acids such as acetic and trifluoroacetic acid used as reagents or solvents for acidolysis can esterify the hydroxyl groups of serine and threonine residues (see Section 6.6). Polar solvents promote stereomutation (see Section 4.12).[60–63]

60. JM Stewart, JD Young. Solid phase peptide synthesis.
61. M Hudlicky. An improved apparatus for the laboratory preparation of diazomethane. *J Org Chem* 45, 5377, 1980.
62. D Seebach. Safe substitute for HMPA. *Chem Brit* 21, 632, 1995.
63. A Ogawa, DP Curran. Benzotrifluoride: a useful alternative solvent for organic reactions currently conducted in dichloromethane. *J Org Chem* 62, 450, 1997.

8.11 ORGANIC AND OTHER SALTS IN PEPTIDE SYNTHESIS

The functional groups of reagents, amino acids, and their derivatives and peptides form salts, with a few exceptions, in the presence of acids or bases. Knowledge of the properties of these salts is of value for facilitating manipulation of the compounds. Carboxyl groups form salts with amines including ammonia. Derivatives containing a carboxyl group that do not crystallize readily are often obtained as the dicyclohexylammonium (Dcha) salts. Acid-sensitive derivatives are also advantageously stored as Dcha salts. The acids are regenerated for use by dissolving them in an organic solvent and neutralizing the base with a dilute solution of a stronger acid. Dcha hydrochloride is insoluble in water, but Dcha sulfate is soluble in water, so sulfuric acid is preferred. Amino groups form salts with acids including acetic acid. Hydrochlorides are preferred to hydrobromides because the latter are more likely to be hygroscopic; hydrochlorides are preferred to trifluoroacetates because the former are more likely to be solid. Benzyl esters that are obtained as the *p*-toluenesulfonates (see Section 3.17) are often converted to the hydrochlorides. Neutralization of an aqueous solution of the salt in the presence of an organic solvent that is not miscible gives the free ester in the organic solvent. Bubbling hydrogen chloride through the solution or evaporation of the solution in the presence of dilute hydrochloric acid gives the amino acid ester hydrochloride. Trifluoroacetate salts of peptides can be converted to the hydrochlorides by lyophilizing a solution of the peptide in the presence of 0.01 *M* hydrochloric acid. Trifluoroacetic acid can be removed from a peptide by depositing a solution of the peptide on a reversed-phase column. A thorough wash with acetate buffer before displacement of the peptide from the column removes the acid. When mixed-anhydride reactions are carried out in tetrahydrofuran, *N*-methylmorpholine hydrochloride precipitates out as a white powder; the salt is very soluble in dichloromethane. Some peptides that are slightly soluble as the hydrochlorides are more soluble as the perchlorates. The bulk of perchloric acid in aqueous solution can be removed by filtration as potassium perchlorate, which is much less soluble than sodium perchlorate. Chaotropic salts (see Section 8.4) disrupt secondary structures and facilitate synthesis on a solid support. Lithium salts facilitate the solution of peptides in polar solvents, increase the rate of swelling of polar resins, and improve coupling yields but are deleterious to couplings not involving preactivated acylators (see Section 8.4). Salts such as tertiaryamine hydrochlorides or *p*-toluenesulfonates promote epimerization (see Section 4.12).

Amino groups form salts with acetic acid, and carboxyl groups form salts with aqueous ammonia. However, amino acids that are not basic or acidic do not form salts with acetic acid or aqueous ammonia, though they do form salts with stronger acids such as hydrochloric and trifluoroacetic acid and stronger bases such as dicyclohexylamine (Figure 8.14). Only lysine and arginine with the extra basic group

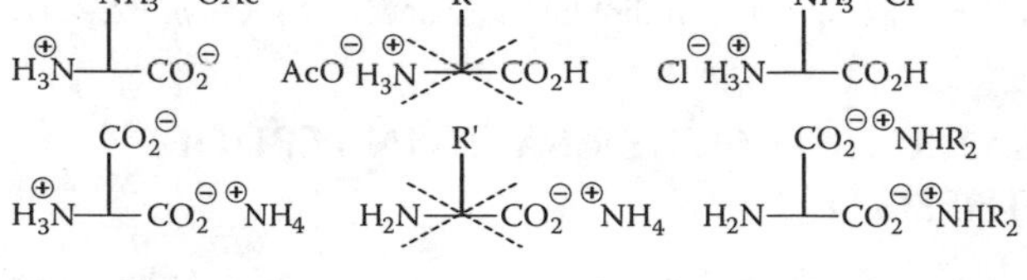

FIGURE 8.14 Salts of amino acids. (A) Acetic acid and aqueous ammonia form salts with the basic and acidic amino acids but (B) not with neutral amino acids. (C) Stronger acids and bases form salts with all amino, guanidino and carboxyl groups of amino acids. Ac = acetyl; R = alkyl or cyclohexyl.

form salts with acetic acid; only glutamic and aspartic acid with the extra carboxyl group form salts with aqueous ammonia. Monoprotected derivatives of these four amino acids also do not form salts with aqueous ammonia or acetic acid. Advantage of this fact can be taken to isolate amino acids from dilute (urine) or salt-containing solutions. Adsorption of all cations from a solution by a cation-exchange resin, followed by displacement of the amino acids with ammonium hydroxide, leaves (after evaporation of the solvent) a residue of pure amino acids with a trace of ammonium ion if glutamic or aspartic acid is present. Similarly, adsorption of anions by an anion-exchange resin followed by displacement of the amino acids by acetic acid and evaporation of the solvent leaves a residue of pure amino acids with a trace of acetic acid if arginine or lysine is present.[64–66]

64. L Benoiton, LP Bouthillier. Synthesis of γ-hydroxyglutamic acid (diastereomeric mixture). (isolation of amino acid with ion-exchange resin) *Can J Chem* 33, 1473, 1955.
65. FMF Chen, NL Benoiton. Hydrochloride salts of amino acid benzyl esters from *p*-toluenesulfonates. *Int J Pept Prot Res* 27, 221, 1986.
66. W König, G Breipohl, P Pokorny, M Birkner. Perchloric acid in peptide chemistry, in E Giralt, D Andreu, eds. *Peptides 1990. Proceedings of the 21st European Peptide Symposium*. Escom, Leiden, 1991, pp 143.

8.12 REFLECTIONS ON THE USE OF TERTIARY AND OTHER AMINES

Amines may possess two chemical properties: they act as bases, which means they combine with protons, and they may act as nucleophiles, which means they react with electron-deficient centers. Their reactivity depends on the strength, basicity, or p*K* of the base (see Section 1.2) and the ease with which the nitrogen atom can come into contact with the proton or electrophile — in common parlance, whether the amine is hindered or not by the presence of substituents on the nitrogen atom. The ability of an amine to bind a proton is hardly affected by hindering groups; however, the nucleophilicity of the amine is considerably affected by hindering groups. As bases, the amines may neutralize liberated protons, or they may deprotonate molecules to initiate reactions such as coupling, deprotection, and methylation. They may also deprotonate molecules when it is not desired that they do so, which usually causes a stereomutation of the residue or may result in a β-elimination. Pertinent to this is the fact that the p*K* of a tertiary amine varies depending on the nature of the solvent that surrounds it. As an example, the relative values of the p*K* of triethylamine and the p*K* of an ester of glycine in dioxane, ethyl acetate, chloroform, and tetrahydrofuran can vary by 1000 times. The p*K* of *N*-methylmorpholine is close to that of the glycine ester in the first two solvents, being higher in chloroform and lower in tetrahydrofuran. As nucleophiles, amines may trap electrophiles that are liberated during deprotections, or they may engender reactions by displacement of protectors or activating moieties. Some amines are basic but not nucleophilic; they do not react with activated acyl groups to form acylium ions or deprotonated 9-fluorenylmethyl to form the adduct. The following (with qualifying remarks) are amines encountered in peptide synthesis. Their structures appear in Figure 8.15. For

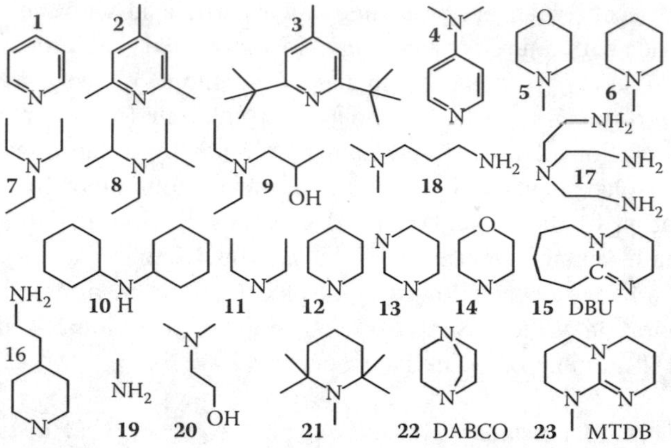

FIGURE 8.15 Alkylamines encountered in peptide synthesis. **1,** pyridine; **2,** 2,4,6-trimethylpyridine **3,** 2,6-di-*tert*-butyl-4-methylpyridine; **4**, 4-dimethylaminopyridine ; **5,** *N*-methylmorpholine, **6,** *N*-methylpiperidine; **7,** triethylamine; **8**, diisopropylethylamine; **9**, 1-diethylaminopropane-2-ol; **10**, dicyclohexylamine **11**, diethylamine; **12**, piperidine; **13**, piperazine; **14**, morpholine; **15**, 1,8-diazabicyclo[5.4.0]undec-7-ene; **16**, 4-(aminoethyl)piperidine; **17**, *tris*(2-aminoethyl)amine; **18**, 3-dimethylaminopropylamine; **19**, methylamine; **20**, dimethylaminoethane-2-ol; **21**, 1,2,2,6,6-pentamethylpiperidine; **22**, 1,4-diazabicyclo[2,2,2]octane; **23**, 7-methyl-1,5,7-triazabicyclo[4,4,0]dec-5-ene.

convenience, couplings effected by phosphonium/guanidinium/methanaminium salts are referred to as o/inium salt-mediated reactions. The amines employed for initiating o/inium salt-mediated reactions and removing 9-fluorenylmethyl-based protectors enantiomerize side-chain-protected cysteine residues that are bound to supports through an ester linkage, but not those bound through an amide linkage (see Section 8.1). Use of the 2-chlorotrityl linker avoids these side reactions.

Pyridine (**1**) is a weak and good base and good solvent for effecting aminolysis of acyl fluorides (see Section 7.12) and for preparing Boc-amino-acid *N*-carboxyanhydrides (see Section 7.14) and activated esters by the carbodiimide method (see Section 7.7), especially the esters of Boc-amino acids, as it prevents decomposition of the activated residue (see Section 7.15). It is the preferred base for aminolysis of acyl fluorides in dichloromethane.

2,4,6-Trimethylpyridine (collidine) (**2**) is more basic than pyridine and is recommended as being superior for minimizing enantiomerization of the activated residue during o/inium salt-mediated reactions of protected segments with an amino group and Fmoc-Cys(R)-OH with the hydroxyl group of a linker-resin (see Section 8.10).

2,6-Di-*tert*-butyl-4-methylpyridine (**3**) is hindered, allows coupling of Fmoc-amino-acid fluorides with weak nucleophiles (see Section 7.12) and is best for minimizing the enantiomerizaton of Fmoc-Cys(tBu)-F (see Section 8.1) during its reaction with the hydroxyl group of a linker-resin.

4-Dimethylaminopyridine (**4**) is a moderately strong base, and as a nucleophile, is a good catalyst for esterifying *N*-protected residues activated as anhydrides (it transfers the activated group; see Section 5.22), but it may cause enantiomerization of the residue (see Section 4.19). It is effective without undesirable effects in catalyzing the esterification of Fmoc-amino-acid fluorides if the solvent is not polar (see Section 4.19).

N-Methylmorpholine (**5**) is a weak, unhindered base traditionally employed for mixed-anhydride reactions because its use gives the best results in terms of chiral preservation during the coupling of segments. It is the base of choice for neutralizing hydrogen chloride during the preparation of Boc- and *Z*-amino acid *N*-carboxyanhydrides and their reactions with the salts of amino-acid esters. It is better than triethylamine for neutralizing the excess of acid after protonation of the side chain of arginyl with hydrogen bromide. However, it is not basic enough to completely neutralize protonated amino groups and strong acids such as hydrogen chloride; its use results in slower aminolysis, leading to more enantiomerization of the activated residues in o/inium salt-mediated reactions of segments than does diisopropylethylamine (see Section 7.18), and much more during aminolysis by a secondary amine of Pht-Asp(OBzl)-Cl (Pht = phthaloyl) than does the use of alkali or 2,6-di-*tert*-butylpyridine. It causes some decomposition of Fmoc-NH- after prolonged contact between the two.

N-Methylpiperidine (**6**) is a strong, unhindered base recommended for mixed-anhydride reactions, in which its use leads to less urethane formation in coupling amino acid derivatives, especially those of *N*-methylamino acids, and less epimerization in coupling segments than does *N*-methylmorpholine (see Section 7.3). It allows the preparation of mixed anhydrides by reaction of *N*-protected amino acids with pyrocarbonates $(ROCO)_2O$ when *N*-methylmorpholine does not and is more efficient than triethylamine for preparing anhydrides R^1OCO-Xaa-O-COPh using benzoyl chloride. It leads to much higher yields than does *N*-methylmorpholine when employed for generating mixed anhydrides of, or as neutralizer for couplings of Boc-*N*-methylamino acids because there is less decomposition of the derivative (see Sections 7.15 and 8.13). It enantiomerizes 2-alkyl-5(4*H*)-oxazolones faster than does *N*-methylmorpholine or triethylamine.

Triethylamine (**7**): is a strong, hindered base originally employed for mixed-anhydride reactions. However, it reduces the rate of anhydride formation if the solvent is dichloromethane or chloroform (see Section 7.3) and promotes disproportionation of the anhydride under conditions in which *N*-methylmorpholine does not (see Section 7.5). It causes enantiomerization of urethane-protected amino-acid *N*-carboxyanhydrides and reaction between two molecules with release of carbon dioxide (see Section 7.14). It is also used in the synthesis of Atosiban (see Section 8.9). There is no reaction for which it is recommended as superior to other tertiary amines, except possibly for coupling employing BOP-Cl (see Section 8.14).

Diisopropylethylamine (**8**) is a very hindered strong base that is appropriate for use in couplings of acyl azides and is the base of choice for initiating o/inium salt-mediated reactions and neutralizing the acid liberated during the aminolysis of Fmoc-amino acid fluorides, whether prepared beforehand or *in situ* using TFFH ($Me_4N_2C^+F \cdot PF_6^-$) (see Section 7.12). Compared to other tertiary amines, it is less conducive to base-catalyzed imide formation, better suited than triethylamine for assisting the reaction of *N*-protected amino acids with the chloromethyl group of linker-resins because it leads to less quaternization of chloromethyl groups, and generally less conducive to enantiomerization of the activated residue during aminolysis, with possible exceptions indicated for the alkylated pyridines. However, it cannot be employed for deprotonation in a mixed-anhydride reaction, because no anhydride is generated, or for reactions of urethane-protected amino acid *N*-carboxyanhydrides in dichloromethane, because it promotes reaction between two molecules of the latter with release of carbon dioxide (see Section 7.14). It causes the enantiomerization of Bz-Leu-ONp under conditions in which it does not enantiomerize *Z*-Cys(Bzl)-ONp and *Z*-Phg-ONp (Bz = benzoyl, Np = *p*-nitrophenyl, Phg = phenylglycyl). Its purity may be crucial. It has been reported that a sample employed for o/inium salt-mediated reactions led to acetylation of amino groups because it contained traces of acetic acid. The same sample employed for neutralization in a carbodiimide-mediated reaction led to no acetylation.

1-Diethylaminopropane-2-ol (**9**) is a good base for use for coupling acyl azides because it is harmless under conditions in which *N*-methylmorpholine, triethylamine, and diisopropylethylamine epimerize the acyl azide (see Section 7.16). It has been employed to neutralize *p*-toluenesulfonic acid in a one-pot synthesis of peptides using activated esters.

Dicyclohexylamine (**10**) is a base employed as counter ion for crystallizing acid-sensitive N^{α}-protected amino acids and *N*-protected amino acids that do not crystallize as the acids. Its use allows removal of unreacted substrate after *N*-methylation of trifunctional amino acid derivatives (see Section 8.13).

Diethylamine (**11**) is an inexpensive base that removes fluorenylmethyl-based protectors without forming an adduct with the liberated moiety, and it can be eliminated by evaporation. It has been employed as nucleophile for palladium-catalyzed allyl-transfer reactions (see Section 3.13).

Piperidine (**12**) is a base employed (20% in dimethylformamide) routinely in solid-phase synthesis for removal of fluorenylmethyl-based protectors; reacting as nucleophile, it binds the liberated moiety. It causes some aspartimide and piperidide formation at the pertinent residues of susceptible sequences (see Section 6.13). It is better than morpholine for deacetylating hydroxyl groups during the synthesis of peptides containing *O*-glycosyl-serine because it does not cause β-elimination or enantiomerization at the substituted serine. It may be that the stronger base stabilizes the residue by deprotonating the peptide bond (–CONH–) (see Section 8.14) to form the aza-enolate [–C(O)=N–].

Piperazine (**13**) is a base and nucleophile employed (6% in dimethylformamide) in solid-phase synthesis for removal of fluorenylmethyl-based protectors. The reaction is slower (three times) than with piperidine, but it leads to minimal aspartimide or piperazide formation.

Morpholine (**14**) is used as base for detaching protected peptides bound to the 9-hydroxyl group of Fm-$NHCO_2$-resin, but it is not compatible with *O*-glycosylated serines.

DBU = 1,8-diazabicyclo[5.4.0]undec-7-ene (**15**) is a nonnucleophilic base employed in conjunction with piperidine in dimethylformamide (1:1:48) for removal of fluorenylmethyl-based protectors. The piperidine is necessary as a nucleophile to trap the expelled moiety that does not react with DBU. DBU has no effect on phthalimido [Pth-NH of –Lys(Pht)–], dialkylphosphoryl [–Tyr(PO_3R_2)–], or Dde-NH [-Lys(Dde)–; see Section 6.4], but it promotes aspartimide formation at the pertinent residues of susceptible sequences (see Section 6.13). In dichloromethane, it promotes a reaction between two molecules of urethane-protected amino acid *N*-carboxyanhydride with release of carbon dioxide (see Section 7.14).

4-(Aminoethyl)piperidine (**16**) is a base and nucleophile employed for removal of fluorenylmethyl-based protectors during synthesis in solution. The adduct formed with the released moiety can be separated from the peptide ester by extraction into a pH 7.4 phosphate buffer (see Section 7.11).

tris(2-Aminoethyl)amine (**17**) is the best base/nucleophile for avoiding emulsion formation following the cleavage of Fmoc-NH during synthesis in solution.

3-Dimethylaminopropylamine (**18**) is useful for converting activated species into stable water-soluble products that allow their removal from an organic solution by extraction with aqueous acid.

Methylamine (**19**) is useful as a nucleophile for generating derivatives from activated acids.

Dimethylaminoethane-2-ol (**20**) is a compound that, by virtue of its nucleophilic center ($Me_2NH^+C_2H_4O^-$), is employed to convert protected segments bound to supports as benzyl esters into acids by transesterification into dimethylaminoethyl esters [$C(=O)OC_2H_4NMe_2$] that are hydrolyzable by a dimethylformamide-water (1:1) mixture. Compound **20** readily forms esters from acid chlorides. The hydrolysis and esterification are facilitated by anchimeric assistance by the adjacent nitrogen atom (see Section 2.10). The amino alcohol also reacts with dichloromethane.

1,2,2,6,6-Pentamethylpiperidine (**21**) is an exceptionally strong base that does not undergo quaternization; hence it is employed as acid acceptor during the quaternization of amines by alkyl halides.

DABCO = 1,4-diazabicyclo[2,2,2]octane (triethylene diamine) (**22**) is a moderately basic amine, removable by sublimation, that causes enantiomerization of amino acid residues that have been reacted to form Schiff's bases ($RR'C{=}NCHR^2CO$–).

MTBD = 7-methyl-1,5,7-triazabicyclo[4,4,0]dec-5-ene (**23**) is a strong base that allows the selective *N*-methylation of *o*-nitrosulfonyl-substituted terminal amino groups of support-bound peptides (see Section 8.13).[67–79]

67. M Bodanszky, A Bodanszky. Racemization in peptide synthesis. Mechanism specific models. (nitrophenyl esters) *Chem Commun* 591-593, 1967.

68. M Loew, Kisfaludy, L. Some observations with N-hydroxysuccinimide esters (3-dimethylaminopropylamine). *Acta Chim Acad Sci Hung* 44, 61, 1965.
69. HZ Sommer, HI Lipp, LL Jackson. Alkylation of amines. A general exhaustive alkylation method for the synthesis of quaternary ammonium compounds. (effects of hindrance) *J Org Chem* 36, 824, 1971.
70. MW Williams, GT Young. Amino-acids and peptides. Part XXXV. The effect of solvents on the rates of racemisation and coupling of some acylamino-acid *p*-nitrophenyl esters. The base strength of some amines in organic solvents, and related investigations. *J Chem Soc Perkin Trans 1* 1194, 1972.
71. MA Barton, RU Lemieux, JY Savoie. Solid-phase synthesis of selectively protected peptides as building units in the solid-phase synthesis of large molecules. (dimethylaminoethane-2-ol) *J Am Chem Soc* 95, 4501, 1973.
72. FMF Chen, M Slebioda, NL Benoiton. Mixed carboxylic-carbonic acid anhydrides of acylamino acids and peptides as a convenient source of 2,4-dialkyl-5(4*H*)-oxazolones. (pyrocarbonates for mixed anhydrides) *Int J Pept Prot Res* 31, 339, 1988.
73. YH Wang, JC Xu. One-pot liquid-phase synthesis of DSIP and 5-DSIP using fluoren-9-ylmethoxycarbonyl-protected amino acid pentafluorophenyl esters. (3-diethylaminopropane-2-ol) *Synthesis* 845, 1990.
74. LA Carpino, D Sadat-Aalaee, M Beyermann. *tris*(2-Aminoethyl)amine as a substitute for 4-(aminoethyl)piperidine in the FMOC/polyamine approach to rapid peptide synthesis. *J Org Chem* 55, 1673, 1990.
75. JD Wade, J Bedford, RC Sheppard, GW Tregear. DBU as an N^{α}-deprotecting reagent for the fluorenylmethoxycarbonyl group in continuous flow solid-phase peptide synthesis. *Pept Res* 4, 194, 1991.
76. R Dölling, M Beyermann, J Haenel, F Kernchen, E Krause, P Franke, M Brudel, M Bienert. Base-mediated side reactions on Asp(OtBu)-X sequences in Fmoc-chemistry (piperazine), in HLS Maia, ed. *Peptides 1994. Proceedings of the 23rd European Peptide Symposium*. Escom, Leiden, 1995, pp 244-245.
77. JA Robl, DS Karanewsky, MM Asaad. Synthesis of benzo-fused, 7,5- and 7,6-fused azapinones as conformationally restricted dipeptide mimetics. (*N*-methylmorpholine and Pht-Asp(OBzl)-Cl) *Tetrahedron Lett* 36, 1593, 1995.
78. P Sjölin, M Eloffsson, J Kihlberg. Removal of acyl protective groups from glycopeptides: base does not epimerize peptide stereocenters, and β-elimination is slow. *J Org Chem* 61, 560, 1996.
79. S Kates, NA Solé, M Beyermann, G Barany, F Albericio. Optimized preparation of deca(L-alanyl)-L-valinamide by 9-fluorenylmethoxyloxycarbonyl (Fmoc) solid-phase synthesis on polyethylene glycol-polystyrene (PEG-PS) graft supports, with 1,8-diazabicyclo[5.4.0]undec-7-ene (DBU) deprotection. *Pept Res* 9, 106, 1996.

8.13 MONOMETHYLATION OF AMINO GROUPS AND THE SYNTHESIS OF *N*-ALKOXYCARBONYL-*N*-METHYLAMINO ACIDS

N-Methylamino acids are of interest for incorporation into peptides because a methylated amide bond has major effects on the properties of a peptide. It eliminates hydrogen bonding at the nitrogen atom, thus disrupting adjacent secondary structures; it changes the conformation of the peptide by introducing a *cis* bond into the peptide chain (see Section 6.19); and it prolongs the life of the molecule in biological systems because the bond is not susceptible to the proteases that usually hydrolyze peptide bonds. Derivatives for synthesis are obtained either by derivatization of *N*-methylamino acids, which does not always proceed efficiently, or by methylation of *N*-derivatized amino acids. Synthesis of *N*-methylamino acids cannot be achieved directly because methylaminoalkyl is more reactive than aminoalkyl (i.e., a secondary amine is more reactive than a primary amine). Methyl iodide or dimethyl sulfate in the presence of alkali reacts with an amino group, producing the quaternized product. Diazomethane reacts with an amino group in aqueous solution, producing a mixture of mono-, di-, and trimethylated products. Reductive methylation of an amino group produces the dimethylated product (see Section 6.20). Hence, monomethylation is achieved by methylating an amino group that is partially protected by a substituent that allows replacement of the remaining proton by a single methyl group. The original substituent for preparing *N*-methylamino acids was *p*-toluenesulfonyl. Methylation of a tosylamino acid produces the tosyl-*N*-methylamino acid from which the *N*-methylamino acid can be generated, using sodium in liquid ammonia. Other methods of removing the tosyl group result in partial enantiomerization of the residue (see Section 8.14). A second protector was benzyl obtained by reaction of the amino acid with benzaldehyde at alkaline pH. Methylation was achieved by the Clarke-Eschweiler procedure, employing aqueous formaldehyde and formic acid. Unfortunately, both steps of this synthesis resulted in partial enantiomerization of the residue. A third protector was alkoxycarbonyl, both Boc and Cbz, with methylation being achieved using methyl iodide and silver oxide. The product of this reaction is the methyl ester of the *N*-methylated derivative. Methylation employing methyl iodide and sodium hydride in dimethylformamide also produced esterified derivatives. Unfortunately, saponification of an ester of an *N*-protected *N*-methylamino acid results in slight enantiomerization of the residue (see Section 8.14). A direct method for preparing Cbz- and Boc-*N*-methylamino acids of acceptable enantiomeric purity was finally developed in the early 1970s, when it was shown that methylation of Cbz- and Boc-amino acids with methyl iodide and sodium hydride generated the products without esterifying the carboxyl groups if the reaction was carried out in tetrahydrofuran at ambient temperature (Figure 8.16). This has been the method of choice for preparing these derivatives ever since. It transpires that under these conditions the reagent methylates weaker acids such as phenol, but not stronger acids such as *p*-nitrophenol, and it methylates sodium carboxylates ($-CO_2Na$) very slowly (33% of R^1OCO-MeAla-OMe after 5 days). This is in contrast to diazomethane, which preferentially methylates stronger acids. For maximum

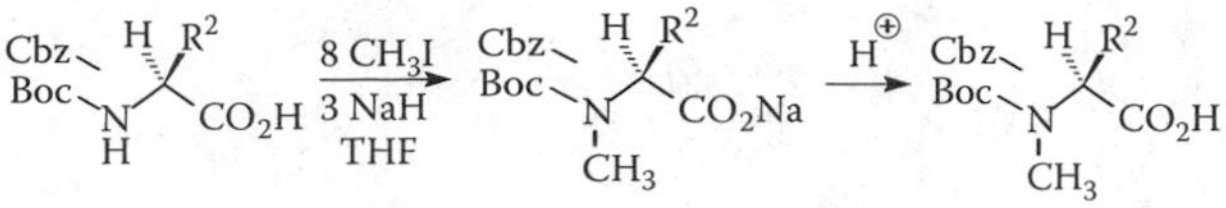

Cbz/Boc–NH–CH(R^2)–CO_2H $\xrightarrow{8\ CH_3I,\ 3\ NaH,\ THF}$ Cbz/Boc–N(CH_3)–CH(R^2)–CO_2Na $\xrightarrow{H^{\oplus}}$ Cbz/Boc–N(CH_3)–CH(R^2)–CO_2H

FIGURE 8.16 Preparation of Cbz/Boc-*N*-methylamino acids by methylation of Cbz/Boc-amino acids. [Benoiton et al., 1972]. Sodium hydride is added to substrate and methyl iodide in tetrahydrofuran, producing the sodium salt of the *N*-methylated derivative.

performance, the sodium hydride must be added to the mixture after the methyl iodide. Methylation of Cbz-NH– is faster than that of Boc-NH–. Most products are crystalline compounds. Attempts to crystallize derivatives of trifunctional *N*-methylamino acids as dicyclohexylamine salts revealed that they are soluble in petroleum ether. Advantage of this can be taken to remove traces of unreacted starting material that crystallizes out of solution as the dicyclohexylamine salt. Because the protector is sensitive to base, Fmoc-*N*-methylamino acids cannot be obtained using methyl iodide and sodium hydride. They are satisfactorily prepared by converting the Fmoc-amino acids into oxazolidinones by reaction with formaldehyde under dehydrating conditions, followed by ring opening and methylation employing triethylsilane and trifluoroacetic acid (Figure 8.17). They can also be obtained by reaction of Fmoc-chloride with the *N,O-bis*-trimethylsilylated *N*-methylamino acid. A more recent approach for incorporating an *N*-methylamino acid into a peptide is methylation of the amino group of a peptide that is attached to a support. In a method compatible with Fmoc/tBu chemistry, the amino group is partially protected by reaction with 2-nitrobenzenesulfonyl chloride. Selective methylation is achieved by deprotonation with MTDB (see Section 8.12), followed by reaction with methyl 4-nitrobenzenesulfonate (Figure 8.18). Use of MTDB is critical, as other tertiary amines failed to provide the required selectivity. The protecting group is removed with β-mercaptoethanol and DBU (see Section 8.12), with the methyl group making it more sensitive to cleavage than when attached to an unmethylated amino group. An analogous approach compatible with Boc/Bzl chemistry involves partial protection of the amino

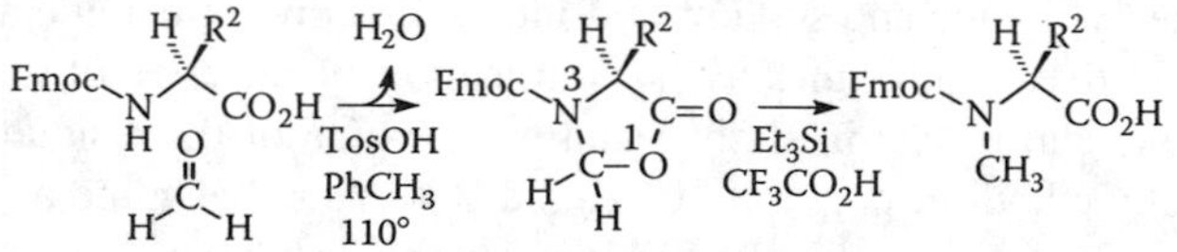

FIGURE 8.17 Preparation of Fmoc-*N*-methylamino acids by methylation of Fmoc-amino acids.[90] Acid-catalyzed reaction of substrate with formaldehyde at elevated temperature with removal of water by azeotropic distillation produces the oxazolidinone, which is then opened and reduced to the *N*-methylated derivative.

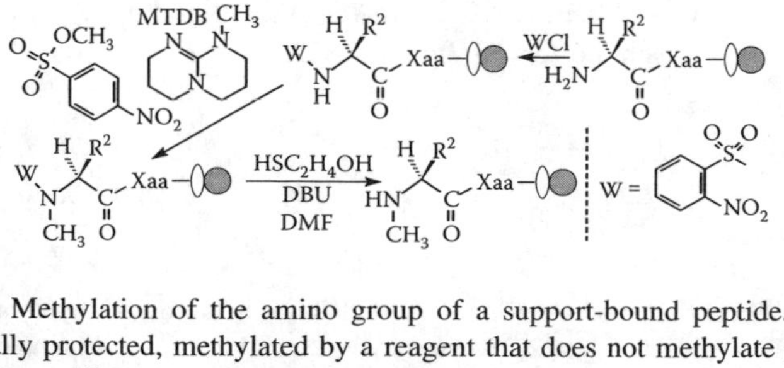

FIGURE 8.18 Methylation of the amino group of a support-bound peptide.[92] The amino group is partially protected, methylated by a reagent that does not methylate peptide bonds and then deprotected.[91] DMF = dimethylformamide.

group by 4,4′-dimethoxydiphenylmethyl [$(4MeOC_6H_4)_2CH$–] and reductive methylation (see Section 6.20) using formaldehyde and sodium cyanoborohydride ($NaCNBH_3$), followed by deprotection with trifluoroacetic acid. An interesting observation emanating from this work is that amino groups can be acylated without affecting *N*-methylamino groups by reaction with the benzotriazolyl ester of trimethylacetic acid [$(CH_3)_3CCO_2Bt$]. This allows removal of unmethylated contaminant from an *N*-methylated compound.[80–93]

80. RL Dannley, M Lukin. Use of sodium hydride in alkylation of urethans. *J Am Chem Soc* 22, 268, 1957.
81. P Quitt, J Hellerbach, K Vogler. Synthesis of optically active *N*-monomethylamino acids. (methylation of benzylamino acids) *Helv Chim Acta* 46, 327, 1963.
82 RK Olsen. A convenient synthesis of protected *N*-methylamino acid derivatives. (silver oxide and methyl iodide) *J Org Chem* 35, 1912, 1970.
83. NL Benoiton, RE Demayo, GJ Moore, JR Coggins. A modified synthesis of α-*N*-carbobenzoxy-L-lysine and the preparation and analysis of mixtures of (ε-*N*-methyllysines). (diazomethane) *Can J Biochem* 49, 1292, 1971.
84. JR Coggins, NL Benoiton. Synthesis of *N*-methylamino acid derivatives from amino acid derivatives using sodium hydride/methyl iodide. (in dimethylformamide) *Can J Chem* 49, 1968, 1971.
85. BA Stoochnoff, NL Benoiton. The methylation of some phenols and alcohols with sodium hydride/methyl iodide in tetrahydrofuran at room temperature. *Tetrahedron Lett* 21, 1973.
86. JR McDermott, NL Benoiton. *N*-Methylamino acids in peptide synthesis. II. A new synthesis of *N*-benzyloxycarbonyl-*N*-methylamino acids. *Can J Chem* 51, 1915, 1973.
87. ST Cheung, NL Benoiton. *N*-Methylamino acids in peptide synthesis. V. The synthesis of *N*-*tert*-butyloxycarbonyl,*N*-methylamino acids by *N*-methylation. *Can J Chem* 55, 906, 1977.
88. ST Cheung, NL Benoiton. *N*-Methylamino acids in peptide synthesis. VI. A method for determining the enantiomeric purity of *N*-methylamino acids and their derivatives by ion-exchange chromatography as their C-terminal lysyl dipeptides. *Can J Chem* 55, 911, 1977.
89. ST Cheung, NL Benoiton. *N*-Methylamino acids in peptide synthesis. VII. Studies on the enantiomeric purity of *N*-methylamino acids prepared by various procedures. Can J Chem 55, 916, 1977.
90. RM Freidinger, JS Hinkle, DS Perlow, BH Arison. Synthesis of 9-fluorenylmethoxycarbonyl-protected *N*-alkyl amino acids by reduction of oxazolidinones. *J Org Chem* 48, 77, 1983.
91. K Kaljuste, A Undén. New method for the synthesis of *N*-methylamino acids containing peptides by reductive methylation of amino groups on the solid phase. *Int J Pept Prot Res* 42, 118, 1993.
92. SC Miller, TS Scanlan. Site-selective *N*-methylation of peptides on solid-support. *J Am Chem Soc* 119, 2301, 1997.
93. JF Reichwein, RMJ Liscamp. Site-specific *N*-alkylation of peptides on the solid phase. *Tetrahedron Lett* 39, 1243, 1998.

8.14 THE DISTINCT CHIRAL SENSITIVITY OF *N*-METHYLAMINO ACID RESIDUES AND SENSITIVITY TO ACID OF ADJACENT PEPTIDE BONDS

It was known in the middle of the last century that coupling of segments at activated proline was not accompanied by epimerization. Epimerization during coupling at other activated residues is a result of formation of the 5(4*H*)-oxazolone, which is initiated by removal of the proton from the nitrogen atom of the activated residue (see Section 4.4). *N*-Substituted proline does not have a hydrogen atom on the imino group, and hence cannot easily form the oxazolone. *N*-Methylamino acids are imino acids similar to proline. On the basis of this chemical similarity it was thought that *N*-methylamino acids were more chirally stable than amino acids. On the contrary, it was shown in the early 1970s that activated *N*-methylamino acids have a similar or greater tendency to enantiomerize during aminolysis than do activated amino acids; moreover, substituted *N*-methylamino acids are more likely to isomerize than the corresponding unmethylated residues in alkaline solution and under some acidic conditions. *N*-Methylamino acids undergo partial enantiomerization under saponification conditions when both functional groups are substituted (Figure 8.19), but not if one of the groups is unsubstituted. A saponifiable residue is subject to enantiomerization until it has been saponified; hence hindered residues that are more difficult to saponify undergo more isomerization during saponification. The greater tendency of substituted *N*-methylamino acids to isomerize is explainable on the basis that the α-proton in these molecules [e.g., Boc-N(CH_3)C*H*R-CO_2R'] is the only one that can ionize in the presence of base. Other molecules such as Boc-N*H*CHR-CO_2R' and Boc-N(CH_3)CHR-CO_2*H* have a second ionizable proton, and it dissociates more readily than the α-proton. The unusual chiral sensitivity of piperazine-2,5-diones containing a prolyl residue is consistent with this hypothesis. It follows from the above that any nonterminal *N*-methylamino-acid residue in a peptide is chirally sensitive to alkali. A carboxy-terminal *N*-methylamino-acid residue is also chirally sensitive to a reagent formerly employed for removing benzyloxycarbonyl groups; namely, hydrogen bromide in acetic acid (Figure 8.19). Isomerization is slower at

Compound	Conditions	% Epimer
Z-Ala-Leu-OtBu	MeOH-aq NaOH (2:1), 4M 15' or 1M 60'	0.35
Z-Ala-MeLeu-OtBu		3.5
H-Ile-OH	5.6 N HBr, AcOH, 4 h	<0.5
H-MeIle-OH	1 h	3
H-Ala-MeIle-OH	4 h	17
Z-MeIle-OH	4 h	34
Z-Ala-Leu-OH	iBuOCOCl/Et_3N 0°, Tetrahydrofuran	2.0
Z-Ala-MeLeu-OH		7.0

FIGURE 8.19 Data showing the relative sensitivities to isomerization of *N*-methylamino- and amino-acid residues under conditions of saponification, acidolysis by HBr (5.7 *M* = saturated) in anhydrous acetic acid and aminolysis of the mixed anhydrides by H-Gly-OBzl·TosOH/Et_3N.[94] Ile = isoleucine.

lower concentrations and nil if water is present. No isomerization occurs in trifluoroacetic acid or in dioxane saturated with hydrogen chloride. Isomerization can be attributed to ionization promoted by protonation of the carboxyl group (see Section 4.1). It was known at the time that *N,N*-dimethylamino acids are enantiomerized by hot dilute aqueous acid. Acid can have another deleterious effect on peptides containing *N*-methylamino acids. Peptide bonds at the carboxyl group of *N*-methylamino acids (–MeXaa–Xbb–) are sensitized to acidolysis and may undergo cleavage by strong acids such as trifluoroacetic acid. A methylated bond at a carboxy terminus (–Xaa–MeXbb-OH) is also sensitive to strong acid. The sensitizing effect of a methyl group is dramatically illustrated by the extreme sensitivity to acid of tBuN(Me)SO_2Ph relative to tBuNHSO_2Ph. It issues from this that for the synthesis of peptides containing *N*-methylamino-acid residues on a solid support, one cannot employ a linker that requires strong acid to detach the peptide from the support.

Despite the absence of an NH-proton, activated *N*-acyl-*N*-methylamino acids undergo enantiomerization during aminolysis by formation of the oxazolonium ion (Figure 8.20). Proof of the existence of the intermediate was established by the addition of dicyclohexylcarbodiimide to a solution of Z-Ala-MeLeu-OH in tetrahydrofuran, followed 10 minutes later by methyl propiolate. The latter reacted by a well-known reaction (Figure 8.20) to give an 85% yield of the pyrrole. The oxazolinium ion is stable for a few hours at 15°C. It has a yellow color with a band of maximum absorbance at 400 nm. The oxazolinium ion does not appear if 1-hydroxybenzotriazole has been added to a dicyclohexylcarbodiimide-mediated coupling. Isomerization that is observed can be attributed to abstraction of the α-proton of the activated ester by the nitrogen atom (N-2) of the amide oxide form of the adduct (–CO_2BtO; see Section 7.17). The lability of the α-proton has been confirmed by studies of deuterium exchange. *N*-Hydroxysuccinimide also eliminates the appearance of the oxazolinium ion in a carbodiimide-mediated reaction. However, the α-proton of the succinimdo ester does not undergo exchange in the presence of deuterium oxide. The difference in the labilities of the α-protons of the two esters is the same for the esters of the unmethylated residues. The stereomutation of *N*-methylamino-acid residues during coupling is not catalyzed by base, with the

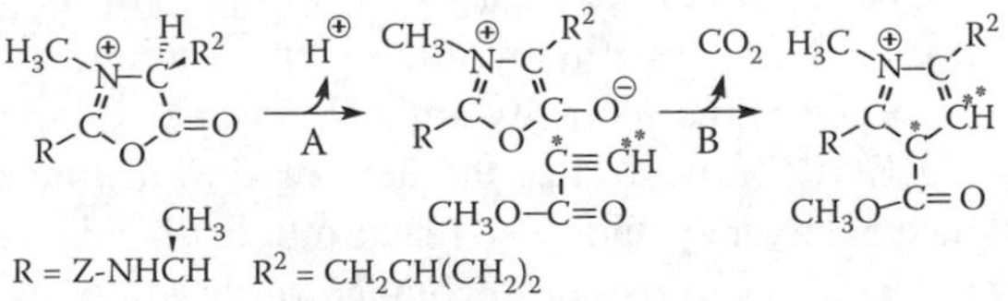

R = Z-NHCH(CH_3) $R^2 = CH_2CH(CH_2)_2$

FIGURE 8.20 Peptides activated at an *N*-methylamino-acid residue are postulated to epimerize because of the formation of the oxazolonium ion. Evidence for the latter resides in spectroscopic studies,[96] and the isolation of a substituted pyrrole that was formed when methyl propiolate was added to a solution of Z-Ala-MeLeu-OH in tetrahydrofuran 10 minutes after dicyclohexylcarbodiimide had been added.[95] The acetylenic compound effected a 1,3-dipolar cycloaddition reaction (B), with release of carbon dioxide, with the zwitter-ion that was generated (A) by loss of a proton by the oxazolonium ion.

latter being unnecessary for cyclization to the oxazolinium ion, and markedly dependent on the polarity of the solvent and the presence of salts, as would be expected because the responsible intermediate is charged. In addition to the tendency to isomerize, there is more time for isomerization to occur during couplings because the functional groups are often less reactive as a result of the presence of the methyl group. It should be emphasized that isomerization during coupling pertains in particular to the coupling of segments, though there may be instances when the aminolysis of *N*-methylamino-acid derivatives is so retarded that enantiomerization does occur (see Section 8.1).[94–97]

94. JR McDermott, NL Benoiton. *N*-Methylamino acids in peptide synthesis. III. Racemization during saponification and acidolysis. *Can J Chem* 51, 2555, 1973.
95. JR McDermott, NL Benoiton. *N*-Methylamino acids in peptide synthesis. IV. Racemization and yields in peptide-bond formation. *Can J Chem* 51, 2562, 1973.
96. JS Davies, AK Mohammed. Assessment of racemisation in *N*-alkylated amino-acid derivatives during peptide coupling in a model dipeptide system. *J Chem Soc Perkin Trans 1* 2982, 1981.
97. J Urban, T Vaisar, R Shen, MS Lee. Lability of *N*-alkylated peptides towards TFA cleavage. *Int J Pept Prot Res* 47, 182, 1996.

8.15 REACTIVITY AND COUPLING AT *N*-METHYLAMINO ACID RESIDUES

The reactivities of the functional groups of *N*-methylamino-acid residues are reduced by the steric bulk of the methyl substituent. The effect is actualized more at the carboxyl group than at the imino group. There are numerous examples of reduced reactivity at the carboxyl group. Hydrazides (see Section 2.13) of *N*-methylamino-acid residues do not form easily, and esterification of *N*-methylvaline employing isobutene and sulfuric acid (see Section 3.18) does not go well, fluorides of Fmoc-*N*-methylamino acids cannot be made, and benzotriazolyl esters of *N*-alkoxycarbonyl-*N*-methylamino acids are so inert to amino groups that they are in fact dead-end products. A striking exception in which good reactivity remains are Fmoc-*N*-methylamino-acid chlorides, which are excellent acylating agents. Triphosgene (see Section 7.13) in the presence of 2,4,6-trimethylpyridine (for 10 minutes) is an alternative for generating the activated Fmoc-*N*-methylamino acids. Coupling efficiency can be ensured by including KOBt to neutralize the acid produced. Fmoc-MeVal-Cl with the catalytic assistance of silver cyanide also reacts efficiently with *tert*-butyl alcohol. Fmoc-*N*-methylamino acids have also been coupled successfully by the symmetrical anhydride method (see Section 2.5). As for coupling-protected *N*-methylamino acids by other classical methods, the efficiency depends greatly on the nature of the aminolyzing residue, with difficulties being encountered with the amino groups of valine and isoleucine and more with methylamino groups (see below). Methylamino groups are more nucleophilic than the corresponding amino groups; however, the steric bulk of the methyl group outweighs the increased nucleophilicity. In situations in which the coupling of protected amino acids to *N*-methylamino acid residues is sluggish, symmetrical anhydrides often acylate methylamino groups efficiently. A

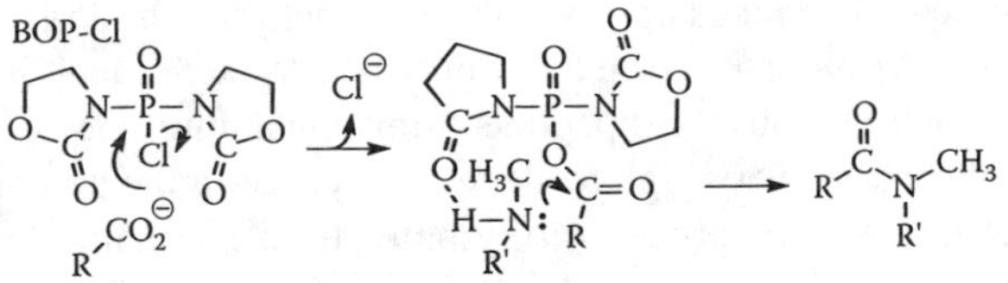

FIGURE 8.21 Proposed mechanism for the BOP-Cl-mediated reaction of a carboxylate anion with a methylamino group. Formation of a mixed anhydride is followed by aminolysis that is facilitated by anchimeric assistance provided by the oxygen atom of the ring carbonyl.[101] (van der Auwera & Anteunis, 1987). BOP-Cl = *bis*(2-oxo-3-oxazolidino)phosphinic chloride.

reagent that is uniquely applicable for coupling to methylamino groups is BOP-Cl (Figure 8.21). It reacts with a carboxylate anion to produce a phosphinic mixed anhydride, which is then aminolyzed to give the peptide. Triethylamine or diisopropylethylamine give higher yields than *N*-methylmorpholine or *N*-methylpiperidine when employed to generate the anion. This indicates that the tertiary amine deprotonates the acid but does not participate further in the reaction, as it does for mixed anhydrides (see Section 2.10). Amino groups react with BOP-Cl, so it is not suitable for reactions involving aminolysis by unmethylated amino groups.

In addition to hindrance caused by the methyl group, the nature of the protector sometimes has a role to play in the coupling of *N*-alkoxycarbonyl-*N*-methylamino acids. Higher yields have been obtained for coupling Z- or Fmoc-*N*-methylamino acids than for coupling Boc-*N*-methylamino acids, and this has been attributed to hindrance by the tertiary alkyl of the Boc group. However, the explanation most likely resides in the tendency of activated Boc-derivatives to decompose in the presence of protic molecules (see Section 7.15). The exceptional lability of activated Boc-*N*-methylamino acids is illustrated by a preparation of the mixed anhydride (see Section 2.8) of Boc-*N*-methylvaline that gave a product containing 20% of *N*-methylvaline *N*-carboxyanhydride. The symmetrical anhydride of the same derivative is too unstable to prepare. Coupling of Boc-*N*-methylvaline as the mixed anhydride with *N*-methylmorpholine as base resulted in 19% decomposition but gave a normal yield when the more basic *N*-methylpiperidine was employed. The base prevents the decomposition that is acid-catalyzed. The 4-nitro-6-trifluoromethyl equivalent of PyBOP [$BtOP^+(Pyr)_3 \cdot PF_6^-$; see Section 2.19] has been recommended as efficient for coupling to methylamino groups.

The greatest obstacle in handling *N*-methylamino acids emerges when two of the residues have to be combined. This became apparent during pioneering work on the synthesis of the 11-residue cyclic peptide cyclosporin A and its analogues, which contain seven *N*-methylamino-acid residues. Couplings were first successfully achieved using mixed anhydrides formed with pivalic (trimethylacetic) acid. Then followed recommendations that BOP-Cl (Figure 8.21) was the reagent of choice for such couplings. Other reagents and methods have since been found to be efficient in some cases. These include the use of KOBt with Fmoc-amino-acid chlorides, the 4-nitro-6-trifluoromethyl-benzotriazole-containing reagent alluded to above, HATU [$OaBtC^+(NMe_2)_2 \cdot PF_6^-$], and diisopropylcarbodiimide assisted by 1-hydroxy-7-azabenzotriazole. The superiority of 7-azabenzotriazole-based compounds over benzotriazole-based compounds may be caused by better anchimeric assistance from the N-7 atom than the N-2 atom (see Section 8.10). An additional obstacle emerges at the dipeptide stage because a dipeptide ester containing one or two *N*-methylamino-acid residues has a very high tendency to cyclize to the piperazine-2,5-dione (see Section 6.19) unless the protector originates from a tertiary alcohol.[98–107]

98. M Zaoral. Amino acids and peptides. XXXVI. Pivaloyl chloride as a reagent in the mixed anhydride synthesis of peptides. *Coll Czech Chem Commun* 27, 1273, 1962.
99. RM Wenger. Synthesis of cyclosporine. Total synthesis of cyclosporin A and cyclosporin H, two fungal metabolites isolated from the species *Tolypocladium inflatum Gams*. *Helv Chim Acta* 67, 502, 1984.
100. RD Tung, MJ Dhaon, DH Rich. BOP-Cl mediated synthesis of cyclosporin A 8-11 tetrapeptide fragment. *J Org Chem* 51, 3350, 1986.
101. C van der Auwera, MJO Anteunis. *N,N′*-bis(2-oxo-3-oxazolidinyl) phosphinic chloride (BOP-Cl); a superb reagent for coupling at and with iminoacid residues. *Int J Pept Prot Res* 29, 574, 1987.
102. KM Sivanandaiah, VV Suresh Babu, SC Shankaramma. Synthesis of peptides mediated by KOBt. (Fmoc-amino-acid chlorides). *Int J Pept Prot Res* 44, 24, 1994.
103. M Angell, TL Thomas, GR Flentke, DH Rich. Solid-phase synthesis of cyclosporin peptides. *J Am Chem Soc* 117, 7278, 1995.
104. JCHM Wijkmans, FAA Blok, GA van der Marel, JH van Boom, W Bloenhoff. CF_3-NO_2-PyBOP: A powerful coupling reagent, in PTK Kaumaya, RS Hodges, eds. *Peptides Chemistry, Structure and Biology. Proceedings of the 14th American Peptide Symposium*. Mayflower, Kingswoodford, 1996, pp 92–93.
105. SY Ko, RM Wenger. Solid-phase total synthesis of cyclosporine analogues. *Helv Chim Acta* 80, 695, 1997.
106. JM Humphrey, AR Chamberlin. Chemical synthesis of natural product peptides: coupling methods for the incorporation of noncoded amino acids into peptides. *Chem Rev* 97, 2243, 1997.
107. M Wenger. The chemistry of cyclosporine, in R Ramage, R Epton, eds. *Peptides 1996. Proceedings of the 24th European Peptide Symposium*. Mayflower, Kingswoodford, 1998, pp 173–178.

Appendices

APPENDIX 1: USEFUL REVIEWS

INTRODUCTORY

E Gross, J Meinhofer. The peptide bond, in *The Peptides: Analysis, Synthesis, Biology,* Vol. 1, pp 1–64, Academic Press, New York, 1979.

JH Jones. The formation of peptide bonds: a general survey, in *The Peptides: Analysis, Synthesis, Biology,* Vol. 1, pp 65–104. Academic Press, New York, 1979.

PROTECTION, DEPROTECTION, AND SIDE REACTIONS

R Geiger, W König. Amine protecting groups, in *The Peptides: Analysis, Synthesis, Biology,* Vol. 3, pp 1–99, Academic Press, New York, 1981.

RW Roeske. Carboxyl protecting groups, in *The Peptides: Analysis, Synthesis, Biology,* Vol. 3, pp 101–136, Academic Press, New York, 1981.

RG Hiskey. Sulfhydryl group protection in peptide synthesis, in *The Peptides: Analysis, Synthesis, Biology,* Vol. 3, pp 137–167, Academic Press, New York, 1981.

JM Stewart. Protection of the hydroxyl group in peptide synthesis, in *The Peptides: Analysis, Synthesis, Biology,* Vol. 3, pp 170–201, Academic Press, New York, 1981.

J-L Fauchère, R Schwyzer. Differential protection and selective protection in peptide synthesis, in *The Peptides: Analysis, Synthesis, Biology,* Vol. 3, pp 203–251, Academic Press, New York, 1981.

JK Inman. Peptide synthesis with minimal protection of side-chain functions, in *The Peptides: Analysis, Synthesis, Biology,* Vol. 3, pp 253–302, Academic Press, New York, 1981.

H Yajima, N Fujii. Acidolytic deprotecting procedures in peptide synthesis, in *The Peptides: Analysis, Synthesis, Biology,* Vol. 5, pp 65–109, Academic Press, New York, 1983.

M Bodanszky, J Martinez. Side reactions in peptide synthesis, in *The Peptides: Analysis, Synthesis, Biology,* Vol. 5, pp 111–216, Academic Press, New York, 1983.

E Atherton, RC Sheppard. The fluorenylmethoxycarbonyl amino protecting group, in *The Peptides: Analysis, Synthesis, Biology,* Vol. 9, pp 1–38, Academic Press, New York, 1987.

JP Tam, RB Merrifield. Strong acid deprotection of synthetic peptides: mechanisms and methods, in *The Peptides: Analysis, Synthesis, Biology,* Vol. 9, pp 185–248, Academic Press, New York, 1987.

H Yajima, N Fujii, S Funakoshi, T Watanabe, E Murayama, A Otaka. New strategy for the chemical synthesis of proteins. *Tetrahedron*, 44, 805–819, 1988.

GB Fields, RL Noble. Solid phase synthesis utilizing 9-fluorenylmethoxycarbonyl amino acids. *Int J Pept Prot Res* 35, 161–214, 1990.

F Albericio. Orthogonal protecting groups for *N*-amino and C-terminal carboxyl functions in solid-phase peptide synthesis. *Biopolymers (Pept Sci)* 55, 123–139, 2000.

R Sheppard. The fluorenylmethoxycarbonyl group in solid phase synthesis. *J Pept Sci* 9, 545–552, 2003.

Coupling

M Bodanszky. The myth of coupling reagents. *Pept Res* 5, 135–139, 1992.

NF Albertson. Synthesis of peptides with mixed anhydrides. *Organic Reactions* 12, 157–355, 1962.

D Kemp. The amine capture strategy for peptide synthesis — an outline of progress. *Biopolymers* 20, 1793–1804, 1981.

M Bodanszky. Active esters in peptide synthesis, in *The Peptides: Analysis, Synthesis, Biology*, Vol. 1, pp 105–196. Academic Press, New York, 1979.

J Meienhofer. The azide method in peptide synthesis, in *The Peptides: Analysis, Synthesis, Biology*, Vol. 1, pp 197–239. Academic Press, New York, 1979.

DH Rich, J Singh. The carbodiimide method, in *The Peptides: Analysis, Synthesis, Biology*, Vol. 1, pp 241–261. Academic Press, New York, 1979.

J Meienhofer. The mixed carbonic anhydride method of peptide synthesis, in *The Peptides: Analysis, Synthesis, Biology*, Vol. 1, pp 263–314. Academic Press, New York, 1979.

TJ Blacklock, R Hirschmann, DF Veber. The preparation and use of *N*-carboxyanhydrides and *N*-thiocarboxyanhydrides for peptide bond formation, in *The Peptides: Analysis, Synthesis, Biology*, Vol. 9, pp 39–102. Academic Press, New York, 1987.

LA Carpino, M Beyermann, H Wenschuh, M Bienert. Peptide synthesis via amino acid halides. *Acc Chem Res* 29, 268–274, 1996.

WD Fuller, M Goodman, FR Naider, Y-F Zhu. Urethane-protected α-amino acid N-carboxyanhydrides and peptide synthesis. *Biopolymers (Pept Sci)* 40, 183–205, 1996.

F Albericio, LA Carpino. Coupling reagents and activation. *Methods Enzymol.* 289, 104–126, 1997.

Stereomutation

DS Kemp. Racemization in peptide synthesis, in *The Peptides: Analysis, Synthesis, Biology*, Vol. 1, pp 315–383. Academic Press, New York, 1979.

J Kovács. Racemization and coupling rates of N^{α}-protected amino acids and peptide active esters: predictive potential, in *The Peptides: Analysis, Synthesis, Biology*, Vol. 2, pp 485–539. Academic Press, New York, 1979.

NL Benoiton. Quantitation and the sequence dependence of racemization in peptide synthesis, in *The Peptides: Analysis, Synthesis, Biology*, Vol. 5, pp 217–284. Academic Press, New York, 1983.

NL Benoiton. 2-Alkoxycarbonyl-5(4*H*)-oxazolones and the enantiomerization of *N*-alkoxycarbonylamino acids. *Biopolymers (Pept Sci)* 40, 245–254, 1996.

A Scaloni, M Simmaco, F Bossa. Characterization and analysis of D-amino acids, in P Jollès, ed. *D-Amino Acids in Sequences of Secreted Peptides of Multicellular Organisms*, pp 3–26. Birkhäuser, Basel, 1998.

Solid-Phase Synthesis

G Barany, RB Merrifield. Solid-phase peptide synthesis, in *The Peptides: Analysis, Synthesis, Biology*, Vol. 2, pp 1–284. Academic Press, New York, 1979.

E Atherton, MJ Gait, RC Sheppard, BJ Williams. The polyamide method of solid phase peptide and oligonucleotide synthesis. *Bioorg Chem* 8, 351–370, 1979.

G Barany, N Kneib-Cordonnier, DG Mullen. Solid-phase peptide synthesis: a silver anniversary report. *Int J Pept Prot Res* 30, 705–739, 1987.

P Lloyd-Williams, F Albericio, E Giralt. Convergent solid-phase peptide synthesis. *Tetrahedron* 49, 11065–11133, 1993.

R Angelleti, G Fields. Six year study of peptide synthesis. *Methods Enzymol* 289, 607–717, 1997.

J Alsina, F Albericio. Solid-phase synthesis of C-terminal modified peptides. *Biopolymers (Pept Sci)* 71, 454–477, 2003.

Synthesis in Solution

TW Muir, PE Dawson, SBH Kent. Protein synthesis by chemical ligation of unprotected peptides in aqueous solution. *Methods Enzymol* 289, 266–298, 1997.

JM Humphrey, AR Chamberlin. Chemical synthesis of natural product peptides: coupling methods for the incorporation of noncoded amino acids into peptides. *Chem Rev* 97, 2243–2266, 1997.

S Aimoto. Polypeptide synthesis by the thioester method. *Biopolymers (Pept Sci)* 51, 247–265, 1999.

K Barlos, D Gatos. 9-Fluorenylmethoxycarbonyl/tbutyl-based convergent protein synthesis. *Biopolymers (Pept Sci)* 51, 266–278, 1999.

S Sakakibara. Chemical synthesis of proteins in solution. *Biopolymers (Pept Sci)* 51, 279–296, 1999.

JP Tam, A Yu, A Miao. Orthogonal ligation strategies for peptide and protein synthesis. *Biopolymers (Pept Sci)* 51, 311–332, 1999.

DM Coltart. Peptide segment coupling by prior ligation and proximity-induced intramolecular acyl transfer. *Tetrahedron* 56, 3449–3491, 2000.

L Andersson, L Blomberg, M Flegel, L Lepsa, B Nilsson, M Verlander. Large-scale synthesis of peptides. *Biopolymers (Pept Sci)* 55, 227–250, 2000.

JS McMurray, DR Colman IV, W Wang, ML Campbell. The synthesis of phosphopeptides. *Biopolymers (Pept Sci)* 60, 3–31, 2001.

Cyclic Peptides

KD Kopple. Synthesis of cyclic peptides. *J Pharmaceutical Sci* 61, 1345–1356, 1972.

JS Davies. The cyclization of peptides and depsipeptides. *J Pept Res* 9, 471–501, 2003.

Other

VNR Pillai, M Mutter. Conformational studies of poly(oxyethylene)-bound peptides and protein sequences. *Acc Chem Res* 14, 122–130, 1981.

APPENDIX 2: YEAR, LOCATION, AND CHAIRMEN OF THE MAJOR SYMPOSIA

A = American, E = European, I = International, J = Japanese; P = Peptide, S = Symposium.

1958	EPS-1	Prague (J Rudinger)
1959	EPS-2	Munich (F Weygand)
1960	EPS-3	Basel (R Schwyzer)

1961	EPS-4	Moscow (Y Ovchinnikov)
1962	EPS-5	Oxford (GT Young)
1963	EPS-6	Athens (L Zervas)
1964	EPS-7	Budapest (K Medzihradsky)
1966	EPS-8	Noordwijk, The Netherlands (HC Beyerman)
1968	EPS-9	Orsay, France (E Bricas)
1969	EPS-10	Albano Terme, Italy (E Scoffone)
1971	EPS-11	Vienna (H Nesvadba)
1972	EPS-12	Reinhardsbrunn Castle, GDR (H Hanson)
1974	EPS-13	Kiryam Anavim, Israel (Y Wolman)
1976	EPS-14	Wepion, Belgium (A Loffet)
1978	EPS-15	Gdansk (G Kupryszewski)
1980	EPS-16	Helsingør, Denmark (K Brunfeldt)
1982	EPS-16	Prague (K Bláha)
1984	EPS-17	Djurönäset, Sweden (U Ragnarsson)
1986	EPS-18	Porto Carras, Greece (D Theodoropoulus)
1988	EPS-19	Tübingen (G Jung)
1990	EPS-20	Platja D'Aro, Spain (E Giralt)
1992	EPS-21	Interlaken (CH Schneider)
1994	EPS-23	Braga, Portugal (HLS Maia)
1996	EPS-24	Edinburgh (R Ramage)
1998	EPS-25	Budapest (S Bajusz)
2000	EPS-26	Montpellier (J Martinez)
2002	EPS-27	Sorrento, Italy (E Benedetti)
2004	EPS-28/IPS-3	Prague (M Flegel, M Fridkin, G Gilon, M Lebl, P Malón, J Slaninová)
2006	EPS-29	Gdansk (P Rekowski, K Rolka, J Silberring)
1968	APS-1	New Haven (S Lande, B Weinstein)
1970	APS-2	Cleveland (FM Bumpus)
1972	APS-3	Boston (J Meienhofer)
1975	APS-4	New York (R Walter)
1977	APS-5	San Diego (M Goodman)
1979	APS-6	Washington (E Gross)
1981	APS-7	Madison (D Rich)
1983	APS-8	Tucson (V Hruby)
1985	APS-9	Toronto (CM Debber, K Kopple)
1987	APS-10	St. Louis (GR Marshall)
1989	APS-11	San Diego (JE Rivier)
1991	APS-12	Boston (JA Smith)
1993	APS-13	Edmonton (RS Hodges)
1995	APS-14	Columbus (PTK Kaumaya)
1997	APS-15	Nashville (JP Tam)
1997	JPS-35/IPS-1	Kyoto (Y Shimonishi)
1999	APS-16	Minneapolis (G Barany)
2001	APS-17/IPS-2	San Diego (R Houghten, M Lebl)

2003	APS-18	Boston (M Chorev, TK Sawyer)
2005	APS-19	San Diego (JW Kelly, TW Muir)
2007	APS-20	Montreal (E Escher, WD Lubell)

APPENDIX 3: ON THE "PRIMARY SEQUENCE" OF PEPTIDES AND PROTEINS

In the early 1950s, KU Linderstrøm-Lang, an eminent Danish enzymologist, delivered a lecture at Stanford University. To facilitate communication, he employed three new terms — primary structure, secondary structure, and tertiary structure — to convey to his audience the notion that there was a structural hierarchy in proteins. These terms referring to the different levels of organization in proteins were gradually adopted by the scientific community and have become part of our scientific language. Primary structure refers to the chemical structure of the peptide chain (i.e., the sequence of amino acid residues linked together by peptide bonds). Folding that is brought about by hydrogen bonding between the carbonyl and imide groups of the peptide chain is referred to as secondary structure. When a globular unit results from the packing of one or more structural elements, the molecule is said to have a tertiary structure. A fourth term, quaternary structure, emerged when it was recognized that a distinct entity may be formed by the association of several separate polypeptide chains. The terms are used routinely in writings and books on enzymes and proteins. They are defined in different ways or are employed without explanation by authors of books. There is never disagreement over their meaning. However, there is another expression that occasionally surfaces in scientific writings: primary sequence. A peptide or protein does not have a secondary sequence, so one might ask, Why is the sequence primary? The meaning of the word "sequence" is clear, so the expression seems to have originated from a confused grasp of the terms originally introduced. The words "primary structure" and "amino acid sequence" have the same unambigous meaning and adequately convey the message that is to be disseminated. These are the words that should be employed. Use of the term "primary sequence" shows a misundertanding of our scientific language and should be abandoned.

REFERENCES

KU Linderstrøm-Lang. *Lane Lectures*, Vol. vi, p 115. Stanford University Press, Palo Alto, CA, 1952.

KU Linderstrøm-Lang, JA Schellman. Protein structure and enzyme activity, in PD Boyer, H Lardy, K Myrback, eds. *The Enzymes*, 2nd ed. Academic Press, New York, 1959, pp 443-510.

Index

A

C

E

H

I

M

N

O

P

R

S

T